Robots in American
Popular Culture

Robots in American Popular Culture

STEVE CARPER

McFarland & Company, Inc., Publishers
Jefferson, North Carolina

ISBN (print) 978-1-4766-7041-6 ∞
ISBN (ebook) 978-1-4766-3505-7

LIBRARY OF CONGRESS AND BRITISH LIBRARY
CATALOGUING DATA ARE AVAILABLE

LIBRARY OF CONGRESS CONTROL NUMBER: 2019942359

The front cover image is from the novelette "The Defenders," the cover
story for the January 1953 issue of *Galaxy Science Fiction*,
illustrated by Ed Emshwiller (© 2019 Wood River Gallery)

Printed in the United States of America

*McFarland & Company, Inc., Publishers
Box 611, Jefferson, North Carolina 28640
www.mcfarlandpub.com*

Table of Contents

Introduction

Robots emerge from myth and legend, born the first time a mind imagined an inanimate object imbued with human form and properties. In the *Epic of Gilgamesh*, Aruru formed Enkidu out of clay; this wild man eventually became the hero Gilgamesh's best friend. In the Torah, Yahweh created Adam from dust and water and Eve from Adam's rib; they were innocently wild until they disobeyed their creator's orders, thereby explaining the rest of humanity's history. Placing one of the names of God onto a figure of clay gave life to the golem. The Cypriot artist Pygmalion sculpted a woman out of ivory (given the name Galatea by later storytellers) so beautiful he fell in love with it, and Aphrodite granted his wish that it become human. The legendary Virgil of Naples made a magic statue that saw all and informed the emperor of plots. Virgil was also the creator of the sexbot. According to robot historian John Cohen, "he is said to have made an artificial prostitute for the use of the Romans, and there is a Rabbinic legend to the same effect."[1]

More directly ancestral are metallic figures, constructed piece by piece and brought to life directly by their creators. Callistratus wrote that Daedalus' statues moved on their own by mechanical means. Apollonius visited India and found "four Pythian tripods glid[ing] in of their own accord, like the moving tripods which Homer describes, and behind them came cup-bearers of dark bronzes resembling the figures of Ganymede or of Pelops among the Greeks."[2] (Homer was writing of the god Hephaestus, from whose forge Prometheus stole fire, the source of all later technology.)

"Like Friar Bacon's brazen head, I've spoken," wrote Lord Byron in *Don Juan*. Brazen (i.e., made of brass) figures were attributed to several of the greatest minds of the Middle Ages, like the astrologer Pope Sylvester II (946–1003), whose creation, *Meridiana*, only prophesied truth; or St. Albertus Magnus (c. 1200–1280), who spent thirty years constructing a figure that answered questions; or Robert Grosseteste (1175–1253), about whom John Gower wrote in the 14th-century *Confessio Amantis: Book 4*:

> For of the grete clerc Grosseteste
> I rede how besy that he was
> Upon clergie an hed of bras
> To forge, and make it for to telle
> Of suche thinges as befelle.[3]

Roger Bacon (c. 1220–1292) followed Grosseteste at Oxford, building on his work. Earlier brazen figures had been individual miracle workers, often aided by the gods. From Bacon we get the concept that scientific and technological advancement are achievements gathered from and spread across entire societies, the works of many adopted for the use of everyone. Bacon saw these inventions not just changing the present but also at the core of a different and transcendent future. The idea that technology was transformative was itself transformative; new technologies broke open a world guarded by elites and offered a pathway other than primogeniture and war. Truly a vision before its time, Bacon's future needed centuries and the Enlightenment to take hold, helped by the introduction of a new version of democracy, invented by America's science-mad founding fathers. However, once this future became reality, human culture could never return to the static oppression of rulers and peasants. Every agile and fertile mind could create, add to, and utilize the bounty of scientific advancement. The future suddenly roiled with ideas and achievements. Bacon's *Discovery of the Miracles of Art, Nature, and Magick* reads like a short history of the 19th century, a view through a futurescope. The chapter titled "Of admirable Artificial Instruments" is the forerunner of all futurism, science fiction, popular science articles, and every exercise in the wonders that human ingenuity will inevitably contrive (including this book).

> That I may the better demonstrate the inferiority and indignity of Magical power to that of Nature or Art, I shall a while discourse on such admirable operations of Art and Nature, as have not the least Magick in them, afterwards assign them their Causes and Frames. And first of such Engines, as are purely artificial.
>
> It's possible to make Engines to sail withall, as that either fresh or salt water vessels may be guided by the help of one man, and made sail with a greater swiftness, than others will which are full of men to help them.
>
> It's possible to make a Chariot move with an inestimable swiftnesse (such as the Currus falcati were, wherein our fore fathers of old fought,) and this motion to be without the help of any living creature.
>
> It's possible to make Engines for flying, a man sitting in the midst whereof, by turning onely about an Instrument, which moves artificiall Wings made to beat the Aire, much after the fashion of a Birds flight.
>
> It's possible to invent an Engine of a little bulk, yet of great efficacy, either to the depressing or elevation of the very greatest weight, which would be of much consequence in several Accidents: For hereby a man may either ascend or descend any walls, delivering himself or comrads from prison; and this Engine is only three fingers high, and four broad.
>
> A man may easily make an Instrument, whereby one man may in despight of all opposition, draw a thousand men to himself, or any other thing, which is tractable.
>
> A man may make an Engine, whereby without any corporal danger, he may walk in the bottome of the Sea, or other water. These Alexander (as the Heathen Astronomer assures us) used to see the secrets of the deeps.
>
> Such Engines as these were of old, and are made even in our dayes. These all of them (excepting only that instrument of flying, which I never saw or know any who hath seen it, though I am exceedingly acquainted with a very prudent man, who hath invented the whole Artifice) with infinite such like inventions, Engines and devices are feasable, as making of Bridges over Rivers without pillars or supporters.[4]

Bacon's brazen head, if anything of the sort ever existed (like the other examples, it's better attested through fictional works rather than any remaining physical evidence), is a disappointment compared to his future visions. Bacon purportedly worked on it for seven years, yet missed its prophecy when an assistant guarding his creation while he slept ignored the mystical head's gnomic words, though that's a wonderful metaphor for trusting the permanence of physical handiwork over relying on ephemeral magic.[5]

For centuries after Bacon, the most spectacular proto-robotic technology was shared and proudly public in the form of clock towers. Townspeople gathered in squares to watch moving figures on the clocks act out complicated playlets, allegorical minidramas of the progression of the seasons and of humans doing Promethean work by capturing time inside gears and levers. Clockwork eventually miniaturized into watches and metastasized into toys. Fiction slowly became fact as technological expertise spread across civilizations and down from a few extraordinary individuals to craftsmen so numerous they could organize into guilds. The rich commissioned artificers to create life-sized toys for themselves and their children to play with. Jacques de Vaucanson (1709–1782), whose place in the history of automatons is similar to James Watt's in the history of steam engines, built "a mandolin-player which sang as it played and kept time with its foot, and a piano player which moved its head and simulated the act of breathing. He is said to have cherished a secret ambition to make an artificial man."[6] And why not? Seemingly everybody else had the same goal.

René Descartes (1596–1650) was known to dabble in constructing automatons, including a dancing man. A story tells of his voyage, late in life, to the court of Sweden's Queen Christina. During a storm, the sailors sought to confirm the philosopher's safety; instead, they found "a living doll, they thought, which moved and behaved exactly like a living being." Superstitiously terrified, they threw it overboard, conveniently removing all evidence of the figure's existence.[7] More pertinently, Descartes provided the philosophical foundation for the next two centuries of thought in his *Discourse on the Method of Rightly Conducting One's Reason* (1637). Garth Kemerling's summary is like a roadmap of the robot works to come:

> Non-human animals, on Descartes's view, are complex organic machines, all of whose actions can be fully explained without any reference to the operation of mind in thinking.
>
> In fact, Descartes declared, most of human behavior, like that of animals, is susceptible to simple mechanistic explanation. Cleverly designed automata could successfully mimic nearly all of what we do. Thus, Descartes argued, it is only the general ability to adapt to widely varying circumstances— and, in particular, the capacity to respond creatively in the use of language—that provides a sure test for the presence of an immaterial soul associated with the normal human body.
>
> But Descartes supposed that no matter how human-like an animal or machine could be made to appear in its form or operations, it would always be possible to distinguish it from a real human being by two functional criteria. Although an animal or machine may be capable of performing any one activity as well as (or even better than) we can, he argued, each human being is capable of a greater variety of different activities than could be performed by anything lacking a soul.[8]

The first two great modern fictional works on robots—one featuring an artificial man, the other a mechanical woman—explore whether creations are indeed soulless. The former is Mary Shelley's *Frankenstein*, the latter E. T. A. Hoffmann's "The Sandman." Both will be examined at length in chapter 1. They are the most useful ur-works for a robot historian because they contain nearly every theme and complication that later writers would borrow, adapt, struggle with, oppose, alter, and resignedly return to over and over again. Throw in Karel Čapek's 1921 play *R.U.R.*, and the few holes that the two earlier stories leave are neatly patched. *R.U.R.* stood for Rossum's Universal Robots, with the neologism "robot" taken from an obsolete Czech term with many connotations: worker, peon, slave. In the play, Rossum factories turn out robots by the millions, grown from artificial protoplasm rather than built, outwardly indistinguishable from humans and— in the 20th century's refutation of Rousseau—possibly inwardly indistinguishable as well.

—꙾—

This base of historical myth and fiction gave rise to six overarching motifs that have been repeated throughout the history of robots, in ways both overlapping and contradictory, giving homage to ancient and modern greats or reintroduced with apparent wonder.

1. Robots as Servants

Machines are Promethean fire, codified as concrete examples of human ingenuity that make tasks easier, faster, more convenient, less time dependent, and often remote and impersonal. Machines do not talk back or require payment or time to sleep; they will perform a function properly and precisely in perpetuity, never abandoning their owners. Robots are machines in human form: we give them our work to do and hate them for taking away our jobs; we ask them to take care of our families and resent them when our families care about them in return; we need them to plan the future and balk when that future is not what we foresaw; we expect robots to act rationally at all times and rue that robots as self-aware as humans will develop human feelings. Robots have grown to be metaphors for human striving against the technology that we so fervently desire and so often can't cope with—the future personified.

2. Robots as Enemies

Machines are designed to be stronger than people. Without careful and constant control, they can crush us without a thought, just as ironclad ships made wooden battleships instantly obsolete. Though robots could misbehave or even run amok in early fiction, it was the lesson of the first fully technologized world war that give rise to robots as the enemies of humanity. *R.U.R.* debuted shortly after the end of World War I, Čapek beating others to the inevitable extrapolated horror of millions of superior manufactured soldiers wiping out all of humanity. Other writers personalized the danger, with individual robots turning on their creators or being usurped by criminals or engineered as ultimate fighting machines. The arrival of the computer age put a robot in every home, in every appliance, in every gadget—closer to us than our fellow humans—and rendering us ever more dependent on robots for our survival. As both palpable enemy and faceless terror, robots are the opposite of docile servants, to be feared in person and in the abstract.

3. Robots as Lovers

Hoffmann's Olimpia is yet another idealized female that men have perpetually dreamed of. (Woman dreaming of idealized men are rarer. In one story, a young maiden named Bertha shapes a paste out of "half a hundredweight of sugar, half of ambrosian almonds, six bottled of scented waters, musk, and ambergris, amber, forty pearls, two sapphires, a few garnets and rubies, [and] gold thread" and asks the goddess of love to bring this tasty confection to life, but her story is a fairy tale written by a man, found in Giambattista Basile's *Pentamerone* [1634].)[9] Hoffmann's innovation is making the love object the creation of another, something distant and misunderstood. Later stories would be split between the loves one makes and those created by others. Many writers give these stories cautionary messages: humans must fall in love with other humans as they are; making beings for our own pleasures too easily slips into slavery. The rebellion of a created slave is the most personal rejection of all.

4. Robots as Children

Men, in the cynical worldview, are said to build imitations of men because they cannot bear children and envy women for that superiority. It's certainly true that few females in history are credited with creating artificial life. *Frankenstein* was published anonymously and the college freshman whose experiment goes awry had to be male, given the times. Nevertheless, Mary Shelley's life informs the work: she was a mother of two children, one a premature baby who had quickly died. *Frankenstein* is a horror novel not because the student created a monster but because a father rejects a "child" who is stronger and wiser and more deeply moralistic than he. Children are symbols of hope: our children will—must—do better than us and be better than us; our children must be our saviors (and they often are in robot stories).

5. Robots as Successors

Machines can go where humans cannot survive: into outer space, to the bottom of the ocean, into fires and chemicals and atomic reactors. They let us see at a distance through camera eyes and hear through microphone ears. Robots are stronger, hardier and more adaptable than humans. As they grow in intelligence, they can take over human functions. If robots are superior to humans in all specific physical attributes, why shouldn't they take over from us, either by force or by default? Short-lived, fragile, fixed biological intelligences may be only a stage in the evolutionary process, with immortal, upgradable, hard-shelled beings taking our place in a larger universe.

6. Robots as Doubles

Robots do not have to resemble machines. Through art and artifice, they can look human, of any age, sex, race, or size. We want children and lovers that look human; we fear Others who might be spies infiltrating our communities. The face of a human robot is malleable; it can be made to appear like anyone, including ourselves. Doubles and doppelgängers are ancient in fiction, instigators of merry hijinks and deadly attacks. If robots look human, do they deserve the rights that humans have? Where is the line drawn? Robots have been asymptotically approaching humanness since writers first dreamed them into existence. If they are us, how do we treat them? Answering that question also addresses how we should treat one another. It is the profoundest of questions, one with a multitude of answers, none of them obviously correct and all of them of deepest importance.

—※—

The entire public history of robotics is a series of thin lines and fuzzy boundaries. Other than power source, do true definitional differences exist between clockwork automatons, steam men, and electric-powered robots? How humanoid must a robot appear before it stops being a mere machine? Is an automaton less of a robot if it cannot walk on its own without support? Are robots that are silent servitors dismissed in favor of those that appear to talk and think and perform independent deeds? When are robots real, and when are they frauds? People often seem not to care. They want the majesty of technological innovation, regardless of whether they see only a magic trick. The line between robots and automatons is blurry but as thick as the Great Wall of China compared to the distinctions among robots, clockwork figurines, steam men, ventriloquist dummies, and humans acting without expression on thousands of stages in thousands of cities over the past two centuries.

Perhaps that's why the coinage of an all-encompassing term for these variants was so eagerly seized upon. The word *robot* popped up in unrelated newspaper articles before *R.U.R.* ended its first New York stage run, a stunning (and possibly unique) example of a word from a foreign art play immediately entering popular culture. No one had complained of a lack in the language; old terms like *automaton* and *mechanical man* remained common referents alongside *robot* for another decade. Yet somehow *robot* filled an unknown need for a condition as new as the technology surrounding it, a metaphor for the servility, repetition, perfection, and lack of individuality endemic in the Machine Age. *Robot* became the default term not just for a humanoid machine but also for any automatic or remote-controlled machine. This development created a definitional quandary, as Čapek's Rossum did not build machines: he created synthetic life. Over time, the older term of *android* came to be applied to non-metallic robots, though *robot* remained the category descriptor: all androids are robots, even if not all robots are androids. To individualize their works, writers have continued to pile on coinages and neologisms and slang and jargon for their creations. A quick pass over the literature drops a thesaurus on the audience: cybernaut, anabot, fembot, sexbot, synthbot, synthopoid, bot, droid, drone, humanoid, mecha, cyborg, terminator, transformer, transmorpher, simulacrum, animatronic, robotoid, replicant, electroid, mandroid, and so forth.

In practice, a far more important and fundamental duality shakes out of the word-salad: artificial creations are either visibly artificial or confusable with humans. Most early automatons were the former. Artificial life that could pass easily for human was a brainstorm of movie and television producers, struck *en masse* in the 1960s by the realization that hiring humans was far cheaper and more reliable than building working humanoid machines. Plotlines followed; a nonhuman slipped into human society offered infinite possibilities. Besides, actors love playing dual roles with themselves. (Chapter 13, titled "Robots as Androids," expands on this historical dichotomy.)

For purposes of clarity, this book uses *robot* as the default term for all of these creations, regardless of the physical process that made them. Other terms will be used in specialized situations. *Mechanical man* and *automaton* are used when older works are discussed as more appropriate for the historical context. Creations that can pass for humans are called *androids* to cue readers to the way they are used. Other terms coined by specific creators will be used to refer to their works for consistency. Writers, naturally, do not always cooperate in placing their creations in neat boxes. In the movie *Bicentennial Man* (1999), the title character starts as an obvious robot and ends as Robin Williams. The "Lovergirls" that Benjamin Rosenbaum invented for his short story "Droplet" (2002) can take on any shape at all—human, post-human, or nonhuman. In "Farewell to the Master" (1940), the Harry Bates story from which the movie *The Day the Earth Stood Still* was adapted, "Gnut" (as spelled there) is the master and Klaatu is the puppet; whether Gnut is a natural being or a creation is never clarified.

Robots are a cloudlike subject—solid and massive at a distance, invisible and ephemeral up close, and always moving. If they cannot easily be defined, they must then be confined in some way. Herein, a robot is an artificially created and artificially brained nonhuman that is humanoid in appearance. Cyborgs—human brains (and sometimes other parts) in otherwise artificial bodies—are not robots. Nor are sentient computers or brains without bodies. The intelligent nonhumanoid machines found in science fiction, such as the "Genuine People Personalities" that Douglas Adams gave to elevators in *The Hitchhiker's Guide to the Galaxy* (1979), are not robots either. Nonrobots, by this definition, have been excluded from the discussion, with a few exceptions addressed on a case-by-case basis for historic or thematic importance. To compensate, what constitutes "humanoid" is given wide leeway.

Robots that are humanoid in appearance almost invariably are assigned a gender appropriate to that appearance. For most of their history, robots were depicted as metallic and brawny, usually larger-than-life. They were assigned culturally male roles and given male voices. Just as "he" once was the default pronoun unless the subject was explicitly a "she," robots were always, by default, male unless created to explicitly mimic females. Except for those few serving maids and love objects, robots were far more often referred to as "he" than as "it." The thorough masculinization of robots became obvious when human actresses started portraying androids. They not only automatically became "she" but also were relegated to stereotypically female roles and professions and attitudes, including the ability to flirt. With great reluctance, this historical quirk will be retained in this book to maintain continuity with the many quotes and references that treat generic humanoid robots as "he." All too few examples can be found of "sexless" robots referred to as "she." They are almost certainly outnumbered by the many robot animals identified as "he," like Sparko, the robot dog of 1940.

Popular culture is immediate and short-memoried. New is good, novel is better. This book lives in a perpetual "now," less concerned with the place of the robot in historic context and more with how the public encountered robots in the pages of fiction or on screens or through breathless newspaper headlines. Popular culture is a constant barrage of contradictory information, exemplified by the way robots have been presented since the 19th century. They have always been, from one story to another, from one page to the next, simultaneously servant and enemy, helpmate and replacement, savior and menace. Today's public image of computers as threats to our privacy while also serving as indispensable links to the world is an outgrowth of these earlier fears. Both are facets of the larger arguments about the value of technology and progress ongoing since the start of the Industrial Revolution, with optimism and pessimism taking turns prevailing. The complexity of the robots' presentations, the dominance of comical goodness or villainous badness, speaks to the tenor of the times as contemporaries saw them, in hindsight often unexpectedly.

One last point: The "popular culture" that examples are drawn from is that of the United States. Robots did not start here, of course, and that history can't be ignored. However, in all cases, the emphasis will quickly shift from those historic precedents to the many imitators they spawned in America. American popular culture is a growth industry. The number and frequency of robot appearances in popular culture per decade in the pre-computer era pales in comparison to the hordes of robots that modern media turns out every year in the 21st century. A hard stop of early 2018 was rigorously adhered to; the number of new robots introduced between then and the time that you as a reader see these pages undoubtedly would require an equal-length book to log and examine. Part One is the chronicle of a futile, if rewarding, attempt to include every major mention of a robot across multiple media. Part Two pretends no such thoroughness. Apologies to anyone whose favorite robot has been stinted or left out entirely.

The Robot Pre-Computer

Robots are now completely identified with the genre of science fiction. Introducing a robot into a story or movie immediately creates the expectation that the work is not part of the here and now, that it is set in the—or at least *a*—future, and that some additional outrageous travesties of the quotidian world will be dragged into the plot to accompany the robot.

Originally, however, no such expectations existed among audiences. For at least the first century in which robots were present in popular culture, none of this was true. The mechanical man, the android, the automaton, the electric man, or the artificial mother would be just another modern marvel in the spirit of the telegraph and the automobile, similarly designed to work alongside (or perhaps replace) human workers.

Various media seized upon different aspects of artificial humanity. Stage performers exaggerated the jerky movements of automatons or their exceptional stillness and imperturbability. Comic book writers emphasized their size, strength, and menace. Exhibitors highlighted the ways that human movements could be mimicked by machinery. Cartoonists found them caricatures of humanity in endless variation. Early science fiction writers were as likely to place their robots in contemporary settings as to include them as automatic background details that any future would surely contain.

Robots never dominated any medium, not even pulp magazines or classic science fiction, nor any particular time period. They appeared in steady but small numbers year after year from the beginning. Their slow evolution from automaton toys and clockwork figures made them familiar marvels, always signaling the new, always presaging the future, yet comfortable to contemplate. That comfort vanished after the technological horrors of World War I.

Machines had always had the capability to kill—newspapers continually editorialized about the appalling number of deaths caused by exploding steam boilers or derailing trains or downed electrical lines—but machines designed specifically for the purpose of killing were the rot in the gleaming utopia promised by technology. Post-1920 robots could not be assumed to be benign. Mad scientists might design them to do their bidding; crooks could steal the controller and bend them to crime. Robots might even rebel against their human masters. If robots were truly the symbol of machines capable of anything, then everything becomes a possibility. The chapters in Part One show the robot evolving with the times. Those in Part Two reveal how the times caught up with and surpassed them.

1

The Robot and the Android
The Origin of the Species

The border between the possible and impossible never seemed thinner than in the immediate post–World War I period. A war to end all wars fought by machines (or men sitting in machines on land, in the air, and under the sea, so as to make them one and the same) made the end of humanity by its own hand believable for perhaps the first time. If humans destroyed themselves, who else would inherit the world but machines? A warning needed to be sounded, and a European intellectual delved deep into his country's history for a term that would resonate, a reminder of the serfdom that kept laborers as quasi-slaves to the "superior" classes for centuries, an obsolete term that would come to define the future: *robot*.

> "In the eastern part of Europe the peasantry are allowed certain quantities of land for their own use, on the condition of giving a certain number of days' service on the lands of their lord. This species of service is termed Robot."—Henry Duncan Macleod, *The Elements of Political Economy* (1858)[1]

The technical term had largely been forgotten by the 20th century because modern reforms in the old Austrian Empire had wiped away the existence of the robot as a form of feudal servitude. Even in Czechoslovakia, the word *robota* had a pleasantly retro feel when Josef Čapek (pronounced chop'-ek) suggested its use to his younger brother and erstwhile collaborator, Karel, for the synthetic men he envisioned. The brothers had been writing partners for years until Josef was drafted into the army during World War I. Frail and sickly his entire life, Karel Čapek was saved from service because of a spinal ailment so severe that he couldn't even handle the book lifting called for as a librarian. Imagining a simplified and perfected human body must have been a natural daydream for him.

The end of the war brought freedom to a country in which political opponents, intellectuals, and writers (often one and the same) had been imprisoned or killed for their beliefs. Though Karel was born in 1890 in the tiny mining town of Malé Svatoňavice, his mother forced the family to move to the larger city of Prague in 1907 to give him better schooling. Although she imbued the family with a love of learning and literature, she doted on her sickly youngest to the point of hysteria, neglecting her older children, including a sister named Helena, until they finally tore him away from her clutches. Karel Čapek received an excellent cosmopolitan education, studying in Berlin, exploring night-clubs with his brother when they both were at the Sorbonne in Paris, and finally returning to Prague for his doctorate—becoming (rare among notables in this history) a doctor of philosophy in philosophy.

Yet Čapek remained as devoted to prodigious writing as William Wallace Cook. Stories, newspaper articles, and plays flowed incessantly as he made a name for himself after the war's end. Freedom to speak one's thoughts was the minimum demand. Karel also felt horror at both the mechanized slaughter of soldiers in the war's trenches and the dehumanizing horror of machines demeaning workers' toil in a modern world in which capitalist excess used them up and threw them away. As early as 1908, a joint story by the brothers Čapek, "The System," featured a factory owner spouting the themes that later plays would dramatize:

> Exploit the entire world! The world is nothing but raw material…. The worker must become a machine, so that he can simply rotate like a wheel. Every thought is insubordination! … A worker's soul is not a machine, therefore it must be removed…. I have sterilized the worker, purified him; I have destroyed in him all feelings of altruism and camaraderie, all familial, poetic, and transcendental feelings.[2]

The above paragraph might serve as a thematic précis for the play Karel Čapek wrote in 1920 and debuted on stage in 1921. The play's title, *R.U.R.* (for Rossum's Universal Robots), contained in itself a level of wordplay that extended throughout the piece, making the bombastic tolerable. Even in the original Czech script the name appears in English, already the cosmopolitan argot of international business. Rossum is derived from *rozum*, Czech for "reason," although Čapek liked to translate it as "intellect" or "brain."[3] The individual named Rossum is so obviously a device that he doesn't even appear in his own play. *R.U.R.* is the ultimate condemnation of the lack of intelligence of collective humanity, satisfying individual needs at the expense of the needs of the whole.

This contradiction sets the tone for a play full of contradictions, starting with the fact that it is a play. Theater is the province of individual humanity, with actors delving into the psychological complexities of character through confrontational wordplay and internal soliloquies. Mass action is the most difficult problem for a playwright to solve. The gruesome slaughter of all humanity seems directly opposed to the intimacy of a bare stage, better suited for media like novels or radio that allow readers to supply the missing horrors in whatever form their brains choose, from smoking ruins to piles of festering corpses. As Heywood Broun wrote of *R.U.R.*, "For years the complaint against theater has been that profundity is impossible because all drama must be developed in terms of action."[4] As in the comedy stylings of H. G. Wells and Orson Welles, wittily scaring the pants off the public with *The War of the Worlds*, the seemingly non-literary genre of science fiction allows—and, in its best forms, demands—ideas to be given life as actors as real and expressive as any stage stars. Is science fiction melodrama when it is not farce? Excellent. Čapek traps an otherwise noncompliant audience in a web of profound ideas by immersing them in farce that fades into melodrama. They arise applauding wildly.

Although *R.U.R.* was an instant worldwide success, and has been revived repeatedly and adapted into other media, its plot is as forgotten as the language of serfdom. The setting is the far future, possibly even the year 2000. As a young scientist many years earlier, Rossum discovered the formula for creating protoplasm that would duplicate life. A Dr. Frankenstein figure, he succeeded in manufacturing only distorted horrors, a reality hidden from history and the biographies of his genius along with the contemptible truth that he was an atheist who wanted to prove that God was not necessary. The true break-through was made by his son, an engineer, who—in an inadvertent but devastating critique of Creationism—deemed humans far too complicated to be manufactured and simplified the process to industrial practicality. Robots are now everywhere, indistinguishable—outwardly, superficially—from humans. That provides the farce of the first act, when the impossibly beautiful 21-year-old Helena Glory arrives by ship on the distant island that holds the factory with the naively romantic hope of forcing the corporate drones to treat the robots with the dignity of the human workers they have replaced in their scorned menial jobs. The few humans on the island, all men, instantly fall in love with her, with the winner being Harry Domain (Domin in some translations), master of the works—a Seinfeld joke waiting to happen.

After ten years, nothing and nobody on the island has changed while Europe, the totality of civilization, devolves into turmoil. Workers displaced by the robots are rebelling, but governments merely order more robots and turn them into the most efficient of slaughterers. Humans cannot compare in this dimension, either. However, the robots do have one fatal flaw: Because of Helena's influence, they are gradually being given more human traits, making them unhappier and angrier. "The era of man has come to its end," their leader explains. "A new epoch has arisen! Domination by robots!"[5] The robots rebel and kill every human on Earth save those on the island. Domain wants to flee on the last boat, but the robots have taken it over as well. Helena cannot launch even one ship. Surrounded by a sea of identical dehumanized beings, the last humans plot fantastic schemes like making robots in national varieties so that they will hate each other, just as humans do. When the power plant providing juice to the electrified fence fails, the humans all die—all but one.

The lone human survivor is allowed to live because he is a maker, a builder, and therefore superior and more robotlike. The only thing he cannot make is more robots, the plans for which Helena burned. Robots, like the replicants in *Blade Runner* (see chapter 13), have limited lifespans and are already dying off. Yet two of the new and improved robots find themselves capable of love. This development apparently gives them the automatic gift of reproduction, and they are sent off, literally named Adam and Eve, to repopulate the Earth. Man has turned into God after all.

With its commentary on every important issue of the decade except Prohibition, *R.U.R.* was an instant sensation, translated and staged in most European countries, the United States, and even Japan, where, under the title *Jinzo Ningen* (*Artificial Human*), it was the first Western-style theatrical production ever staged.[6] Most of the productions took some liberties with the script and the staging, often omitting the magical Adam and Eve epilogue, so it is difficult to know exactly which version was seen by whom. Those who saw the epilogue could understand that it follows directly from earlier acts, in which reproduction is made the central glory of humanity. Nana (sometimes Emma), Helena's maid, whose sole function is demonstrating the stupidity and religious bigotry of the lowly servants that robots could and should replace, notes that even dogs and cats can reproduce, unlike robots. (Helena is herself still childless despite her many years of marriage, probably to reflect the worldwide sterility of women because it no longer makes sense for humans to reproduce.) Emma, despite her ignorance, states truth. All the characters do, as mouthpieces for various sides in the argument. Čapek believed completely in every one of his ideas and became intensely frustrated that commentators saw only farcical contradiction. For him, *R.U.R.* was "a comedy, partly of science, partly of truth."[7] Science, progress, is inescapable, noble, and dooming, so it must be avoided. Humans are ignorant fools too stupid to accomplish anything but an imitation of God. Atheism is ridiculous because God is obvious, but religious belief is ridiculous because it negates the wisdom of science. Love is both cheap lust and Edenic beauty. Workers can be given utopia by machines only if machines render them obsolete. If by the end audience members were not hopelessly confused at the meaning of it all, they would go home with pleased vindication of the correctness of their position. *R.U.R.* is the ultimate triumph of the playwright's art.

Reviewers in New York, where the play was staged by the Theatre Guild, an outlet for serious (therefore noncommercial) works, staggered to their typewriters awed and confused, but in a state very near rapture. "The mark of genius, or thereabouts, is also on it," wrote Heywood Broun of the debut on October 9, 1922.[8] The poet Carl Sandberg allowed no small qualifiers in a letter he wrote to the *Times*: "*R.U.R.* is significant, important, teasing, quizzical, funny, terrible, paradoxical. It has its kinship with the strongest plays of Henrik Ibsen, who fought many years against the view that his dramas set forth propaganda, his own thought being that his plot and characters ask only big and terrible questions, leaving the answers to those who chose to fathom the depths of their minds for answers."[9] William J. Perlman similarly understood its scope: "[Čapek says] 'No sooner the present life will become extinct than new life will take root.' To be sure this is little enough comfort to gain from the play, but what is a civilization or two compared with the infinite chasms of time and space!"[10]

One of the mysteries of the English language is the way certain terms sweep through popular speech without any obvious cause. Within a few years the word *robot* could be found in newspapers across the country, referring to any mechanical gadget, from an automatic vending machine to a traffic light regulator. Yet Čapek's robots were the opposite of mechanical. He hammered home again and again the point that his creatures of artificial protoplasm were indistinguishable from people—that they were human in all ways save for the lack of a soul. He was a philosopher flogging the message of a parable. He was also a theater's art director. Characters indistinguishable from humans are cheap and easy to create on stage. For modern eyes, the best early productions gave the (relentlessly logical and unemotional) robots identical Mr. Spock haircuts plus uniforms that would get them past the guards of Starfleet. By 1938, however, in the first ever televised science fiction program, the BBC would swathe them in Monty Python–esque medieval helmets and riveted tinfoil outfits.[11] The confusion began immediately. Broun's first night review said flatly that "a Robot is a mechanical man."

While a play about the end of the world undoubtedly garnered smaller total audiences than *Abie's Irish Rose*, the smash hit of 1922, *R.U.R.* had an entirely respectable run of 184 performances in a slightly larger theater. Most Americans would never have a chance to see the original play run for themselves, but New York dominated the cultural world, and reviews of New York plays were syndicated to smaller papers in every state. So were the bright sayings of columnists and intellectuals, who, like modern late-night hosts, seized upon the unusual and buzzworthy for fodder. In the *New York Evening Post*, Christopher Morley wrote, "We are hoping that the intellectual phrase mongers, who got hold of the word moron a few years ago and worked it to death, will now make use of the robot instead. It seems to us a delightfully applicable term and is quickly adhesive to any vocabulary. It is much better than the deplorable epithet dumb-bell."[12] The *New York*

Tribune's "Conning Tower," a column run with genial majesty by Franklin Pierce Adams, laden with contributions from the Algonquin Round Table crowd, ran the following poem:

> To make life more pleasantly
> A robot I should choose
> To do such tiresome things for me
> As reading movie news.[13]

Both comments were quoted in syndicated articles in November 1922, along with one from the press agent of the New York Theatre Guild, who wrote that "robot" seems here to stay.... 'Robot' applies to all of us in so far as we do only what we are wound up to do."[14] That was the meaning that most quickly and thoroughly caught on. By Christmas of that year, James W. Dean, whose theater and movie reviews were widely syndicated, could confidently expect that a casual reference to the word would be understood by every reader: "The more I see of film comics the more I am convinced that the most useless thing in the cinema world is the leading lady in the comics in which the burden of action is carried by a male star.... Mildred David was an essential character in *Doctor Jack* and *Grandma's Boy*, but in his other comedies Harold Lloyd could well have had a robot in her place."[15]

Robot as a word and a concept had—remarkably—captured the country before the end of *R.U.R.*'s first theatrical run. No newspaper or magazine could consider itself *au courant* without including a mention of the Theatre Guild. Comments appeared by the hundreds, not just in the expected run of intellectual and worldly magazines but also in special interest publications ranging from the *Single Tax Review* to the *Monthly Bulletin* of the Massachusetts Society for Mental Hygiene. *R.U.R.* was "the most talked about play of the year," opined the *Oakland Tribune*, which decreed that it was beyond doubt the "strongest piece of writing that has graced the American stage in years and there is small wonder than Čapek, a young Czecho-Slovakian, awoke to find himself famous after The Theater Guild's presentation of the piece."[16] The *Mexia Evening News* even romanced the word in rhyme:

> When I and you will be robots two
> In the automo-clock work age,
> And gay and free with mechanical glee
> We'll strut on the manikin stage—
>
> When you be nutty to pal with me
> And I'll make a bolt for you—
> Our hearts will lock with a tick and a tock
> And we'll stick like a bottle of glue.[17]

Americans wrenched the term from the gloomy predictions of its creator. As productions of the play popped up in various cities for the next several years, the word *robot* grew in popularity as the perfect metaphor for the modern condition, whether for workers becoming robots or humans surpassing them. "The man who wants to arrive as a business executive, the woman who wishes to be more than a 'Robot' secretary, must get the best our excellent business and efficient secretarial schools provide," editorialized the *Woodland Daily Democrat*.[18] The *Salt Lake Tribune* dubbed Bobby Jones, then the world's best golfer, "the remorseless robot of the links."[19] Still, for all the otherworldly excellence of the long-distance Olympic medalist Paavo Nurmi, a dispatch from Paris maintained that while he had "a machine-like precision in his movements, [he was] precise like a watch, not like a machine. He is no robot."[20] Stanley E. Babb, in his review of Romer Wilson's novel *The Grand Tour*, caught the meaning of the vogue word when he declared, "So cunningly has Wilson done her book, so surely has she created the sculptor, that he lives and strides from her pages not as a Robot, but as a very intelligent and as a very interesting Parisian."[21] And the reduction of the worker to a nonentity was made stark in 1927's *Matinee Ladies*, when a college student condescendingly notes that a chauffeur "was an automaton, a Robot, and hardly a thinking or interesting human being. He was a perfect chauffeur."[22]

Robot remained a word more often seen than heard. Newspapers felt readers needed instruction on the proper pronunciation. Wrote one in 1929, "[T]he first letter 'o' is the same as in 'no.' The last syllable 'bot' is the same as the 'b-o-t' in the word bottle."[23] As late as 1934, an exhibitor kept being corrected by people who thought the "t" on the end should be silent, as if the word were French.[24] *Robot* wouldn't completely take over from *mechanical man* until about 1940, but anyone familiar with modern language knew the word and all that it implied. Far from a device limited to the pages of science fiction— not coined as a term in its modern sense until 1927—robots were ubiquitous in popular culture. The world awaited them.

—⚏—

The term *android* did not get as neat and dramatic an entrance. It was used in French as *androïde* or *andréide* since the 18th century to indicate "an automaton with a human figure which, by means of an internal mechanism [i.e., clockwork], reproduces certain movements of a human being." Mentions of androids are common in French literature in the 19th century, but much less so in English.

The first story to feature the word in its title was "L'Andréïde Paradoxale d'Edison" by Jean-Marie-Mathias-Philippe-Auguste, Comte de Villiers de L'Isle-Adam. Villiers (the shortened version of his name that can fortunately be used) published that story in 1878, a follow-up to 1877's "Miss Hadaly Habal," in which the title figure is referred to as a *sosie*,[25] the name of a character in Molière's play *Amphitryon*, which had entered the French language as meaning *double* or *Doppelgänger*.[26] Those stories were incorporated into serializations in 1880 in *Le Gaulois*, then in *L'Etoile française* (1880–1881), before being published in book form in 1886 as *L'Eve nouvelle* (reprinted in 1891 as *L'Eve Future*). The early dates are critical to understanding the way that real-world technology, especially electricity, appeared to embody the truly miraculous, a break with the meager output of the past, and a certain route to a transformed future in which any impossibility suddenly became possible. The "Edison" in the 1878 title was *the* Thomas Alva Edison, the Wizard of Menlo Park, the inventive genius who created a factory for invention in the wilds of New Jersey in 1875 and in immediate succession produced the phonograph and the light bulb, both of which play prominent roles in *L'Eve nouvelle*. Villiers had accidentally created the Edisonade, "any story dating from the late nineteenth century onward and featuring a young US male inventor hero who ingeniously extricates himself from tight spots and who, by so doing, saves himself from defeat and corruption, and his friends and nation from foreign oppressors," a huge and lucrative subgenre in dime novels, story weeklies, and young adult books[27] (see chapter 2).

Villiers makes the connection explicit in the first paragraph:

> Twenty-five leagues from New York, at the heart of a network of electric lines, is found a dwelling surrounded by deep and quite deserted gardens. The doorway looks out across a grassy lawn crossed by sanded paths and leading to a kind of large isolated pavilion. To the south and west two long avenues of ancient trees bend their shadows in the direction of this pavilion. This is Number One Menlo Park; and here dwells Thomas Alva Edison, the man who made a prisoner of the echo.[28]

Can anything be more amazing than capturing an echo? Perhaps, if the Edison whose hagiography continues for pages is involved.

Making a pilgrimage to Menlo Park comes Lord Ewald, the quintessential English aristocrat, as blindingly rich as he is handsome and manly. Yet he is going to kill himself and wants only to say goodbye. He has fallen hopelessly in love with the 20-year-old Alicia Cleary, a poor girl who happens to look exactly like the Venus de Milo, complete with arms. She has a voice as extraordinary as her features but disdains both. She has no soul, says Lord Ewald. Horror creeping into his voice, he tells Edison that she wishes to marry only for money. Even after he dragged her around Europe to visit cultural capitals, her lack of education prevented her from worshipping them as he did. What else can he do with a being who lacks the exquisite aesthetics of the aristocracy?

The modern reader might find Alicia's attitude thoroughly understandable and wonder whether Villiers is subtly satirizing the English class distinctions that would allow Lord Ewald's blindness. However, evidence suggests that Villiers wrote himself into the character. An 1890 newspaper quotes the late impoverished nobleman expressing these exact sentiments:

> I, the heir of all these titles, I do not possess the $1,000 required to make my titles recognized in European Courts of Chancery. Many a time I have been advised to marry a rich woman—advised to sell my name for gold. Always I have answered: "God shows how little He regards riches by the merit of those to whom he has given them. Ah! yes; I am too poor to claim my rights, but that is not a reason to neglect my duty."[29]

Edison has already created the perfect android he calls Hadaly, a being composed of constrained electricity. "The vision appeared to have a face of shadows, phantomlike. In the center of the forehead a network of pearls caught together and held in place folds of black gauze which completely hid the rest of the head. A suit of armor, fashioned of leaves of burned silver, which were molded with a myriad of perfect shadings, covered her girlish form."[30] Edison also has ready artificial flesh, and he makes Lord Ewald a proposition: If the aristocrat refrains from killing himself, Edison will present Hadaly with the flesh of Alicia but a true soul. He needs but three weeks. Ewald must ask Alicia to come to Edison under the guise of being prepped for the stage, and he will duplicate her to the last molecule and store away sufficient phrases on his phonographs to last Lord Ewald a lifetime.

(The stories by Jerome K. Jerome, Alice W. Fuller, and Ernest Edward Kellett mentioned in chapter 2 all use this device, with Kellett's "A Lady Automaton" being a near retelling of *L'Eve Future*. There is no way of determining whether any of them read the original book in French, or whether one did with the others copying, but Villiers had a strong reputation in England and the United States, one that might attract those with a knowledge of French. A posthumous biography in 1894 enthused that with this book, "Villiers was forthwith consecrated a great writer, his renown crossed the Channel, and penetrated across the frontier, causing much preoccupation in Belgium, that literature-loving country, always on the watch for whatever succeeds in France," giving him temporary monetary success until cancer destroyed him in 1888, leading to an agonizing death the next year.[31] And *The Athenæum* of London said [with period understatement] that the "exceedingly strange book 'L'Eve Future' made some stir in literary circles last summer [i.e., 1887]."[32] *L'Eve Future* also falls under the umbrella of stories dealing with the Pygmalion myth. Since Kellett has been suggested as a predecessor to George Bernard Shaw's play, Villiers may be a likelier source. Shaw boasted that he could read French as easily as English, after all.)[33]

The final "andraiad" (as she is called in the first English translation) has four layers to her essence: the vital system, including voice and movement; the incarnation, the plastic mediator (or armature); the incarnation, the artificial flesh; and the epidermis, the pliable skin that mimics the fluidity and expression of human skin, all powered by the literally miraculous electricity. "She is an angel," Edison says. "If theology informs us aright, angels are creatures of fire and light, and that is what Hadaly is made of."[34]

And yet, though Villiers spends pages on the details of Hadaly's construction, he, like Čapek, cannot help but add magic to the mixture. Just as Rossum's robots miraculously gain the power to beget a new race, Hadaly's soul is powered by a mystic connection to a spirit world via a clairvoyant in a trance. She fools Lord Ewald with her humanity and thrills him to his toes with her soul. She is everything a real woman should, in his estimation, be but is nowhere found in a human shell. She will never age, never need the concealment of cosmetics, never deviate from this perfection and grow into her own person. She is too perfect for this world and is as doomed as Čapek's Helena. A shipwreck (that useful deus ex machina no different from "it was all a dream") destroys the android in the end, an early example of the soon-to-be ubiquitous fate of devices that man—here specifically *man*—was not meant to know.

Perhaps because of Villiers' swift decline, no English translation of *L'Eve Future* appeared until 1926. The novel was serialized in six parts, from December 18, 1926, through January 22, 1927, in *Argosy All-Story Weekly*. *Argosy* may have been a pulp magazine, but it was easily the pinnacle of the field, publishing 100,000 words of fiction every week that guaranteed readers a mixture of every possible type and genre of story for a mere dime. The editors even induced Florence Crewe-Jones, a noted French interpreter, to do the translation. Crewe-Jones made many changes, most notably replacing Edison with Professor X, even though Edison was still alive at the time of the serialization. Edison might have been one of the most famous and honored people in the world, but the frail man of almost eighty could no longer be expected to be the glamorous wizard of his peak years. Anonymous mad scientists had stolen his thunder, impossibilities sitting better in their hands than those of a known hero. Aside from that, the timing was auspicious. The first issue of *Amazing Stories*, the earliest true science fiction pulp, had appeared on newsstands in early 1926, providing the inchoate genre with a fixed home in which robots and androids could flourish by the dozens. Genre writers read Villiers' tale in *Argosy* more surely than earlier British writers read the original. The happy coincidence that the movie *Metropolis*, starring a luminously perfect female robot, appeared in 1927 pinned the thread of idealized artificial women to the core of science fiction.

—〰—

The unusual 40-year gap between *L'Eve Future*'s original release and its publication in English opens the door to a footnote about android history. Because of that gap, William Douglas O'Connor's "The Brazen Android" (1891) appears to have been the first story to use the word in its title in English. O'Connor (later to become better known to history as Walt Whitman's friend and champion) probably wrote the work between 1860 and 1862, predating even Jules Verne's works. At the time he wrote mainly ghost stories, with this new story a close cousin. In a January 1861 letter, he agonized about the project:

> For the last few days I have been vainly trying to psychologize [*sic*] myself into commencing my story for the *Atlantic*—"The Brazen Android"—which is the one I told you of. I have decided to make the Brass Head of Roger Bacon an automaton, as Friend [*sic*] in "History of Physic" [*sic*] suggests or says it was, and then I can insinuate that it was something else—something daemonic—and spread the dark wings of my imagination in my picture of the satanic Paduan. I am also going to lug in DeMontfort and the politics of the thirteenth century to help out the tale.[35]

Today the notion that ancient civilizations possessed advanced technologies that have since been lost is confined almost entirely to the fringiest pseudoscientists and archaeologists. That stance is a complete reversal of the attitudes of Western writers for a thousand years, when the study of Greek and Roman literature and still-standing astounding relics of the past gave credence to the idea of classical superiority. The poet Virgil, transformed into a sorcerer, had in legend built a head of brass that could talk and prophesy. "The motif of the metallic talking head seems to have become especially popular between the fourteenth and sixteenth centuries, when scholars and scientists became associated with such devices," wrote Roger Mills.[36]

Doctor Malatesti, the evil Paduan of O'Connor's letter, informs Roger Bacon of the occult possibilities, which Bacon, true to his reputation as the founder of the scientific method, insisted upon believing only after experimentation made it real "that articulations, to a great extent, can be effected by simply natural means, so that a machine may be made to utter certain sentences. This machine, compact in form, placed with a bust of brass and set in motion, and lo, you have a brazen android which seems to speak of itself what by means of art it uttereth!"[37] *Brazen* here is used in its original meaning "of brass," though the plan—to use the android to frighten the superstitious King Henry III into democracy—may well be considered brazen in the later sense of being shameless.

The long story was printed in two parts in the April and May 1891 issues of *The Atlantic Monthly*, two years after O'Connor's death.[38] Contemporary reviewers explained that the lapse occurred because the magazine had rejected the story as written, calling for extensive revision.[39] This is understandable, for the story is a densely written and near indigestible historical sludge, the literary equivalent of pease porridge in the pot nine days old. O'Connor resubmitted it untouched; this time the magazine accepted it, probably because his was a more recognizable name a generation later. Lightning, another useful deus ex machina, destroys the android in the end, another independent early example that keeps the author from dealing with the costs of artificial humanity.

Čapek, writing in the 20th century, had many predecessors to draw from when he acknowledged (or at least prophesied) that superior beings would replace humanity. The threat of the robot replacement never lifted after Čapek; it would return in all fictional media representations of the robot as a major thread. Robots resemble humans too closely not to make superb enemies.

2

The Heimlich Maneuver
Robots in Early Fiction

Two stories, published almost simultaneously in the early 19th century, contain not merely the seeds but almost the entirety of the attitudes that writers have since taken toward artificial re-creations of humanity via science and technology. Both tales of beauty, obsession, madness, and death (one subtle, one blatant), they would later be transformed into every possible form of visual media as well as mined by others for use, reuse, and transformation.

January 1818 saw the publication of an anonymous three-volume novel titled *Frankenstein; or, The Modern Prometheus*. Five years later, a French edition revealed the author to be Mary Shelley, wife of poet Percy Bysshe Shelley and a mere eighteen years old when she began a year of work on it in June 1816. That month, husband and wife, along with Lord Byron and John Polidori, had read aloud from a German work of ghost stories to while away long rainy hours during a stay in Switzerland, playfully challenging one another to write ghost stories of their own. Polidori started what became *The Vampyre*, the first modern vampire story; Mary literally envisioned *Frankenstein*. Her whole short life, filled with births and deaths of a gothic hue, pointed her toward this culmination.

Slow at first to find a workable idea, Mary on June 22, 1816, had a vision of a "hideous phantasm of a man stretched out, and then, on the working of some powerful engine, show signs of life."[1] The subtitle "The Modern Prometheus" gave away the plot to contemporary readers, who were certain to be familiar with the Greek myth that the Titan Prometheus had been the creator of humanity. Mary's conceit gave modern humans a similar power through the gains made by natural philosophers (what we now call scientists). In her story, a professor at the Swiss University of Ingolstadt, where a youthful Victor Frankenstein goes to study, lectures the incoming freshman, firing his blood and intellect with the cause of science. "These philosophers," he said, "have indeed performed miracles. They penetrate into the recesses of nature, and show how she works in her hiding places. They ascend into the heavens; they have discovered how the blood circulates, and the nature of the air we breathe. They have acquired new and almost unlimited powers; they can command the thunders of heaven, mimic the earthquake, and even mock the invisible world with its own shadows."[2]

The "miserable monster" that Frankenstein creates is a horror; yet it is bigger, stronger, faster, and more capable than he. Frankenstein recoils from its mere presence, as does every other human it encounters—a father committing the ultimate sin of rejecting his own child. (Mary had borne Percy's illegitimate child in 1815 and was deeply wounded when he dismissed the premature infant's death.) Buried deeper is the implicit question of whether a man can ever usurp woman's role of giving birth—whether any (necessarily artificial) offspring of a man must be, as the monster in the end describes himself, "an abortion, to be spurned at, and kicked, and trampled on."[3] When they meet for the first time, the "wretched devil" berates his creator/god: "Remember, that I am your creature; I ought to be thy Adam; but I am rather the fallen angel, whom thou drivest from joy with no misdeed. Every where I see bliss, from which I alone am irrevocably excluded. I am benevolent and good; misery made me a fiend. Make me happy, and I shall again be virtuous."[4] Unfortunately, Frankenstein has no such power. At the end of the book, the monster flees to the North Pole to die without further danger to himself or others, sacrificing himself for humanity, a trope that would almost come to define the ending of robot narratives.

In a posthumously published review of the novel, Percy Shelley seized upon this conclusion as the great moral meaning of his wife's writing: "Treat a person ill, and he will become wicked. Requite affection with scorn—let one being be selected, for whatever cause, as the refuse of his kind—divide him, a social being, from society, and you impose upon him the irresistible obligations—malevolence and selfishness. It is thus that, too often in society, those who are best qualified to be its benefactors and its ornaments, are branded by some accident with scorn, and changed, by neglect and solitude of heart, into a scourge and a curse."[5] The original sin of the monster is its failure to correspond to society's preconceptions of the bounds of humanness. Unlike Adam, he has been cast out of Eden without cause. The uncanny valley, the seemingly innate revulsion that humans feel for beings that almost but not quite properly resemble them, is inherent in *Frankenstein* and would be the basis for hundreds of future explorations of artificially made creatures.

Yet it is the other seminal work from which the uncanny explicitly emerged. "Der Sandmann" ("The Sandman") appeared in an 1817 collection of short stories, *Die Nacht-stücke* (*The Night Pieces*) by E. T. A. Hoffmann. Hoffmann showed talent as a musician, music critic, and composer from an early age, leading to a career as an itinerant (and often failed) performer further hindered by the Napoleonic Wars, which ravaged that section of Europe and kept driving him into exile. Life didn't turn around for him until he started writing the stories about ghosts, apparitions, obsession, madness, and doom that we now celebrate as the light-hearted German Romantic tradition. For his written work, Ernest Theodore Wilhelm Hoffman dropped the Wilhelm from his name in favor of Amadeus as a tribute to Mozart. This slight recasting made his signature a mellifluous E. T. A. Hoffmann, and that is how history remembers him.

In "The Sandman," another university student, this one German, is ostensibly in love with a girl at home when he spies an impossibly lovely face through the window of his professor's house. Olimpia is a recluse and the subject of rumors, supposedly imbecilic or even deranged. Nathanael, our putative hero, obsessively watches her from his window and tries to make contact without any success until the day that the professor arranges for her debut. "Olimpia played on the piano with great skill; and sang as beautifully an aria di bravura aria, in a voice which was, if anything, almost too brilliant." Her dancing is equally perfect, putting Nathanael's to shame. When he takes her hand, "it was cold as ice; he shook with an awful, frosty shiver." After the dance, they sit while Nathanael pours his heart out to an unreceptive audience: "'Oh, you glorious heavenly lady! You ray from the promised paradise of love!' ... But Olimpia only continued to sigh 'Ah Ah' again and again."[6]

Nathanael is alone in these feelings. Others are repelled by the beauty's muteness and frigidity. Hoffmann prefigures our expectations with an earlier discussion that sets the bar low when a fellow student says, "We feel quite afraid of this Olimpia, and did not like to have anything to do with her; she seemed to us to be only acting like a living creature, and as if there was some secret at the bottom of it all."[7]

Nathanael is horrified beyond measure when he learns that Olimpia is indeed an automaton, built up over twenty years by the professor and a strange foreign tinkerer named Coppelius. They destroy their creation in a jealous fight. "Nathanael was stupefied—he had seen only too distinctly that in Olimpia's pallid waxed face there were no eyes, merely black holes in their stead; she was an inanimate puppet."[8] The professor throws the eyes at Nathanael when he tries to intervene, a shock he cannot recover from, the voyeur undone by eyes. The particulars of this destruction of his beloved are exacerbated by childhood trauma from a nursemaid who told the boy frightening tales of the Sandman, a horror who stole little boys' eyes (not the sleep-inducing friendly fairy of Hans Christian Anderson). The clockmaker likewise disappears until he shows up in Nathanael's hometown, his mere presence driving the young man to more melodrama and eventual suicide, an ending that parallels Frankenstein's pursuit of his monster into oblivion.

Gothic horror demands constant suffering and hideous death; the fates of Frankenstein and Nathanael were predetermined. The instruments of their destruction—artificial, mechanical mockeries of humanity—struck deep nerves that begged to be explained in psychological terms when that discipline emerged.

The first to do so was German psychologist Dr. Ernst Jentsch. His 1906 paper "Zur Psychologie des Unheimlichen" ("On the Psychology of the Uncanny") delved into the phenomenon using automatons as a core device.

> It is thus comprehensible if a correlation "new/foreign/hostile" corresponds to [and therefore subverts] the psychical association of "old/known/familiar." ...
>
> Among all the psychical uncertainties that can become a cause for the uncanny feeling to arise, there is one in particular that is able to develop a fairly regular, powerful and very general effect: namely, doubt as to whether an apparently living being really is animate and, conversely, doubt as to whether a lifeless object may not in fact be animate—and more precisely, when this doubt only makes itself felt obscurely in one's consciousness....
>
> A doll which closes and opens its eyes by itself, or a small automatic toy, will cause no notable sensation of this kind, while on the other hand, for example, the life-size machines that perform complicated tasks, blow trumpets, dance and so forth, very easily give one a feeling of unease. The finer the mechanism and the truer to nature the formal reproduction, the more strongly will the special effect also make its appearance.[9]

Sigmund Freud drew heavily on Jentsch in 1919 when he wrote "Das Unheimliche" ("The Uncanny"). He unpacks the word through etymology, *Heimlich* meaning "familiar," "native," and "belonging to the home," a word that might be translated as "homelike" or, in an old-fashioned sense, "homely." A haunted house therefore would be an *unheimlisches* house—something uncanny that is unsettlingly strange. Unlike Jentsch, Freud finds most of his examples in fiction, in myths, fairy tales, and ghost stories. Jentsch merely mentions Hoffmann; Freud spends pages retelling "The Sandman" and other Hoffmann stories. Freud being Freud, the loss of eyes is read as a fear as castration: Oedipus, he notes, blinded himself for his crime.

Both discussions of the familiar made strange emphasize that virtually everybody has had an experience of the uncanny, although the particulars vary so greatly from person to person and from incident to incident that no thorough discussion can handle all cases. Writers of fiction have the privilege of picking and choosing from among the universe of examples to focus on those that appear to produce near-universal *frisson*. Freud expands on Jentsch by dealing not merely with an artificial humanoid but a near-exact copy of a human, a double. Using Otto Rank's 1914 essay "Der Doppelgänger" ("The Double"), Freud provides a plotline to generations of future writers: "[In mythology] the 'double' was originally an insurance against the destruction of the ego.... [In modern society, f]rom having been an assurance of immortality, it becomes the uncanny harbinger of death"[10]—death not just for the original but also for the double. Frankenstein and his creation, Nathanael and Olimpia all die at the end of their tales. For decades, robot stories followed this model, until the technophiliac optimism of the 20th century demanded happier endings. Machines can always be rebuilt, writers realized, and a well-cared-for machine is effectively immortal. Ordinary parents want their children to be better than them; god-like creators want their creations to be better than anybody.

Tragedy being endlessly attractive to 19th-century dramatists, "The Sandman" dazzled with its possibilities. *Les Contes d'Hoffmann*, an 1851 Parisian drama by Jules Barbier and Michel Carré, wove together several of Hoffmann's stories. In this opera, an actor playing young Hoffmann tells his fellow students tragic tales of his misbegotten loves, starting with Olimpia, while quaffing a sad glass of beer. Lacking the true Hoffmann's exquisite talent for atmosphere, the authors omitted the necessary touches of melodrama. These were supplied thirty years later by Jacques Offenbach, the king of comic operettas, turning at the end of his life to a serious opera. Keeping the drama's name, known in English as *The Tales of Hoffmann*, he gave each of the ladies a star turn in song, using musical highs and lows as a foundation for the love and loss inherent in the stories. Debuting at Paris' Opéra-Comique in February 1881, the show was an immediate success, if a puzzling and qualified one: Offenbach had died the previous October without receiving either his due praise or finishing the work. Despite its many faults, the opera was soon deemed a classic and revived over and over, a tribute to stagers' egos, each hoping that a new production could finally fill in the original's missing pieces.

The one constant in all incarnations of the opera is "Olimpia, the charming automaton, fashioned by the skillful craftsman Coppelius into the semblance of a living creature. Struck by her beauty, paying no attention to her want of intelligence, Hoffmann is enamored of the doll, and his heart breaks when the puppet falls to pieces," as the *New York Times*' Paris correspondent described it in a long but mixed review.[11] The opera received far better notices the next year when it opened at the Fifth Avenue Theater in New York and even better ones for an 1887 performance at New Orleans' French Opera House.[12] *Coppélia*, a ballet by Arthur Saint-Léon and Léo Delibes, happened to make its American debut that year. Coppélia was the name given to the automaton invented by Coppelius, who intends to use a magic spell on the man who is infatuated with the doll to transfer his life essence into her clockwork body. The National Opera Company toured the country from Boston to San Francisco to rave reviews. "The transformation of the automaton into the living 'Coppelia' was a wonderful piece of pantomime acting," said a Philadelphia reviewer.[13] The effect always killed: a whole subgenre of vaudeville acts developed from it (see chapter 4).

—⁓—

Writers are inherently indisposed to actual work; even in 19th-century Britain few rose from the working class. They were accustomed to hiring servants, so much so that one of the poorest writers, Karl Marx, had a live-in housekeeper, Helena Demuth, who moved in with Frederick Engels after Marx's death. Larger houses required a squadron of specialized servants, all of whom came with human wants and needs, competencies (or lack thereof), and often burning resentment of the inequities that the class system imposed. If machines served humans in industry, why not in the household, already invaded by machinery to reduce or improve household chores?

Over the rest of the 19th century in England and in America, numerous writers were inspired to transform the figurative into the literal and devise automaton servants more efficient, less temperamental and far better mannered than real-life servants. The hugely bestselling technological utopia *The Coming Race*, published anonymously in 1871 but soon acknowledged to be by Edward George Bulwer-Lytton, first Baron Lytton of Knebworth, includes multiple references to automaton servants merely as background, because, like flying machines and electricity, any advanced society would naturally include them. Contrarily, a more ambiguous utopia, Samuel Butler's *Erewhon*, published anonymously in 1872, banishes all machines entirely, for otherwise their superiority would displace humanity. These contrasting views of the future—which have persisted until the present day—were both logical extrapolations from contemporary discussions of evolution, as Butler made clear in an essay, "Darwin Among the Machines," whose essence he incorporated into his book:

> What sort of creature man's next successor in the supremacy of the earth is likely to be. We have often heard this debated; but it appears to us that we are ourselves creating our own successors; we are daily adding to the beauty and delicacy of their physical organisation; we are daily giving them greater power and supplying, by all sorts of ingenious contrivances, that self-regulating, self-acting power which will be to them what intellect has been to the human race. In the course of ages we shall find ourselves the inferior race. Inferior in power, inferior in that moral quality of self-control, we shall look up to them as the acme of all that the best and wisest man can ever dare to aim at. No evil passions, no jealousy, no avarice, no impure desires will disturb the serene might of those glorious creatures. Sin, shame and sorrow will have no place among them. Their minds will be in a state of perpetual calm, the contentment of a spirit that knows no wants, is disturbed by no regrets. Ambition will never torture them. Ingratitude will never cause them the uneasiness of a moment.... We take it that when the state of things shall have arrived which we have been above attempting to describe, man will have become to the machine what the horse and the dog are to man.[14]

Most satiric works were decidedly lighter in tone. In 1865, Charles H. Bennett, a noted British writer and illustrator of children's books, joined the staff of *Punch*, the humor magazine, as a cartoonist. His book *The Surprising, Unheard of and Never-to-Be-Surpassed Adventures of Young Munchausen* also appeared that year in London and New York to continue the series of over-the-top tall tales attributed to the titular German nobleman. Bennett's book is barely coherent, a mere compilation of ideas and notions flowing past the reader like the crawl at the bottom of a cable news show. At one point he stops the book to provide a description of Munchausen's ingenuity. As a little boy (with emphasis on "little"), he invented a cast-iron schoolteacher who could not only instruct in every field but also maintained such order that the floggings every schoolboy dreaded were never required. A few years later, he furnished an elaborate upper-class household with a flotilla of automated servants whose virtues lay in what they *didn't* do:

> A BUTLER who bottled in, laid down, took up, and put on the table at the right moment the most delicious wines known in Europe, without ever tasting or spilling a drop.

A FOOTMAN, with calves and irreproachable manner, who waited at table, cleaned plate, and adorned the hall-door without either making love to the cook or flirting with the housemaid.

A COOK, equal to Francatelli and modest as Soyer, without a trace of bad temper.

A HOUSEMAID, could sweep, scrub and dust without insolence or disdain.

A SCULLERY MAID who was clean.

A PAGE who never cut off one of his hundred buttons.

A GROOM who used good language, and never stole the oats, because there were none to steal.

A HORSE which never ate, drank, jibbed, reared or stumbled.[15] [capitals in original]

Frederick Beecher Perkins, of the celebrated New England Beecher family, stayed busy as lawyer, librarian, editor, bibliographer, and writer. He also abandoned his wife and children, leaving them impoverished (no doubt one of the reasons his daughter Charlotte Perkins Gilman became an ardent feminist and wrote *Herland*, an all-female utopian novel). *Devil-Puzzlers and Other Studies* was a collection of short stories Perkins hadn't been able to sell elsewhere, though he was in 1877 enough of a name to get the prestigious firm G. P. Putnam's Sons to publish it. "The Man-ufactory" is perfectly described by its title. Here, as with Joseph Marie Jacquard's punch cards, automatic type-setting machines use a series of perforations—in this case, notches—that can be grabbed or not by a preset code, another form of machine programming. A Washington inventor adapts this notion and installs it into gutta-percha (rubber) manikins to control their speech and movements. His factory then churns out ministers of every possible denomination into which canned sermons can be loaded like tunes into a player piano. More insidiously, he copies the heads of notable lecturers onto his bodies and sends them out, as many as five at a time, to deliver the same insipid speech in different cities simultaneously, multiplying his earnings. His ultimate plan is to automate Congress; politicians do nothing other than speechifying, and his tireless creations can deliver a whole year's worth of speeches to one another in a closed back room and not inflict them on anyone else.[16]

The first use of Edison's phonograph in an English-language automaton story parallels Perkins' story. Published anonymously in the *Detroit Free Press* on July 20, 1884, and then widely syndicated, "The Clericomotor" is the perfect answer to a small church's need for an up-to-date replacement pastor. The Reverend Dr. Dummeigh stands, gestures grandly, and speaks with "a powerful, but well-modulated voice." All that is needed is to swap out the phonos with his sermons and have a small boy constantly crank the apparatus, which he can do hidden behind the altar. Small boys are symbols of eternal mischievousness in 19th-century stories. This one decides to find out what happens if he turns the crank backward. Time doesn't reverse, but the order of the sermon does until the reverend tries to go before the beginning and explodes.[17]

Speaking of Putnam's, its then owner, George Haven Putnam, indulged himself in 1894 by publishing in hardcover a short story of his own (as by "G. H. P.") titled *The Artificial Mother: A Marital Fantasy*. A father whose wife has just had their eighth and ninth children, twins, laments that his brood demands every minute of his wife's time. All the infants really need is to be gently rocked, he reasons, and a machine that looks like a mother could do that just as well. He buys a manikin, has a painter make her look real, and fits in clockwork to make her move. When the mother finds her babies in the automaton's arms, she is not pleased. But it is all a dream, one that nevertheless should be applied to the real world.[18] (Though rare in robot stories, the dream device truly was overused in fiction of the period.)

For those not as global in their thinking, the gap between the human and the mere machine remained a sting in the tail. Women perhaps saw this more clearly than men, since their contributions to society had been deprecated by men throughout the Victorian era. They struck back with stories that spoofed the sheer folly of the substitution with an amused condescension that foretold a million male schemes in a million sitcoms to come. M. L. Campbell's "The Automatic Maid-of-All Work: A Possible Tale of the Near Future" appeared in the July 1893 *Canadian Magazine*. Nothing is known about the writer, although the story is told from a woman's point of view, a wife whose inventor husband has created a maid that will take over all the household chores. Eschewing the uncanny valley, Campbell embraced farce: "It was a queer-looking thing, with its long arms, for all the world like one of those old-fashioned wind-mills you see in pictures of foreign countries. It had a face like one of those twenty-four hour clocks, only there were no hands; each number was a sort of electric button."[19] (Push-button preprogramming appeared far earlier than computers did.) A whiz at every household chore, the maid has a fatal flaw: after being programmed to do a task for a set time, the machine will continue to do that chore with superhuman strength until the time has elapsed, no matter what. The entire house is nearly destroyed by the end of the story.

A nearly identical tale appeared in the December 1899 issue of *The Black Cat*. "Ely's Automatic Housemaid," one of the last stories by the veteran Southern writer Elizabeth W. Bellamy, addresses the servant problem, that bane of *fin de siècle* upper-middle-class Americans who could no longer find an endless supply of cheap, yet highly capable, labor to act as household help. Like Campbell's family, Bellamy's replaces inferior (i.e., ethnic) servants with supposedly superior mechanisms, the ElectricAutomatic [*sic*] Household Beneficent Geniuses of an eccentric inventor. The new cook is better than the old cook, and so is the housemaid, but they, too, are pre-timed and cannot be turned off, which leads to a disaster as they battle to the death for the one broom in the house. Bellamy had the lightest touch and best comic imagery. So popular was her story that *The Black Cat* printed a direct rip-off the next year, awarding W. W. Stannard a minor prize in a short-story contest along with publication of "Mr. Corndropper's Hired Man" (October 1900), whose only difference was that its automatons replaced inefficient farm laborers.

—⁊⁊—

Similarly, satirists told stories about robots replacing humans in social settings, a few programmed words and skills being all that were needed to get by in the formalized social structures of the Victorian age.

> [The business of Nicholaus Geibel was mechanical toys.] He made rabbits that would emerge from the heart of a cabbage, flop their ears, smooth their whiskers, and disappear again; cats that would wash their faces, and mew so naturally that dogs would mistake them for real cats, and fly at them; dolls, with phonographs concealed in them, that would raise their hats and say "Good morning, how do you do?" and some that would even sing a song.
>
> But he was something more than a mere mechanic; he was an artist.[20]

This cautionary tale by Jerome K. Jerome, later retroactively titled "The Dancing Partner," appeared in the English magazine *The Idler* of March 1893 and was incorporated later that year into the book *Novel Notes*. Jerome lived at a time when the near-perfection of imitation mechanical figures was a standard tourist attraction. The fictional Geibel's inventions were at most only slight exaggerations of a multitude of brilliantly crafted toys, the late 19th century being known as the Golden Age of Automatons.

When the girls in Geibel's town jokingly call for a clockwork dance partner who will never tread on their toes, or tear their dresses, or mop his face with a handkerchief (again a robot prized for lacking human failings)—and whose sum total of stock phrases can fit onto a phonograph record—he rises to the challenge. The resulting automaton is the ultimate gentleman, tireless, superhumanly strong, perfect in every way ... except that he lacks an off switch, an oversight that neither Geibel nor the doomed dancing partner ever considered.

Either by a remarkable coincidence or a quick journey of the magazine to America, a columnist in the *New Orleans Times–Democrat* for March 19, 1893, called for automatons to act as harmless male escorts for unaccompanied women, since they, "fitted with a phonograph 'filled' with small talk from some society chappie, would satisfy all the demands of conventionality and be just as entertaining as the real thing."[21] Similarly, a story about a wholly fictitious Neptune Novelty Company's Automatic Hugger appeared in newspaper syndication in 1896, touting its phonograph-equipped product as the perfect gentleman to lure women to one's establishment for a delightful novelty.[22]

A feminist tale by Alice W. Fuller, "A Wife Manufactured to Order," in the July 1895 issue of *The Arena*, starts with a typical forty-ish bachelor of the day. Though growing older, he immediately dismisses the thought of settling down with Miss Florence Ward. "She unfortunately had strong-minded ways, and inclinations to be investigating woman's rights, politics, theosophy, and all that sort of thing. Bah! I could never endure it."[23] Better by far is a waxen figure imbued by an inventor with beauty that entrances him, Hoffmann-like, at first viewing and a set of phonos with stock phrases of love barely more articulate than Olimpia's monosyllables. Males are barely above robots in their fixed simplicity.

Fuller's story reinforces the contemporary view that machines are not as adaptable as people and will repeat themselves mindlessly. When the man's business is wiped out by a financial panic, he discovers that he is an emotional infant without his place as a success in male terms. He contritely rushes back to the patient Florence and pleads with her: "I have learned my lesson. I see now it is only a petty and narrow type of man who would wish to live only with his own personal echo. I want a woman, one who retains her individuality, a thinking woman. Will you be mine?"[24] She will, although the reader is free to wonder about the long-term viability of that match. Walker undoubtedly hit a contemporary nerve. *Stone* magazine reviewed the story in its August 1895 issue, saying, "Since so many confirmed bachelors seem to demand something of this sort Miss Fuller's [bright and amusing] story will obtain a wide reading among the ladies."[25]

The senior English master at Leys School in Cambridge would today be an unlikely candidate to write a more melodramatic version of Fuller's story; at the time, however, this form of science fiction was fit for mainstream publications. Ernest Edward Kellett wrote serious essays and criticism but in 1900 burst out with a collection of fantastic short stories, titled *A Corner in Sleep and Other Impossibilities*. One of those impossibilities was "A New Frankenstein," reprinted in the June 1901 *Pearson's Magazine* under the title of "A Lady Automaton."

Arthur Moore is a young inventor as brilliant as Young Munchausen, or at least Thomas Edison, whose phonograph he has perfected as both a recorder and a playback machine ("Fancy what a preventive of crime a phonograph fastened on every lamp-post would be!"). He accepts a challenge to build a sort of thinking phonograph, one whose prerecorded talk could serve as an unchallenged substitute for conversation. Before you can say "Pygmalion," Moore introduces the narrator to "the most beautiful girl I had ever seen: a creature with fair hair, bright eyes, and a doll-like childishness of expression."

Amelia is, of course, an automaton that, like Shaw's Henry Higgins, the inventor wants to introduce to society, where a few small skills make a young woman more than acceptable: "I have taught her French—drawing-room French, I mean—and three songs. She can enter a room, bow, smile, and dance. If with these accomplishments she can't oust the other dolls and turn them green with jealousy, I am much surprised." Amelia's shortcoming is the inverse of the one Perkins identified: his clerics needed cranking but could say anything recorded to disc, whereas Amelia's limited repertoire forces her to answer similar stimuli with similar responses. When two men propose to her in the same terms, she says yes to both. The result is disastrous.[26]

—〰—

Lyman Frank Baum led a life absurdly stereotypical of the striving American of the 19th century, with temporary successes in half a dozen businesses, followed by devastating failures, after which he would pick himself up and try again. His turning point came as a designer of store window displays. Never content to work on only one project, in his spare time Baum dabbled in writing stories and poems for children. Illustrated by artist William Wallace Denslow, *Father Goose: His Book*, a collection of nonsense rhymes, sold 75,000 copies in 1899, a marked success. Baum's publisher, George M. Hill, wanted more of the same, so when Baum pitched a novel instead, he was told that he'd need to supply all the printing plates—a huge expense. To ensure success, Baum and Denslow worked together to create the equivalent of the brightly lit, infinitely enticing, cram-packed, riot-of-color shop windows that Baum advocated, calling it *The Wonderful Wizard of Oz* (1900). Twenty-four color plates and 130 text illustrations (each tinted to match the color scheme of the Ozian area they were located in: blue, green, or yellow) set an almost unbreakable standard for children's books of the era. The pictures flowed across pages and crept behind text, making the volume a forerunner of the graphic novel. Baum collected 202 reviews of the book, only two unfavorable, yet it didn't come close to matching *Father Goose*'s sales from the previous year. Then something clicked. *The Wonderful Wizard of Oz* sold 780,000 copies in the next few years, defining it as the archetypal American fairy tale and spawning an Oz industry of merchandised products and media offshoots like nothing seen before in American publishing.

Oz clung to Baum like an albatross. Though suddenly famous and wealthy, his auctorial half had ideas for multitudinous fairylands that he plunged into writing, to little response. Readers wanted Dorothy, in Oz, with her friends, and he yielded. Only Baum's early death from a lifelong heart condition at the age of 63 stopped the Oz series at fourteen novels, two of which appeared posthumously.[27]

Two of the major recurring Ozian characters are metal men, as existentially different in their relationship to humanity as Dorothy and Toto. The first and most familiar is the Tin Woodman. Nick Chopper (his name as of the second book) is cursed by the Wicked Witch of the East for wanting to marry her servant girl. She enchants his axe so that he keeps nicking himself, lopping off in succession his legs, his arms, and his head; he ends by cleaving his torso in two. Each time he hies to the talented tinsmith (whom we later learn is named Ku Klip) to replace the lost part with an excellent substitute made of tin. Ku Klip has no problem keeping Nick's identity and memory intact in his long-nosed, bald dome, but he has lost his heart and begs the Wizard for a new one.

In *Ozma of Oz* (1907), the third book, Dorothy again comes across a frozen metal figure in her wandering, this one made of copper. He is explicitly a machine, crafted from scratch, as an advertising card attached to him proclaims:

SMITH & TINKER'S
Patent Double-Action, Extra-Responsive,
Thought-Creating, Perfect-Talking
MECHANICAL MAN
Fitted with our Special Clock-Work Attachment.
Thinks, Speaks, Acts, and Does Everything But Live.[28]

Throughout the Oz books, the Tin Woodman is treated exactly like a person who happens to have his body transformed (a fate repeatedly suffered by Ozians who, after all, live in a fairy land full of beings who know magic), while the clockwork man is not alive at all and must be wound like a grandfather clock to function, the sound made by his key giving him the name of Tik-Tok. Both Denslow and John R. Neill (who replaced Denslow as illustrator as of the second book) drew the Tin Woodman as a series of vertical lines, a cylinder head over a battery-like torso, with thin arms and legs; his hat is an inverted funnel leading to a point. By contrast, Tik-Tok is all spheres, a spherical head over a huge spherical torso with a hat like a brimmed beanie. Tik-Tok, oddly, is the one with hair on his head as well as a huge mustache. They are a physically mismatched comic team, like Laurel and Hardy or Abbott and Costello. The point is rammed home in *The Road to Oz* (1909):

> You could love the Tin Woodman because he had a fine nature, kindly and simple; but the machine man you could only admire without loving, since to love such a thing as he was as impossible as to love a sewing-machine or an automobile. Yet Tik-Tok was popular with the people of Oz because he was so trustworthy, reliable and true; he was sure to do exactly what he was wound up to do, at all times and in all circumstances. Perhaps it is better to be a machine that does its duty than a flesh-and-blood person who will not, for a dead truth is better than a live falsehood.[29]

Once wound up, Tik-Tok talks in what would become known as the robotic voice (seemingly a Baum invention), his words "uttered all in the same tone, without any change of expression whatever," indicated in the text by syllable breaks: "From this time forth I am your o-be-di-ent ser-vant. What-ev-er you com-mand, that I will do will-ing-ly."[30]

Ozians are notoriously cantankerous and independent. Many of the problems Baum creates for his wanderers everywhere in the series are sheer contrivance based on the mulishness of his characters. Tik-Tok himself (always referred to as "he" rather than "it") is not even from Oz. After a shipwreck, Dorothy was washed up in the Land of Ev, across the desert from Oz, where she first encountered Tik-Tok. They think differently in Ev; it wouldn't occur to an Ozian to manufacture machinery. Oz doesn't have sewing machines or automobiles either: those were advances from Dorothy's America left behind as irrelevant in an unchanging pastoral fairy world. Ozians cannot die, except by extreme violence. Dorothy and the other visitors from America are mortals and so always in peril. Robots like Tik-Tok inhabit a third world, outsiders that may be immortal but depend on others for life and motion. Robots are always destined to live outside of human societies as well. Living, thinking machines may be useful, but they cannot find a social niche of their own.

Tik-Tok of Oz (1914) features a giant that Smith & Tinker made out of plates of cast iron, built for the Nome King as a guard on the road to his kingdom. Baum's first novel, written as *Adventures in Phunnyland*, published in 1900 as *A New Wonderland*, and retitled in 1903 *The Surprising Adventures of the Magical Monarch of Mo and His People*, also features a cast-iron giant mechanical man built for a king. Both guardians fail because they cannot think. They are programmed to do one task and cannot adapt when change is needed. People are more adaptable.

In *The Tin Woodman of Oz* (1918), Mrs. Yoop, a giantess who knows Yookoohoo magic, transforms the Ozian boy Woot into a green monkey, the Scarecrow into a stuffed toy bear, the fairy princess Polychrome into a canary, and the Tin Woodman into a tin owl. They do not like their new shapes, but they talk, think, and behave exactly as they did before. Later in the book, Baum presents us with the ultimate test of personhood: The Tin Woodman is on a quest to find the servant girl he abandoned when he lost his heart. On the way he encounters another rusted tin man who looks exactly like him, except that he wears a military cap. This other tin man also loved the Witch's servant girl; he, too, was enchanted so that his sword cut him to pieces; and he likewise went to Ku Klip for a replacement tin body. Captain Fyter (the pacifistic Baum had always before made his soldiers buffoons, but this was 1918, the United States was at war, and his son was in the army) may have been constructed in an identical fashion, but his mind is forever his. Or is it? When the two rival suitors finally find Nimmie Amee, the servant girl rejects them both. She is married now—to Chopfyt. If the name sounds like a portmanteau, it's because Ku Klip, remarkably careless with spare parts, threw the cut-off pieces of the tin men into a common barrel and years later (nothing can die in Oz, not even parts) patched them together. Chopfyt is surly, perhaps because his pieces do not sit comfortably together. Chopfyt has the Captain's head; Nick Chopper's resides in a closet at Ku Klip's smithy and is perfectly satisfied to sit in the dark and not think. This situation violates all we know about the continuity of personality in the tin men, but the story hurriedly continues on and existential conundrums are quickly forgotten.

—⟁—

These early stories all share a similar comic premise: that a humanized automaton would be a direct replacement for a specific type of perceived inferior. As social satire, they could be set in the here-and-now even if the technology behind the machines was not yet feasible. Machines had dueling allegorical aspects as the century turned. For all their transformative possibilities, despite their symbolic nature as daily improving representations of progress in an era of humanity wrest from farms into enormous industrial landscapes, machines were often humorously balky, inefficient, badly designed, mindless, cantankerous, and maddening—yet capable of feats of strength, endurance, and tirelessness that outshone the John Henrys who died striving to compete. Mechanical men—none of the early stories used the term *mechanical women* even when the robot was female—provided the ultimate symbol of the future of machinery, and they could be manipulated by proponents and antagonists alike to bring to life visions of those futures. Just as the 19th century, the age of the individual, morphed into the 20th century, the age of mass humanity, the image of what we now think of as robots expanded from replacing the individual laborer into replacing all human labor—even all humanity itself. Though the subject seems tailor-made for a socialist futurist like H. G. Wells, history made one of its frequent zigs and handed the honor off to a pulp writer extraordinaire, the self-proclaimed "Fiction Factory," the mind behind *Plotto: The Master Book of All Plots*, so prolific that he was termed "the man who deforested Canada."

William Wallace Cook was born in 1867 and ran through the usual miscellany of jobs before selling his first story at the age of 22. In a few years, he had doubled his mundane paycheck by prodigious application to the earliest of the 25 primitive typewriters he would experiment with. In *The Fiction Factory*, a how-to manual/self-congratulatory hagiography he wrote under the name of John Milton Edwards, he boasted that he had "written two 30,000 word stories a week for months at a time," scattering at least 18 pseudonyms across the pages of magazines, newspapers, and nickel and dime novels.[31] That level of production broke his health—rheumatism and typing don't mesh—and he shared in the ubiquitous 19th-century writer's malady of investing in mines that produced no lode. Like Baum, he bounced from luxury to destitution and back and finally found a field in which he could be original, the one endeavor to which he signed his real name so he could bask in the deserved fame that it brought him. He invented American science fiction.

Cook's novels (most serialized in popular magazines so that his name appeared in tables of contents for years on end) included *Adrift in the Unknown* (1904), actually set on Mercury; *A Gift from Mars* (1906), featuring a stone that transmutes iron into gold; *The Eighth Wonder* (1906), describing a giant electromagnet that straightens the Earth's tilt; *Around the World in Eighty Hours* (1920), about a super-airplane race; and *Marooned in 1492* (1905), possibly the first story about traveling to the past to change history.

Cook's first, most reprinted, and arguably most influential work in the field splashed across the pages of *The Argosy*, the first all-fiction pulp magazine, from July through November 1903. *A Round Trip to the Year 2000; Or, A Flight Through Time* reads like nothing Wells ever wrote, or Hoffmann, or Jerome, or any of the genteel story writers of the 19th century. It is completely modern, a breathless adventure romp chronicled via immediate humorous banter rather than a third-hand story told through exposition from a later, safe, vantage point. Cook wrote entirely in cliffhangers, with ridiculous coincidences, last-minute escapes, chases across rooftops, a sublimely evil villain, loyal friends, a besotted female, and a knack for whisking his hero into a new scrape just before he had to explain away the last adventure. *Round Trip* would have been converted into a weekly silent serial if only it hadn't appeared a decade too early.

Briefly, Emerson Lumley has been hypnotized into committing a crime, for which he is being hounded by Detective Kinch, a bulldog with far more stick-to-itiveness than Inspector Javert. Miraculously, Lumley stumbles across an inventor's Time Coupé and is whisked to the year 2000. There, in a delicious bit of satire, he finds a colony of refugees from 1900, part of a vast army of time travelers dotted across the planet, like Edward Bellamy's Julian West, whose awakening day is a subject for public acclaim. Despite the ever-present threat to haul Lumley before the Explainer-General for a 50-page lump of exposition about how the future works (the bane of didactic utopias like Bellamy's *Looking Backward: 2000–1887*), the reader is saved by Cook's inventiveness in showing, not telling, us the future as background to Kinch's appearance and continued maniacal pursuit of Lumley. We get all the standard miracles of future societies—meals delivered as vapors, dragonfly-winged air transports, unisex clothing, instantaneous communications—plus the force that powers the future society: the all-purpose and ubiquitous mechanical workers known as muglugs.

Muglugs pilot the airships, clean the houses, staff the factories, haul the goods, and act as butlers to shield humans with iron umbrellas (debris dropping from the air ships being the everyday danger of 2000). There's no complaining: the Air Trust—trust being the then-common word for monopoly—owns everything, including the air itself, for whose use people are metered and charged accordingly. The million muglugs in Manhattan are controlled by thought transmission from the brainiest of all modern men, known as the Head Center. When one Head Center is run down by overexertion, a new one is chosen in the only feasible pulp magazine fashion: the two contenders thought-power twelve-foot gold- and silver-plated muglugs in a fistfight to the death.

One of the contenders for the title, named Tibilus, has strong opinions about the value of the muglugs, with the story turning serious for a moment.

> "[The Head Center] saves labor for countless thousands of human beings, but the saving of labor results in a demoralizing expenditure of idleness." …
>
> "Every man ought to do a certain amount of labor," commented Lumley, stifling a yawn.…
>
> "Without a certain amount of labor, Lumley, there can be no real happiness; and I aver that the muglug, as a labor-saver, is the greatest success as well as the greatest misfortune, ever flung in the face of a civilized people. Does that sound like a paradox? Listen: Any machine that lightens a laborer's toil is a blessing, but any machine that eliminates the laborer is a curse."[32]

We get the point, sufficiently so that we are saved the labor of reading chapters' worth of exposition on machines putting workers out of their jobs. Lumley simply falls asleep as Tibilus talks.

Cook may have sympathized with timely political causes, but he was merely a pulp dilettante, who buried his message under ludicrous action. The winning Head Center carries out his threat of destroying the evil muglug-based culture by using the muglugs to destroy all modern conveniences and then each other. However, we don't see the consequences because Lumley manages to jump into the Time Coupé and return to 1900. Adventure triumphs.

For better or worse, Cook's style would set the template for American science fiction. Immensely readable, the definition of "page turning," pulp fiction emphasized play, movement, color, exoticism, and imagination over character, theme, naturalism, and depth. The futures of American science fiction were the contemporary world hypertrophied. Whether utopian or dystopian, earth-bound or cosmic, the worlds of the future were those of the American middle or upper-middle class. While sympathies abounded for the worker, the working man made for a poor adventure hero, being unable to afford to take off on adventures and lacking the education to operate muglugs or the income to afford the parts for a Time Coupé. Workers nevertheless seemed to relish reading about the experiences of the worldly and well-to-do, and they flocked to the pulp magazines. While intellectuals from the worlds of writing, unions, and politics would tout the danger of mechanical workers replacing human workers for the next century, the public loved muglugs, automatons, and clockwork figurines too much to worry about their far distant replacement. Let the intellectuals lament. A future containing mechanical men stayed a triumph of human ingenuity, a possibility consummately to be desired.

—〰—

The 20th anniversary issue of *ESPN The Magazine*, cover dated April 23, 2018, spiced the articles with bold predictions about the future of sports in 2028—except where robots were concerned. Considered as a possible time-saving measure for overlong games, robots replacing players weren't forecast until at least 2048.[33] That's only 30 years from the time the article was written. Writers for the *Pittsburg Press* in 1905 were less optimistic; they foresaw robot teams battling each other, but not for another century.

> Chicago, January 1, 2005—It is estimated that 100,000 spectators witnessed the second game of the series between the Chicago and London teams of the International Baseball League. London won, 2 to 0.

The victory was due chiefly to the superiority of the new pitching device invented by T. A. Edis-onon, the greatest inventive genius of the century. The Chicago automatoms could not decipher the riddle propelled by the machine.[34]

Part of a New Year's Day spoof page, the reporter took the huge gains in interest and attendance—and salaries—in baseball since the start of the American League in 1901 and extrapolated them as far as the imagination could stretch. The "automatoms" (an unusual alternate spelling) looked like cylinder pumps with baseballs for heads, the bodies resting on four-wheel chasses for speed. A faux Thomas Edison inevitably got the credit for these marvels, although it's odd to see him inventing for the London team. The writer should get credit for being the first to give robots machine names, anticipating R2-D2 and C-3PO by almost three-quarters of a century.

The Londoners bunched their hits in the second inning and made their runs. R-Q, the first batter, connected with one of the Chicago auto's fast ones and reach reached first before M-B in center field could head it off. L-D which followed sacrificed. C-D doubled to left and both machines scored a moment later when K-V ripped off a hard three-bagger over M-B's upper mechanism.[35]

Transportation was no problem, either.

The teams will be shipped from Chicago Wednesday night via the Chicago and New York Pneumatic Tube Transit Co., beginning a series of three games in the English metropolis Thursday afternoon.[36]

The New York Yankees and the Boston Red Sox will fly to London to play official games in the 2019 season, the first ever in Europe, and the flight will take much longer.[37] Predicting futures is easy; it's the timing that's hard.

3

Is It Mechanism or Soul?
Robots on the Stage

On December 22, 1795, *The Times* of London ran an advertisement declaring that at the Mechanic Theater on the Strand "will be exhibited the ANDROIDES" (all typography as in originals). "These much admired Pieces of Mechanism … not only appear to imitate human actions, but appear to possess rational powers." As the wording implies, androides (also referred to in the piece as automatons) were not new, either in scientific society or on the stage. "MR. HADCOCK [*sic*] flatters himself the Androides will be found more curious than any thing of the kind ever before offered to the public, as the Principles of Action are entirely new." Three years later, Mr. Haddock (the spelling that appears in all later advertisements) announced the closing of his current exposition, also at the Mechanic, of androides, "so well known that comments are unnecessary; suffice it to say, that the first scientific judges of mechanical power who see them, are not alone entertained, but astonished."[1]

Europe was far ahead of America when it came to automatons in the 18th century; the young country had had its industry deliberately suppressed by Britain to create a greedy market for English manufactured goods. Acts were imported from many countries, but an innate familiarity with English gave London acts an edge: performers in close, small theaters well knew the power of audience interaction. Automatons who would appear to engage in call and response with audiences made English a professional necessity. The *New York Evening Post* printed advertisements for such acts as early as 1803, featuring "*Robertson, so much applauded in London & Paris.*"[2] Interestingly, the May 26, 1820, *Evening Post* announced a Mr. Haddock bringing his act, "well known in the capitals of Great Britain," to America for what is suggested to be the first time.[3] Some performers spent their entire lives in the profession; this might be the same Haddock from 1795, or perhaps his son, a successor using the famous name, or else a blatant fraud trading on Haddock's fame (all possibilities occurring in stage history). In any event, Haddock's androides continue to pop up in newspaper databases for yet another decade.[4] The 1820 ad refers to him later in the same paragraph as Mr. Maddock. Getting the name of the act correct should be an advertisement's minimum responsibility; such typos are a remarkably common researcher's bane in sources from the 19th century.

Clockwork figurines in the shape of humans received the name *androides* (singular, despite the plural form; later backformed to *androide* and *android* for the singular, with *androides* and *androids* as plural) from the French *andréïde*. Chambers' *Cyclopedia* included the term as early as the 1727 edition as a synonym for *automaton*, both usages expanding from historic usages to contemporary popular culture well before the end of the century. Lifted out of the province of mechanical novelties for the very wealthy, automatons needed to pay their way. Acts competed against one another for novelty and spectacle. "Hadcock" boasted of half a dozen androides acts to fill out his bill. One had a skill remarkable even among its fellows: "the LITTLE CHIMNEY SWEEPER will … give the usual cry of 'Sweep!' several times." Robertson promised even more: "A wonderful *Speaking Automaton*, suspended by a string in the middle of the room, which will make suitable answers to the questions that may be proposed."

Massive amounts of skill, ingenuity, mechanical genius, and contempt for the suckers produced an advanced seminar in artful lying for such acts. Until the invention of the phonograph, we can assume that all talking automatons were frauds, as skilled in plying the audience as their mechanisms.

Audiences as far back as Haddock's time surely suspected a ventriloquist in the room and just as surely admired a good one. Ventriloquists emerged as stars in their own right, traveling with a troupe of characters that included automatons. In New York in 1831, audiences could have marveled at Mr. Seaman, "the celebrated ventriloquist from England," whose performance promised that "WONDERS WILL NEVER CEASE!" when he engaged in humorous dialogue with "Tommy Rhymer, the Automaton Figure."[5] An 1840 ad has him adapting his act for the natives, with an "automaton revolutionary soldier" and his "Yankee lady wife."[6] Down in Natchez, Mississippi, the juvenile ventriloquist Master Platt proposed to introduce "*Little Hercules*, the Speaking Automaton" in 1838.[7] Mr. Wyman (who by the end of his long career had been promoted to "Professor Wyman, Ventriloquist and Wizard") introduced his "much admired Automaton Speaking Figure" to the big time in New York in 1840.[8] He was accompanied in New York by Miss Wyman, "the celebrated Lady Magician."

Signor Blitz—the mysteriousness of the actor's craft was apparently emphasized by the lack of a first name—dominated the field, which continued to spawn references after his death. An anecdote that appeared newspapers in 1900 mocked the rough handling that baggage received at the hands of train porters. When one such worker placed a long box wrong way up, a stifled boy's voice came from inside, pleading to be placed on his feet. They instead laid it flat on the ground. This time the voice complained that it was lying face down. A great to-do ensued, with the trainmen berating the box's owner for trying to cheat the company by not paying for the boy's fare. The punchline was that the owner was the famed ventriloquist Professor B, traveling with his automaton Bobby.[9] The story is obviously meant to be a joke, as no one attempts to open the box and rescue the trapped boy. How old the joke was, the editor probably didn't know. A newspaper favorite, that version was a line-by-line retelling of the same story found in the Wilmington, North Carolina, *Tri-Weekly Commercial* in 1853, giving proper credit to the still touring Signor Blitz.[10] Blitz, a ventriloquist and magician who enjoyed a fifty-five-year career on the boards and didn't retire until 1868,[11] had been publicizing himself with variations of this story for decades. One from 1840 has a horse playing a joke on his jockey to drum up business for an act that included "Automata Black Rope Dancers and Speaking Figure."[12]

―᠁―

Automatons as automatons were popular acts throughout this period. Peale's Museum & Gallery of the Fine Arts, which also booked ventriloquists, sat on Broadway, then as now the home of New York's theater district. In 1835, it advertised a spectacle typical of what audiences might hope to see imported from Europe's advanced mechanics:

> Just received from London, the most splendid exhibition of Automaton figures ever seen in this country before. It was known in England, by the title of *Maillardett's Mechanical and Musical Museum*. It consists of a musical lady, who plays on a finger organ a great variety of airs; Tight Rope Dancer; Necromancer, or Fortune Teller; Juvenile Artist, who writes in English or French, and draws most beautifully; Walking Figure; Gold Serpent; Spider; Lizard; and Siberian Mouse; Humming Bird, in a snuff box that warbles melediously [*sic*]; Napoleon Vase, of exquisite workmanship; Self-Acting Piano and Organ.[13]

The Olympic, also on Broadway, boasted of the "Extraordinary and truly WONDERFUL AUTOMATON! whose musical performances (in the opinion of the most distinguished Musicians of the city) may put at defiance, without any fear of rivalry, all the *Professors of the Trumpet* ever yet heard in this or any other country."[14]

Some European artisans certainly made extraordinary clockwork figures that played instruments, but the odds of finding them playing nightly on Broadway were small. Just as with ventriloquists and magicians, audiences willingly suspended disbelief for good acts while reserving the right to jeer at those that did not live up to the extravagant publicity. A revealing notice placed in the *North-Carolina Star* of Raleigh on November 27, 1812, makes this point. "STOP THE VILLAIN!/Fifty Dollars Reward," screamed the headline. A certain William Johnson had stolen John Smith's horse while skedaddling out of town in a hurry. Why did he need to leave? He had come to town under the pretense of exhibiting an automaton figure, which he could no longer do because his partner, "a Fellow by the name of Bud," had left him.

Fraudulent automaton acts must have overcrowded stages, for at least two authors used them for inspiration. "Mosco's Automaton" appeared without credit in *Chambers's Journal of Popular Literature, Science, and Art* on July 17, 1869. (*Chambers's* was a London magazine, but the tale sounds totally American and was quickly reprinted around this country.) The story is told from the viewpoint of Bill, who is hired by a traveling magician to pretend to be an automaton. Bill is an exemplar of the unreliable narrator, by nature so dim and emotionless that he doesn't get the joke of his deception: he thinks he's acting. London magazines definitely published American authors; *Belgravia: An Illustrated London Magazine* ran in May 1878 "An Automatic Enigma" by Julius Hawthorne, Nathaniel's prolific son. A suitor jilted after his girl accuses him of being "an owtomaton" leaves town, seemingly for good. Then "the miraculous, the mysterious, the supernatural, the incomparable Automaton!" appears to an overflow crowd, nods to the stunned young lady in the front row, and disappears along with the girl. Contemporary reviewers were cruel only because they hated the story rather than finding the premise incredible.[15]

When an act becomes such a public cliché that writers feel free to mock it, it usually sinks under its own weight. Automaton acts instead built on the publicity; theater-goers wanted to check these marvels out for themselves—audience participation at its finest. The golden age of automaton acts lasted from about 1888 through World War I, with three main categories of effects: ventriloquist acts with full-size dummies, live actors imitating performing automatons, and stunt people whose specialty was standing still and unblinking for inhumanly long periods. They were perhaps prodded by the enormous success of *The Tales of Hoffmann* and *Coppélia*, an opera and ballet based on E. T. A. Hoffmann's "The Sandman" (discussed in the previous chapter). After decades of strutting automatons ringing bells on the hour and miniatures dancing in music boxes—machines playing humans—humans now played machines, finding new graces in their stiff movements.

In "Aladdin," automatons "are placed before the footlights, their faces are expressionless and their postures awkward. When the music starts they begin to dance and when it stops they stop, leaving them in the most awkward attitudes imaginable."[16] "Around the World in Eighty Days" burlesqued famous actors of the day in a comic ballet, in which "the dance of the Automatons and the automatic prize fight are received with uproarious laughter."[17] Kate Castleton publicized her collection of skits and music, "A Paper Doll," with the sort of press agent invention that newspapers found irresistible:

> Her husband has had a large doll made for use in the show and on a recent evening it was sent home and left on a table to be examined by daylight. When the servant arose in the morning she removed the doll and placed it in a chair, when the thing squeaked the refrain of one of Miss Castleton's songs: "For goodness' sake, don't say I told you." The servant packed her belongings and left the house, which she believes to be an abode of devils and witches.[18]

Not formally based on Hoffmann but unmistakably descended from his work (and Shakespeare's) was *The Toymaker*, a comic opera by Edmond Audran. The plot makes use of the ever-popular confusion of doubles, specifically between the titular toymaker's daughter and Baby Doll 84, an automaton he has made in her likeness. He wants to marry off the daughter, and a novice at a monastery wants the dowry she comes with. Naturally, the young man falls for the automaton, but the real-life girl breaks the doll and marries him herself. The opera was a West Coast favorite for a decade starting in 1901. (Even lighter was "The Automatic Servant Girl," suitable for clubs and schools, a farce to the extent that one of the characters was named "Bob Funneigh." Written by Amelia Sanford in 1905, the one-act burlesque popped up repeatedly for two decades.)

—〰—

Back in vaudeville, ventriloquism experienced one of its periodic up cycles. Lieutenant Nobel, either a Swedish ventriloquist celebrated across Europe or a Dane from the Circus Varlete in Copenhagen (or maybe just a guy from Brooklyn: never believe a theatrical ad), toured the United States in 1896, wowing audiences with his ability to walk back and forth across a stage, arm in arm with a life-sized automaton, "placidly smoking his cigarette during all the performance."[19] Possibly some of the automatons that looked human were in fact human; who was going to tell the audience? The same may be said of William Young, who intended to tour with nine automatons, providing not just voices but song for all of them.[20] And while Trovollo started a year after Nobel with his walking, talking automatons, his act lasted until at least 1906. In the same 1896–1897 seasons, Charles Colby and Allie Way combined all three categories into a single act:

> Mr. Colby gives a clever ventriloqual performance, while upon the stage is apparently a life-sized doll, as motionless as an Indian cigar sign but as pretty as any baby's adored treasure. After the little colored figures have amused the audience for five or ten minutes, the doll performs a clever dance and as a finale falls upon the stage just as an automaton would fall. In fact Miss Way does the doll dance so cleverly that many in the audience are not quite positive that she is alive until she steps to the footlights to bow.[21]

The "New Acts" column in the August 22, 1908, issue of the weekly *Variety* (then one of several competing papers trying to cover the stupendously varied world of show biz) carried a review of a between-pictures "try-out" act at the Manhattan movie theater. William Gane's Automatic Minstrels, wrote Sime Silverman, *Variety*'s energetic founder, consisted of nine figures—an interlocutor (the straight man who sets up the jokes) and eight sidemen who responded with jokes and music. Gane's Manhattan Theatre, at Broadway and Thirty-Third, was no hole-in-the-wall nickelodeon—at 700 seats, it was large for its time and had recently been a legitimate theater featuring stage royalty like Sarah Bernhardt. Sime, who slipped into the back of the theater upon hearing of the new act, couldn't tell whether the joke tellers were real. It was "not yet known," he wrote, "whether the others were human or merely dummies." The real joke was on the audience, because "phonographs situated behind each of the figures supplied a couple of jokes and all the songs.... If the figures were 'dummies' they were very lifelike."[22] Sime thought that Gane might be the genius of all geniuses if the latter supposition were true: a nine-person performance for the cost of one actor!

The next week, *Variety* printed an outraged letter from "Bryon" Monzello accusing Gane of stealing his act. Monzello claimed to have originated "The Mechanical Minstrels" in Indianapolis in 1904. He also took out an elaborate half-page ad detailing the act:

> ACTION: As the curtain goes up all are standing. "Gentlemen be seated" figures, and men and interlocuters are seated. Usual "gags" by end men. Then in rapid succession follows tenor, baritone bass, character, coon and quartet numbers intermingled with "end men gags," giving a beautiful minstrel first part in 22 minutes or less.
> Each figure when introduced arises, bows, looks over the house, orchestra chords. Figure breaks forth in sound just as loud and louder than the human voice; just as clear and sweet as the world's greatest singers; when song is completed figure seats itself; in event of encore figure arises, bows and repeats. [capitals in original][23]

All this entertainment could be had for a mere $5,000—perhaps not so cheap after all, and probably too costly for anyone to snap the acts up. The ad is signed "Byron" Monzello—the only time that either Monzello or his Mechanical Minstrels appeared in *Variety*. Nor did Gane's Minstrels have a second act. In a nostalgic history published in 1951, then *Variety* editor Abel Green, with vaudeville historian Joe Laurie, Jr., would be able to refer to the "first (and last) All-Automatic Minstrels."[24]

Mechanical acts seems to have been code for marionettes, a close cousin to the dummies used by ventriloquists, but with a separate tradition. Like automatons, they came over from Europe. In 1877, advertising drew New Yorkers to their summer playground with notices like "REDMOND'S LONDON MARIONETTES. THE MECHANICAL MINSTRELS. SONG AND DANCE PERFORMERS. No visit to Coney Island complete without seeing these little wonders."[25] Till's Original Royal Marionette Troupe appeared two years later, featuring "an entire Automatic Minstrel Company, performing Pantomimes, Plays, &c, like human beings."[26] The Mexican Automaton Company, Bell's Royal Marionettes, Professor Zera Semon, and DeEsta's Mechanical Minstrels copied them.[27] Abraham and Strauss, the giant department store that was the gem of Brooklyn (then still an independent city), sought to outshine the Manhattan competition for the Christmas toy season in 1895. "THE LARGEST TOY EVER BROUGHT TO AMERICA," their ad screamed on the third page of the *Brooklyn Eagle*. A front page ad boasted, "An Automatic Ethiopian Minstrel Troupe made especially for us in Paris at great cost ... from Interlocutor to Brudders Bones and Tambourines."[28]

As each novelty spiced the acts with a bit of innovation, they necessarily expanded in number, range of instruments, and size, with the figures inevitably becoming life-size automatons before William Gane had his inspiration. The *Washington Post* for October 28, 1906, announced that "J. M. Leavitt has invented an automatic minstrel 'first part' with life-size figures. It will not be hard to mistake it for the real thing."[29] The *Oakland Tribune* described the act as "consisting off eight life-sized figures seated in conventional fashion under elaborately wrought and illustrated throne-like structures.... By ingenious mechanism the figures are made to go through all the motions of the old-time minstrel performer, while at the same time, by means of phonographs connected with them, they are made to sing and talk in a manner that is surprisingly lifelike."[30] (There is no indication of whether this is Monzello's act or yet another independent invention.)

To complete the circle, ventriloquists were inspired by the competition to make their figures get up and move. Alf Caum, "the Yankee ventriloquist," updated the Punch and Judy show with "twenty-four singing, talking and dancing automatons."[31] Jay W. Winton, fresh from five years in Australia, was "a ventriloquist whose figures do more than talk—they act and take part in his work with him."[32]

—⁓—

F. Howard Hill also crossed categories with his performance as Psycho. For more than a decade, starting in 1898, advertisements screaming in large type "A Man or an Automaton?" appeared across America, a sure testament to the public's gullibility: the earliest findable article on Hill gives the game away.[33] Snellenberg's department store in Philadelphia had a display of several full-size wax figures in a window (then the preeminent showcase for stores in an era when everyone strolled downtown streets to peer into the increasingly elaborate displays), one of which would "stand for half an hour on one foot, holding his features absolutely motionless and without winking his eyes or turning the eyeballs" before jerkily smoking a cigar or otherwise acting like a wax figure come to life. The article revealed the actor to be a young Englishman who had spent more than half his life as an artist's model, learning to keep his movements to an absolute minimum. A coat of waxy makeup helped to hide his features. According to a *Wilkes-Barre Times* reporter, "Men and women have stood for hours in front of the window and have made grimaces and gesticulated wildly in the hopes of throwing the party off his guard.... Women have made bets of money.... Men have argued and nearly come to blows." This article also introduced the trope that someone "ran a pin into his flesh without causing him to move a muscle," a yarn that would be repeated regularly.[34] When the appeal of this act trailed off—or possibly his muscles ached too much—Hill used his musical talents to imitate an automaton playing a pianola (i.e., player piano). He appeared again in shop windows, this time of piano stores, drawing crowds with hours of uninterrupted playing (and presumably selling pianolas by the score). As late as 1910, he was "revealed" again as "a Man—Not an Auto Man."[35]

As with many show business names, Hill's was stolen from an earlier and more famous act. John Nevil Maskelyne, the leading English magician and father of a line of stage magicians, introduced Psycho in 1875. A seated half-man automaton raised by a transparent glass cylinder played the card game whist with the audience. This itself was a takeoff on the 18th-century chess-playing automaton called "The Turk," a clever fake that concealed a dwarf in its undercarriage. People a century later knew all about the Turk and its many successors—hence the glass cylinder. As is usual in magic, what made it seem impossible was what made it work: compressed-air tubes run up through the glass powered the automaton's movements from under the stage.[36] American magician

Harry Kellar duplicated Psycho and later gave the apparatus to Houdini. Chess, checkers, and card-playing automatons became common circus and fairground attractions. One called Ajeeb started in Britain in 1865 and toured America until World War II, its success coming from its ability to hide full-sized chess masters.[37]

A more animated automaton act was offered by George H. Webster, supposedly from England but really from McKeesport, Pennsylvania.[38] In his first season on the boards as "Phroso," Webster dropped the pretense, as in New Orleans in 1902, when he finished by addressing the audience "in the most natural voice in the world."[39] He quickly learned that audiences wanted to be fooled (or at least consider the illusion a real possibility). For the next several years, he advertised the act under such headlines as "Is It a Man or a Mechanical Figure?/Phroso Has Set Temple Theater Patrons Guessing."[40] Phroso looked like a 6′4″ man in evening clothes but dragged an electrical wire behind him that served to warn off audiences from getting too close. The *Indianapolis News* looked askance at the performance:

> The "turn" it does is not an elaborate affair. It struts around with the stride of an especially tragic tragedian, and mouths words from what is purportedly a phonograph in its interior. But it answers its purpose well in puzzling the public and bringing many coins to its lord and master.[41]

Webster undoubtedly thought notices like this one gave the act little future, and in 1903 articles were referring to a Frederick Trevaillon as the act's inventor, who costumed himself in an elaborate soldier's outfit with a sword supposedly gifted by King Edward of England. Webster, meanwhile, improved the mystery with a bit of magician's trickery. Unveiling the act in London in 1904, probably under a non-compete agreement with Trevaillon, and then triumphantly taking it throughout the British Empire, Webster unpacked "Zutka" onstage from a box apparently 24 × 18 × 16 inches.[42] Zutka's miraculous appearance seemed to be sufficient as an act; no paper reports it doing more than swinging gymnastically from rings.

Imitators naturally followed, unless Webster simply changed the act's name again. Zutka lasted until 1905, but in 1907 "Yuma," a contortionist, emerged from a box 16 × 22 inches.[43] However, another "alleged automaton" named Yuma appeared in 1911 in London, where it was described as being "on the lines of 'Enigmarelle,' 'Zutka' and others of the 'Phroso' brand." Enigmarelle had the best name and probably the best act:

> Enigmarelle is a life-sized figure of a man that does everything but talk, and those who have seen the automaton, including many scientists of note, have pronounced it one of the wonders of the 19th century. The figure, under Frederick J. Ireland's direction, will walk, dance, write its name on a blackboard, smoke a cigarette, go through calisthenics and do other things almost exactly as a human being would, and were it not for a slight whirring sound, caused by the rapidly revolving motors, it would indeed be hard to believe that the figure was not that of a human being.[44]

Audiences might have expected Enigmarelle to be touted as a *20th*-century wonder, given that the extravagant notice appeared in 1904, but perhaps Ireland had stacks of old publicity notices in a warehouse.

Ireland provided superior stagecraft for Enigmarelle, designing the housing so that the arms and legs could be removed and the head opened. The *Washington Post* noted that clowns had long been dismantled in the same way and said it would only fool "the willingly credulous."[45] Among these were the editors of *Scientific American*, who produced a lengthy article in 1907 with the embarrassing title of "A Clever Mechanical and Electrical Automaton" before tacking on a note that a vaudeville manager had revealed the truth about there being a man concealed inside.[46] With the imprimatur of the scientific community, Ireland did better business than Webster until he, too, sold off the act. In 1910, "D. M. Rhoades, constructor of the first electric chair ever used in the world for the execution of condemned criminals … at the state institutions at Auburn, N.Y.," turned his electrical expertise to running Enigmarelle and kept the act humming along with new feats like driving cars until World War I brought it to an end.[47] (The actual electrician was Edwin F. Davis, but who would be so unkind as to check?)

Of all the imitators of imitators, Fontinelle may be the most derivative. Its press releases boldly copied Enigmarelle's look, act, and language. In its first season, 1905–1906, potential audiences read that the device "walks upstairs, writes, rides a bicycle, and does almost everything but think."[48] Just as trustworthy was the notice in 1907 describing the inventor, Dr. Joseph Farrell, as English and having presented Fontinelle to King Edward in 1903 (apparently the British monarch had little to do besides watching automaton acts).[49] No record supports an English start. More tellingly, Iowa City would welcome its hometown boy back with a hearty greeting in 1909, calling the automaton (which was played by William Sutton) "truly astonishing and perplexing." Sutton, no slouch himself at telling tales, repeated Hill's story about a woman testing him with a pin almost verbatim.[50] *Variety* reported on "Erco," "Araco," and "Mr. Reded," all in 1906 and all in the style of Phroso, panning each and claiming that "the novelty has worn out,"[51] undoubtedly explaining why Farrell updated the act for the fall season with some modernistic abracadabra:

> The interior of the body is a mass of electrical mechanism. To show the audience that he is full of machinery, Dr. Farrell removes a portion of the coat and reveals whirring wheels. Incandescent lamps attached to the back of the figure throw a reflection clear through the body, and the X-ray could disclose nothing more perfect.[52]

—⁂—

Show business was the most egalitarian profession or industry in 19th-century America. Women acted, sang, danced, and entranced audiences from the opening act to the headliners. (Egalitarian does not mean not fully equal; men still ran the theaters, the booking agencies, and the publicity apparatus, and they commanded first billing in mixed acts.) Acting's disreputability ironically helped: a business full of thieves, scoundrels, sharpies, and con men suffered no loss of prestige by allowing women entry. Vaudeville was an equal-opportunity purveyor of bunkum. Numerous women took the opportunity to stage similar acts as "mechanical dolls." "Is she woman or machine?" asked the *Pittsburg Daily Post* in 1902[53] about the Moto Girl, whose act was blandly described in a Chicago ad:

> The Moto girl she (or it) is called, and appears dressed as a huge doll, standing on a pedestal. There is an electric apparatus which is apparently turned on, and a clock-like affair that sounds as though the machinery was clicking inside the figure. After walking jerkily into the audience and back, the Moto girl doffs her hat and smiles blandly at those she has mystified. The act is a good one, and compares favorably with those from which it is copied.[54]

Little wonder that the teenage Marx Brothers conquered vaudeville if this was their competition.

The Moto Girl—variously spelled as Moto-Girl, Motogirl, or La Motogirl (often mistyped as "La Motogril") after the act toured Europe—was Doris Chertney, herself a mere 13-year-old "strip of a girl" when she started. A Frederick Melville got the idea for the act while operating a merry-go-round; a central figure broke, and his wife suggested having a live child pose motionless. Over the next nine years, they toured the world and appeared before the crowned heads of Europe. Their best performance took place offstage: Melville got a receipt from Russian customs when he passed Chertney off as "dutiable" and paid the fee for importing a doll.[55]

One of the acts Moto Girl might have copied was Ruth Everett, who billed herself as "the original mechanical doll," though this act is traceable back only to 1903. Called "the nearest delineator of a human automaton yet to be seen on the American stage," Everett evidently put on more of an act, doing "clever impressions."[56] She also performed as herself singing songs, and the whole act made her a living through 1918.[57]

Another possible predecessor to Moto Girl might have been Ver Valin, whose ventriloquist act appeared in 1898 and who advertised "life-sized automatons" as early as 1900.[58] "A Mechanical Lady Turns Out to Be Real" was the headline over a gushing *Minneapolis Journal* article in 1905. "Seldom are audiences so deceived by cleverness in makeup and acting as those that have witnessed Florence Ver Valin's imitation of a mechanical walking woman at the Unique theater this week."[59] The ventriloquist, who would eventually expand the act to become the Great Ver Valin and His Wood Family, would stay a trouper until 1923.[60]

Ver Valin overlapped "Creo, the Unsolvable Mystery." Professor Boehkle, starting in 1909, would build Creo onstage from "a wooden tripod, a plaster-paris bust, some garments and a 'hank of hair.'" The management of the Bell Theater in Oakland offered three prizes for the best explanation of the miracle.[61] Mechanical women outlived even Ver Valin. Miss Terla, who "plays a drum and zylophone and closes by singing 'A Kiss in the Dark,'" wowed audiences in 1924.[62]

—⚹—

Men rode the diminishing returns of playing dolls for decades longer. Edward R. Davidson claimed in 1930 to have begun "his mechanical man act in 1903 whe nas [*sic*] a boy he watched the originator of the stunt and decided he could go him one better. He has since played in every state of the union as well as in Canada, Mexico, England and Cuba."[63] So long was Davidson's career that when the play *R.U.R* put *robot* into common usage, it rebooted his act, allowing him to adopt "Robota" as his stage name through 1935. By then, "More than half of those who have seen me have read about robots and they insist I am an electrically-operated mechanical device. They come a second time to get another look at me."[64]

The 1933–1934 Chicago Century of Progress Exposition (to give the event its formal name) was America's largest and grandest world's fair since Chicago's 1893 World's Columbian Exposition. Like the earlier fair, the promotors in 1933 strove to put a high technological gloss on an exhibition of the world's wonders, with an official motto of "Science Finds, Industry Applies, Man Conforms" promising Americans that the current Depression was a mere bump on the road to a glorious future. Yet, also like the earlier fair, the crowds were thickest on the Midway, a supersized carnival of freaks, wonders, and bawdiness. Sally Rand's nude fan dance shocked visitors so much that they couldn't stay away. The Sky Ride boosted thrill seekers to a height of 219 feet in "rocket cars."[65] And Robert C. Ripley of "Believe It or Not" fame imported human oddities from around the world to stun gawkers. The Man Who Speaks Without Vocal Cords, the Mole-Face Woman, the Man with an Ostrich Stomach, and the Eye-Popper were joined in the

Odditorium by John R. Stone, "The Miracle Man—Human or Robot?" A publicity post-card shows a heavily made-up man in a Lincoln-esque stovepipe hat and frock coat, wearing oversized gloves, his eyes so wide open that a fringe of white encircles his irises.[66]

Makeup hides expression, tiredness, boredom, and sweat, but it also conceals identity. Acts appeared before new crowds every fifteen minutes for a whole day throughout the fair's duration. Did they see the same John R. Stone at 11:00 and 4:30, on Tuesday, and Thursday, and Saturday, in 1933 and 1934? Nobody at the time knew for sure. A squad of Stones may have come and gone, in the same fashion as an endless procession of chess masters hidden in the innards of Ajeeb, the chess-playing automaton. Or one inhumanly persistent man named Stone may have had his identity stolen in the grand theatrical tradition over the next decade by acts who made their claims secure in knowing they couldn't be proven wrong.

Anonymity suited Monsieur X, "the mechanical man who proved to be a sensation at the Century of Progress in Chicago."[67] Confidence oozed from ads that promised a free car to anyone "who makes him smile." His manager claimed that "he has held a rigid face for seven hours without blinking an eye." The notorious hot summer of 1936 proved to be his undoing, as Monsieur X fainted in 115-degree heat in Mexico, Missouri. That was the last straw even for a seasoned trouper. In Beatrice, Nebraska, "he" threw off his cap and revealed the long blonde hair on the head of Mable Crouse. Monsieur X disappears after that.[68]

Nerv-o, the "World Famous Mechanical Man" from Ripley's Odditorium, popped up in 1938 thrilling those who went to the New Market Basket Stores in East Lansing, Michigan. His management made the lesser claim that he could stand still for five hours.[69]

The best hint that people moved in and out of the role is that competitors appeared while the fair ran. Although he didn't start touting his work in Ripley's Odditorium until 1935, R. H. Harris can be found performing from 1933 to 1942 under a variety of names, including Gloomy Gus and Gloomy Harris, the Gloom Chaser—presumably because the audience entertained itself into fits of laughing while he resisted their antics. The Gloomy man didn't do much. His dare to the audience was that they couldn't make him do *any-thing*—not laugh, or smile, or break his pose: a veritable man of still. He appeared in venues large and small, car dealerships, pharmacies, and department stores, with prizes as small as $10 or as large as a new car to the heckler who made him react.[70]

Confusingly, a "Waxo Gus" also appeared in 1933 doing the same act and paralleled Harris for years billed as "'WAXO' The Mechanical Man from the World's Fair, Chicago."[71] A 1933 ad asked, as always, "Is he human or mechanical?" and then added, "WAXO was on exhibition in Ripley's Odditorium at the World's Fair and attracted a lot of attention and comment."[72] A 1938 article outed Waxo as Lee W. Sessions, a clown who had been performing the act all over the country for eight years (i.e., starting before Chicago). In fact, in 1926 a "Waxo the mechanical man" popped up at a roller-skating rink in Akron.[73] Waxo's name can be found for another generation, in 1951 and 1961, until in 1977 the *Battle Creek Enquirer* wrote a feature about the 50-year career of … Edward Dotson. An orphan, in classic fashion he had run away from a foster home to join a carnival. There a Don Quinn had trained him for two years until he had the act down pat. Dotson claimed to have a book of clippings and a picture of himself at the Chicago Exposition. He also claimed to have retired as Waxo in the early 1940s, getting a series of regular jobs until he retired in 1973 and resumed his old act, this time from the safety of a store window. "I've been pushed, stuck with a pin, and had things thrown at me," he said. His biggest challenge came when a woman opened her coat and flashed him.[74]

Dotson said he changed the name of his act to "Mr. X" after he retired Waxo. This would have been in the 1940s, not to be confused with Monsieur X. Or with Ted Muralt's "Mr. X," also from 1935, a home-made radio-controlled robot that could "wiggle his ears, move his head from side to side, shakes, hands, point with his left hand or move his arm almost straight over his head." Muralt exhibited him at fairs, sideshows, carnivals, and department stores.[75]

Claiming to be at the world's fair in Chicago was doubly confusing because the Odditorium had competition right at the fair. Jack "Heim will interpret a 'mechanical man' in such a manner that until he breaks his character, spectators believe him a machine shop product."[76] Dressed in the clean, crisp, pressed uniform of a Firestone service station attendant, Jack stood ready to aid every potential motorist, giving 7,800 shows during the fair. Heim may have been the most watched fake mechanical man of all time. In September 1934, a publicity photo showed him with a pretty co-ed who, surely fortuitously and not at all handpicked by the press people, happened to be the 13,000,000th visitor to Firestone's exhibit. "Is he wax or human?" the caption read. "Others are baffled, but Miss Betty discovered a clue—he blushes."[77]

Others claimed to have been performing the act since long before the Exposition. Renaud LeClaire (a.k.a. Automa) boasted of fifteen years' experience.[78] Baron "Q," "The Robot Man," having "entertained all over the world," landed in Berkeley in 1936.[79] He was sponsored by a Nash auto dealer. Only the name changed in 1957 when "San-Velo" appeared on behalf of Chrysler Plymouth dealers in Los Angeles.[80]

The sad fate of the fake automaton was to sink lower and lower in prestige until it hit rock bottom with "Claudo." When Claudo first commenced standing stock still in store windows in the 1920s, he was a passable follow-up to Psycho and might have had even better motor control since the Baltimore Gas Range Co. offered a free stove to anyone who could make him smile.[81] The Depression temporarily cut the prize money to $25 in 1935, increasing to $100 in 1937, but in 1956 Furniture City in Camden, New Jersey, offered up to $1,000 in prizes "if you can make him smile."[82] Whoever was using the name was still going in 1963 in the metropolis of Chester, Pennsylvania.[83]

With so many robots competing for the same venues, performers had to keep upping the ante to attract the crowds. Waxo, whoever he was at the time, knew how to play that game. Whenever a mere mechanical man wasn't enough for audiences, they could see him share the bill with "Florine, the human icicle," whose specialty was the old magician's trick of seemingly being frozen into a block of clear ice.[84] They made a perfect couple. Reader, she married him. In 1938, a jaundiced reporter caustically hyped the act:

Footnote to civilization.

Thursday will be children's day at the grave of Mrs. Florine Williams, wife of Waxo, the great, the mechanical man.... Mrs. Williams is buried alive, and will stay buried alive until some time during the American legion dance.... Anyone under 12 years of age can look through the periscope and talk with Mrs. Williams for five cents ... as a special feature of children's day, Mrs. Williams' 5-month old baby will be lowered into the grave so that it can be with its mother. [ellipses in original][85]

Match that, David Blaine.

4

The Wonderful
Walking Mechanical Men

Due to a nearly unprecedented half-century of peace in the Western world, the American Civil War was the Age of Steam's first fully mechanized large-scale war. Factory mass production enabled the North to clothe, feed, and arm two million men. The new telegraph system transformed communications, balloons were used to spy on encampments, and ironclad ships and submarines revolutionized sea battles. The use of steam locomotives forever changed the movement and supply of troops across vast distances. The North emerged from the war a nearly untouched technological powerhouse that inspired thousands to work on creations of their own.

In Newark, New Jersey, the heartland of invention during the war, a young would-be mechanic and inventor daydreamed of creating something great that would see fruition in the winter of 1868. He proposed to build a mechanical man, a giant thing with an iron skeleton that would do the work of three horses or a platoon of men. After six years of effort, the 22-year-old had found a way to use a standard steam boiler to provide propulsion. Somehow he amassed the equivalent of two years' good salary, enough to hire a team of workmen to put the complicated figure together at a staggering cost of $2,000. The mechanical man stood seven feet, nine inches tall and weighed five hundred pounds.[1] The precocious inventor never gave his creation a name, but the workmen nicknamed it Daniel Lambert, after a 700 pounder who gained notoriety for his nine-foot, four-inch girth.[2]

Daniel's chest was a boiler, with a fire-box in the center and the water-jacket around it. A piston rod attached to its back provided power to the articulated legs through two rods. According to the patent, "As the two sets of rods turn, therefore, upon different centres, the foot is turned down at the toe as one leg falls behind the other, and the knee bent, so that as the foot is thrown forward, it is raised by bending the knee, to step over any obstacle, the foot being turned downwards at the toe before being placed on the ground."[3] There was no danger of the figure falling over, for a pair of iron shafts at its center connected it to a trailing wheeled carriage. This structure did double duty as a stabilizer and a platform in which the driver could sit and steer and goods could be loaded. Years before the automobile, here was a complete system for a moving vehicle that could house passengers or freight and move at startling speeds. The three-horsepower engine, like those used in steam fire engines, was said to be capable of driving the mechanical man forward at sixty miles per hour, though its speed would be cut in half for safety on cobblestoned streets (still as fast as most trains of the day traveled). The small engine was remarkably efficient. The fire needed coaling only every two to three hours. The driver just needed to stop, remove some coal from the rear of the carriage, unbutton Daniel's vest, and shovel in the fuel.

Daniel's face was "molded in a cheerful countenance of white enamel, which contrasts well with the dark hair mustache." Its top hat was hollow to serve as a smokestack. The vest covered felt and woolen undergarments to resist scorching by the boiler. The steam cocks and gauges set on its back were to be covered by a knapsack. Pantaloons that covered its iron legs completed the illusion, ostensibly intended to give the figure the appearance of a human lest it frighten the horses it hadn't yet replaced, although making it picturesque for the crowds and reporters is a far more likely reason.[4]

The news of the mechanical giant, initially carried in a single anonymous story printed in the young mechanic's hometown newspaper, the *Newark Advertiser* (dated January 8, 1868), spread rapidly across the country, becoming front-page news in papers in Pennsylvania, Illinois, Indiana, Kansas, South Carolina, and Ohio. Naturally, reporters from the big New York City papers rushed to cover the story only to find that mobs of their countrymen had preceded them, an almost literal embodiment of Ralph Waldo Emerson's adage that "if a man … can make better chairs or knives, crucibles or church organs, than anybody else, you will find a broad hard-beaten road to his house, though it be in the woods."[5] The *New York Tribune* chronicled the frenzy:

> When a description of Mr. Deddrick's steam man was published, not only the Newarkers, a goodly number of whom, like the Athenians, 18 centuries ago, "spend their time in nothing but either to tell or to hear some new thing," all rushed to the shop where, under the hands of skilled mechanics, he was slowly but surely assuming the "human form divine," and so thronged the doorway and darkened the windows of the shop that his completion was at one time made doubtful, but scores of gentlemen from other cities ventured into Jersey, and all, men, women, and children who could not go wrote, inquiring about this new wonder. An enthusiastic Committee of Five traveled all the way from Albany, one day last week, to decide a bet that the whole thing was a "newspaper story."[6]

What we today term *fake news* or *clickbait* is not a modern problem. Articles and headlines that contained embellishments or outright inventions caused readers to flock to newspapers in an Emersonian way throughout American history. A cautious reader might note that not a single article actually quoted the inventor, leaving doubts about whether the reporters had talked to him or were relying on secondhand accounts. Not knowing how to spell his name might be another clue. The original article cites a Zadock Deddrick, but other accounts spelled his name as Zadoc or Zaddock and his last name as Dedrick or Dederick or Drederick or even Derrick. The patent application seems to credit a suitably official Zadoc P. Dederick, but the name on the attached drawings is spelled Drederick, and the 1850 Census listed his first name as Zadock. Today's consensus is that the inventor was Zadock Pratt Dederick, named after the inventor Zadock Pratt, who founded the town of Prattsburg, New York.[7] The patent for the mechanical giant (along with the original article) also bears the name of Isaac Grass; unlike Dederick, whose story has been traced to a career in manufacturing and an obituary for his death in Houston in 1921,[8] Grass has no historical record other than his name appearing on an 1890 census form, implying (if it's the same person) that he was a mere 15 or 16 years old in 1888.[9]

Did Daniel the steam giant ever put his footprints in history? Dederick (and possibly Grass) exhibited him in public, first at Crump's Garden in Newark in late January and February, and then later in February and March in the big leagues of New York at 538

Broadway nearly across the street from P. T. Barnum's legendary museum of bunkum.[10] The fanciful *Tribune* article quoted above is written to suggest more: "The old spiral springs have been replaced by stronger ones, so that the steam man is no longer weak in the knees, and upon steam being generated on Thursday, he stumped off like a live Trojan. In the evening he appeared on Broad st. [*sic*], at Crump's Garden, where he will exhibit himself next week." Although the reader is led to believe there is a connection between the two statements, Daniel did not parade down the street. At most, Dederick got the giant's legs to move before packing him up for transfer.

It's doubtful that those spiked feet ever touched the ground, however. When it got to Crump's Garden, there was not "room for it to walk, its feet have been lifted from the floor, and the giant 'goes through the motions' in the most satisfactory style."[11] Nor did he do better in New York, where something unfortunate always prevented Dederick from displaying his creation walking. "[O]wing to some objection on the part of the owner of the hall," said *Scientific American*, "he is not permitted to 'travel on his muscle' but is hung in slings and merely 'marks time' as our military friends would say."[12] The *New York Express* sent over a reporter who described Daniel in detail but was told that "full running motion … would not be permitted by the insurance company."[13]

The only positive evidence of the mechanical giant working as intended appears in a 1931 article in the *Newark Evening News* quoting the memories of a Mr. Hunt, who "testified that as a boy he saw the robot tried out and that there was great excitement among the equine population and great objection on the part of drivers."[14] Sixty-year-old memories are normally suspect, and Hunt may be confusing this event with the arrival of a steam carriage invented by Joseph Battin of Elizabeth, New Jersey, which was also tried out in Newark around that time, according to another 1931 *Newark Evening News* article. A young carpenter in Battin's employ remembered that "after only a few exciting trips to Newark, during which startled horses tried to climb fences and little children fled to the protection of mother's knee, it was laid up in an Elizabeth shed."[15] The latter explanation seems more plausible simply because the idea that Daniel could work defies mechanical reality.

By late March, the nine-day-wonder hit its expiration date. Stories far shorter than the original front-page headline sensations, but also far more pungent, declared Daniel a fake in the same newspapers that had originally acclaimed him: "It turns out there was a great deal of humbug as well as steam about this man. It was so weak in the knees that whilst it swung its legs in the air, it was unable to 'stand on its pins,' and never walked a step on terra-firma. An attachment has been got out against it, which now prevents it walking at all."[16] "The financial obstacle against it may only be a sham, but the physical difficulties are real. It will tumble over as fast as it can be set up."[17] "That which is called 'the steam man' never did, and in all probability, never will, walk the length of his nose."[18]

Dederick had fallen into a trap. Human inventors cannot easily devise systems that rival evolution's several million years of combining primate erectness and the ball-and-socket joint. Locomotion looks easy to emulate, yet the smooth gait of upright bipeds has proved unachievable even today. Despite the cuteness of experimental robot systems like ASIMO (Advanced Step in Innovative Mobility), whose speed is limited to 9 kilometers per hour (5.5 mph), the internet is filled with videos of robots falling, collapsing, crashing, or bumping into walls. It's not clear today whether Dederick and Grass understood that they had attempted the impossible (especially considering that the first report had Dederick originally contemplating a perpetual motion machine for propulsion).

Filing a patent at a cost of $35 indicates that they were sincere, if two centuries ahead of themselves technologically. Far more certain is that they knew that the public wanted a humanlike machine that could walk.

In truth, Daniel was a ridiculous object for its intended purpose. Propelling a cart with an engine shouldn't require the intricate and failure-prone mechanism of fake legs. As shown by Battin and a large number of earlier and later dabblers, the wheels of a cart could be driven directly and far more efficiently by attaching the axle or wheels to a motor. Motor-carts were the superlatively successful outcome. In their stead, mechanical walking men had no technological value, though they filled a cultural gap previously unrecognized.

The jokes started immediately: political, cultural, and social jokes that would be reinvented again and again every time a robot made the news. An election loomed in the fall of 1868 to replace the greatly disliked Andrew Johnson. Just a few days after the announcement of Daniel's existence, the *New York Herald* put the steam man up for president: "[W]e consider that his possibilities, from the mere ideas his existence involves, must be very great.... Put him in the White House, and let the Senate have the appointment of a stroker and engineer, and there never can be any more trouble in the great republic."[19] Another wit, recognizing that war hero General Ulysses S. Grant had a lock on the Republican nomination, extolled the steam man as a running mate: "He can't talk, and has no political opinions.... In the event of any accident happening to Grant, there would then be no material change in the head of the Government."[20] Southerners still smarting about the war and those whites who had joined "Loyalty Leagues" against the rebel cause saw robots as the ideal replacements for lost slave work, as in this editorial sneer from the *Native Virginian*: "A steam slave, equal to ten mutes and that can't vote or jine 'Loil Leegs,' all for $300. Think of it. Whoop!"[21] Presumably not by coincidence, $300 was also the fee that wealthy northerners had paid to buy a substitute if they got drafted during the war, making it the very literal price of a man.

It was the deliberate cultural tease of forming a machine into the shape of a man that drew the deepest reaction. A Washington newspaper reported that "the Rev. Dr. A. A. Willets is lecturing on 'The Model Wife.' She is said to resemble the 'Steam Man'—stays at home and does all the work for a family of fifteen."[22] "Five women write, ordering cast iron husbands," claimed a syndicated article, "and one gentleman sends for a steam wife, provided the machine does not *talk* [italics in original]."[23]

Daniel's fame meandered around the world. As late as June 1868, a Sydney newspaper reprinted a London article reprinting a New York account of an Englishman viewing the steam man, describing a miraculous contrivance as both a parody of humanity and yet greater than any human.

Physically he is grand, gloomy, and peculiar to the last degree. The iron cast of his castiron features imparts a look of singular determination to a face which might otherwise leave an impression of slight deficiency in mobility. It bears, moreover, the marks of a hard morning's work in the shape of four streaks, of a strange grimy hue, down its broad brow, which realise our conception of Pittsburg perspiration. His steam wash basin and steam towel are probably at Newark for repairs. The chest is wonderfully full and deep, as a steam chest ought to be, and covered with a stylish robe of superior ferruginous cassimere, which our patriotism forbids us to call an English shooting-jacket, and which we suppose must be an American steaming jacket.... [His topper] is a stove-pipe hat, as no one of any style need be told. On the street it is worn quite plain, with only the usual ventilator, like other good hats. But our friend has a queer habit of smoking through this hat, as other gentlemen of accomplishments one degree lower do through their noses, in which he takes great pleasure, and which, to be candid, is known to his selecter friends to have become an inveterate and chronic affection, like opium-eating or impecuniosity.[24]

—⟋w⟍—

Although sophisticated Europeans mocked the naïveté of the brash Americans, that same brashness had an electrifying effect on Dederick's contemporaries. At 28, Edward Sylvester Ellis was only a few years older and an odd specimen of an eastward-moving emigrant, his father having transplanted him from Ohio to Rhode Island and then northern New Jersey as a boy. Ellis was as bookish as Dederick was mechanically inclined. He attended the State Normal School, the teacher's college in Trenton, and taught there for years before moving up the ranks of various institutions of learning until he became Trenton's chief commissioner of schools upon earning his master's degree at Princeton. After retiring in his mid–40s, he started a new career penning more than 50 large volumes of history.[25]

Northern New Jersey was as mechanically civilized as any region in the 1860s United States, perhaps explaining Ellis' obsession with romanticizing the great American West as filtered through stories and memories from his frontiersman father. As with Dederick, he recognized his calling as a teen, serializing a novel (*Dick Flinton; or Life on the Border*) in the *New York Dispatch* at 19 and becoming a national bestseller at 20 with the Beadle and Adams dime novel *Seth Jones; or, The Captives of the Frontier*, said to be a favorite of Abraham Lincoln.[26] Ellis contracted to continue to write four novels a year for Beadle, a number too piddling to contain him and his more than a dozen known pseudonyms, with as many as 300 volumes for numerous publishers attributed to him over 30 years of fiction writing.[27]

Ellis certainly read the newspaper articles on Dederick's steam man and might well have taken the short train trip into Newark or New York to see Daniel on display. By August 1868, he had checked off one more box on his contract with *The Steam Man of the Prairies*, number 45 in Beadle's American Novels dime novel series. Ellis copied the newspaper descriptions faithfully, down to the smallest of details—with one exception. Recognizing that Daniel's spindly legs couldn't possibly have stood up to the stress of pulling the heavy wagon, Ellis massively upgraded them, an indication that he might have seen the unworkable steam man in person.

> Perhaps at this point a description of the singular mechanism should be given. It was about ten feet in hight [*sic*], measuring to the top of the "stove-pipe hat," which was fashioned after the common order of felt coverings, with a broad brim, all painted a shiny black. The face was made of iron, painted a black color, with a pair of fearful eyes, and a tremendous grinning mouth. A whistle-like contrivance was made to answer for the nose. The steam chest proper and boiler, were where the chest in a human being is generally supposed to be, extending also into a large knapsack arrangement over the shoulders and back. A pair of arms, like projections, held the shafts, and the broad flat feet were covered with sharp spikes, as though he were the monarch of base-ball players. The legs were quite long, and the step was natural, except when running, at which time, the bolt uprightness in the figure showed different from a human being.
>
> In the knapsack were the valves, by which the steam or water was examined. In front was a painted imitation of a vest, in which a door opened to receive the fuel, which, together with the water, was carried in the wagon, a pipe running along the shaft and connecting with the boiler.
>
> The lines which the driver held controlled the course of the steam man; thus, by pulling the strap on the right, a deflection was caused which turned it in that direction, and the same acted on the other side. A small rod, which ran along the right shaft, let out or shut off the steam, as was desired, while a cord, running along the left, controlled the whistle at the nose.

The legs of this extraordinary mechanism were fully a yard apart, so as to avoid the danger of its upsetting, and at the same time, there was given more room for the play of the delicate machinery within. Long, sharp, spike-like projections adorned those toes of the immense feet, so that there was little danger of its slipping, while the length of the legs showed that, under favorable circumstances, the steam man must be capable of very great speed.[28]

Ellis generally wrote boys' fiction, so Dederick's conception of the steam man as the product of a teenager suited him perfectly. He made his inventor, Johnny Brainerd, fifteen years old, and a dwarf hunchback to boot, yet gifted "with a most wonderful mind." (Could this conceivably be a reference to a teenaged Isaac Grass?) Agile and hardworking, the fictional hero is a genius in the workshop, making clocks, telegraphs, and hot-air balloons from scratch. The steam man is suggested by his mother as a long-term project to keep him occupied and happy.

In the kind of coincidence that propels dime novel plots, Baldy Bicknell stumbles across Johnny's hidden workshop. Bicknell is a master outdoorsman and Indian fighter, nicknamed Baldy due to having been scalped. He knows of a ravine in the West that contains a fortune in gold washed out of higher rocks and wants the steam man to get back to it. Baldy and Johnny strike a deal that Johnny will go along to control the mysterious workings of the steam man. Baldy is the ideal westerner and Johnny the ideal easterner; together they are unstoppable. They create the steam man and travel up the Missouri River to reach the prairie of the title, along with Irish-dialect stereotype Mickey McSquizzle and Yankee-dialect stereotype Ethan Hopkins. The endless flat grasslands allow them to travel hundreds of miles in record time without needing roads or trails. Even so, the rest of the book is one thrilling near-death experience after another at the hands of raging rivers, vicious animals, and implacable bands of Indians who want nothing except to kill white men.

It's tempting to apply a modern eye to the narrative and make the Indians the metaphorical agrarian, pre-mechanized past that is being overtaken by the startling, futuristic technology of the 19th century, although there is no evidence that Ellis meant them as anything more than bad guys. This theory is also countered by the way the Indians immediately lose their terror at the loud apparition when they realize it's just another iron mechanism like their rifles. The Indians are, at least, brave, tenacious, and ingenious. They corner the returning party and their gold by rolling large boulders into another ravine to trap them until daylight enables a swift massacre. Only by superheating the steam man and sending it into the band of Indians to explode into devastating shrapnel do the adventurers escape one final time, while also conveniently destroying any evidence of the steam man's existence.

American Novels #45 is one of the rarest dime novels, but the story is easily available today because it was reprinted half a dozen times from 1869 to 1904. Some of the subsequent releases were published in *Beadle's Half-Dime Library*, a weekly magazine whose price was within even a child's grasp.[29]

The notion that Ellis may have spent some of his money on viewing the steam man in person is bolstered by a suggestive conversation between Baldy and Johnny in chapter VIII:

"Why, you can make more money with him than Barnum ever did with his Woolly Horse."
"How so?" inquired the boy, with great simplicity.
"Take him through the country and show him to the people. I tell yer they'd run after such things. Get out yer pictures of him, and the folks would break thar necks to see him. I tell yer, thar's a fortune thar!"
The trapper spoke emphatically like one who knows.

There speaks a man whose money has been taken for a less than worthy return.

(The "Woolly Horse" Baldy speaks of was real, a mutant with woolly hair but no mane or tail. Barnum kept the animal hidden until the public's attention was caught by the plight of Colonel John C. Fremont, lost in the "trackless snows of the Rockies." Barnum planted stories in the newspapers that Fremont was returning with a creature never before seen by humans. "[T]he public appetite was craving something tangible from Col. Fremont," Barnum wrote in his autobiography. "They were ravenous. They would have swallowed any thing, and like a good genius, I threw them, not a 'bone,' but a regular tit-bit, a bon-bon—and they swallowed it at a whole gulp!")[30]

By far the reprint of *The Steam Man of the Prairies* with the greatest impact on American popular culture appeared as a Beadle's Pocket Novel. This series of reprints of earlier Beadle titles used the latest technology to add color to the covers, making them stand out, as indicated by the advertisement announcing, "A Marvel of Beauty! A New Series by the New Art!"[31] Volume number 40 is dated January 4, 1876, and contained Ellis' story under the title *The Huge Hunter; or, The Steam Man of the Prairies.*

This time, however, Beadle's competitors weren't content to let it lie. Starting on February 28, 1876, *The Steam Man of the Plains; or, The Terror of the West* began a nine-part serialization in *The Boys of New York*, subtitled "A Paper for Young Americans." As the subtitle indicated, this was not a true dime novel but an example of what are now called story papers, extremely cheap nickel (later dime) weeklies printed in black and white on newsprint. Why the sudden rush to compete? *The Boys of New York* was then part of the publishing empire run by Norman Munro, a former Beadle employee, and a rival to the story paper publisher founded by his brother George with Beadle's brother Irwin.[32] Each apparently strove endlessly to outdo the others and, in an indication of the enormous number of five- and ten-cent publications sold over a half-century, each claimed to end up multi-millionaires, although it is more likely that they undercut one another's profits.[33]

The real winner was to be Frank Tousey, who had started working with Munro in 1872 at the age of nineteen. He had the idea to redo the Steam Man and hired an equally young (yet almost as experienced) Harry Enton to write the story, released as *Steam Man of the Plains; or, The Terror of the West.* Enton slavishly copied Ellis' original work. Frank Reade was a carbon copy of Johnny Brainard—sixteen, a city boy, with "a studious nature, and quite a thinker ... pale, slim, and not overstrong."

> The figure was about twelve feet high from the bottom of the huge feet to the top of the plug hat which adorned the steam man's frame. An enormous belly was required to accommodate the boiler and steam chest, and this corpulency agreed well with the height of the metallic steam chap. To give working room to the very delicate machinery in his interior, the giant was made to convey a sort of knapsack upon his shoulders. The machine held its arms in the position taken by a man while he is drawing a carriage.
>
> [Frank said,] "the lamp will be in his head, and his eyes will be the headlights. His mouth holds the steam whistle. Here, in the belly we open a door and put in fuel, and his ashes fall down into his legs and are emptied into the movable knee-pan, and without injury to the oiled leg-shafts, for they are inclosed in a tube. That is why the feller's limbs are so large. Those wire cords increase the power in one leg, and cause that leg to go so much faster, and in that manner we get a side movement and can turn around." ...
>
> "The hands of the man will hold the shafts of the wagon. The vehicle will carry two or three persons and hold my food and water, sufficient for several days.... Here in the knapsack are my steam valves: the top of the hat is only a sieve, and the smoke will come out of that."
>
> "It can go fifty miles an hour.... On a level road I should not hesitate to run at thirty or thirty-five miles an hour."[34]

Frank has the above conversation with someone who, "breathing the free air of the great prairies, had grown strong and robust in form, a splendid hunter, a dead shot, and a lover of wild adventure"—namely, his sixteen-year-old cousin Charley. They crate the Steam Man and take him out to Missouri, where they meet up with an Irish-dialect caricature, Barney Shea, and a Cockney-dialect caricature, George Augustus Fitznoodle, as well as grizzled veteran Indian fighters, an emigrant wagon train, a troop of Secret Service manhunters, a wild buffalo stampede, a wild horse stampede, a maddened grizzly bear, a prairie fire, and troops of white renegades in league with Osage and Sioux Indians. A series of near-death experiences follow, ending with the Steam Man and his companions seemingly doomed after having been trapped in a ravine. Enton would later tell his successor (presumably with a straight face) that he had been inspired by seeing steam men at the 1876 Philadelphia Centennial Exposition.[35]

Enton was a better writer than Ellis, with a mind that would have been perfect for a modern action movie. His coincidences are jaw dropping, his dooms certain and colorful, and his escapes delightfully preposterous, with the pageantry of the settlers trapped in battle with a stampede bearing down on them from one side and a fire from the other designed for a sweeping helicopter shot. He knew how and when to add dollops of humor to move the reader through the connecting scenes and, as an "Easter egg," gives his Indians names that are Yiddish puns, such as Sholum Alarkum, Schorumanolis and Schlenter (authentic since "Harry Enton" was actually the Jewish Harold Cohen of Brooklyn).[36] In addition, his Indians are given a modicum of character as followers of their ancient ways, in contrast to the white renegades who are purposefully but needlessly evil. The Steam Man is oddly played down, however—a mere device for transportation that occasionally frightens the unwary, simply one of the many astounding inventions of young Frank.

Two more Frank Reade serials followed: *Frank Reade and His Steam Horse* and *Frank Reade and His Steam Team*. However, Tousey had already angered Cohen by misspelling his pseudonym as "Entom" on the first story and fully enraged him by crediting the newer stories to the mysterious "Noname" (always placed in quotes). Cohen quit in a huff, though whether before or after writing 1881's *Frank Reade and His Steam Tally-Ho* (a coach pulled by a team of steam horses) is not known.

Tousey didn't care. He had found yet another youngster to grind out material for him. Also a Brooklynite, Luis Senarens started writing for Tousey before he was 16 and claimed that "every week for 30 years I turned out a story averaging 35,000 words."[37] His serial in *Boys of New York*, starting on February 4, 1882, was titled *Frank Reade, Jr., and His Steam Wonder*, covering the adventures of the 16-year-old son of inventor Frank Reade and also credited to "Noname."[38] Over the next decade, twenty-eight more Frank Reade, Jr., titles followed either in *Boys of New York* or in the companion *Wide-Awake Library* series. Then Tousey promoted Frank Reade, Jr., to star in his own magazine, the *Frank Reade Library*, a thicker story paper of 16, 24, or 32 pages that could contain a complete Frank Reade, Jr., novel or a reprint of a Frank Reade novel. The first adventure in volume 1, number 1, dated September 24, 1892, brought the series back to its first glory: *Frank Reade, Jr., and His New Steam Man; or, The Young Inventor's Trip to the Far West*.[39] Senarens' prolificacy was tested mightily, as ten more novels about the Steam Man or Steam Horse appeared before the end of the year. *Frank Reade Library* reached issue #191 before dying in October 1898. However, Frank Reade, Jr., reprints continued, with the *Frank Reade Weekly Library* running for 96 issues from 1901 to 1904.

Senarens could write quickly but not well. His steam man was a copy of a copy, and it showed. The characters in his stories are one-dimensional clichés who speak at length in clichés. The stereotyped assistants speak in strained dialect, with the "darky" named Pomp, though brave, capable, and loyal, being nearly unintelligible: "Golly, Marse Frank, amn't youse gwine to let this chile go wif yo'?"[40] The evil villain is baser than the earlier renegades; the Indians are killers without any redeeming dialogue or personality. Escapes from certain doom litter every chapter but are often relegated offstage. No camera eye directs the sweeping majesty of huge forces swarming the wide-open plains. Senarens' tales are the stultifying fictions that adults denounced as despoilers of children's minds. So, naturally, they sold by the millions and later became much-loved touchstones of adolescents, as demonstrated in this anonymous 1920 newspaper tribute to Senarens:

> The nickel novels that in our youth we were compelled to sneak off in the attic to read safely out of sight of father or mother, with the consciousness that we were committing an unpardonable sin for which we were likely to call down the wrath of the parental rod with stinging emphasis, have been vindicated in full, and instead of the brain-twisting they were supposed to cause, they were, indeed, the early-day prophets we were unable to comprehend or interpret, because they were beyond the ken of man's blunt, show-me-in-advance vision....
>
> It would seem that Senarens anticipated all of the important inventions of these later days, and one re-reading his stories, and seeing that in thirty or forty years many of the most improbable of them have come true makes them almost uncanny.[41]

Senarens' power lay in his mastery of the imaginative vocabulary of science fiction, introducing futuristic technology into current-day adventures just at the moment that technological advancement started its exponential growth. Frank Reade, Jr., traveled the world with the inventions proudly announced in his titles: Steam Man, Electric Man, Electric Coach, Air-Ship, Flying Ice Ship, Bicycle Car, Sky Scraper, Electric Submarine Boat, Cruiser of the Clouds, and more. His adventures paralleled (if not stole from) Jules Verne, making Senarens the first American writer to make a career of having his heroes using technological marvels to travel on, over, and under the earth. And beyond: *The Sinking Star; or, Frank Reade Jr's Trip into Space with His New Air-Ship "Saturn"* takes young Frank off-planet. The dozen or so steam-powered inventions showcased in that short 1892 spurt probably were ordered by Tousey because steam gadgets had long been abandoned by Senarens in favor of the more than three dozen titles focusing on the new wonder of electricity. He had already written *The Electric Man: or, Frank Reade, Jr., in Australia*, which debuted in *Boys of New York* #588 back on October 10, 1886. Frank describes the Electric Man as

> a Samson in physical strength—in fact, the limit of his strength is unknown even to me, as every bone and muscle in him in made of the finest steel. The machinery that works his limbs is inside of him, and he can walk, run, jump, and kick forward or backward with wonderful agility. The motive power that moves him is in a powerful electric battery inclosed in yonder box, under the floor of the carriage, and is communicated to him through wires inside the shafts.... That globe on his helmet gives forth a light that equals the noonday sun, and his eyes do the same. At will I can extinguish all the lights, or only one at a time, just as I may elect. Then another knob starts him going, and then another will turn him to the right and another to the left—just as a faithful horse obeys the rein and the bit—and all, too, without my being exposed to any danger from without.[42]

Senarens must be given credit for independently inventing the genre of young inventor stories now called Edisonades, after Thomas Edison, already a legendary figure (see chapter 1). One of the earliest appropriated Edison's name for a series of humorous adventures featuring the character Tom Edison, Jr. The first was the delightfully titled *Tom*

Edison, Jr.'s Sky-Scraping Trip; or, Over the Wild West Like a Flying Squirrel, published in *The Nugget Library* #102, dated July 16, 1891, from Street & Smith, which would later become one of the biggest publishers of pulp magazines. This work parodied Senarens' output, as did the overt *Tom Edison, Jr.'s Electric Mule; or, The Snorting Wonder of the Plains*, published in *The Nugget Library* #128 of January 14, 1892, under the name "Philip Reade." The mocking stories only ran to eleven titles over a year, but they may have led Tousey to strike back with the *Frank Reade Library*. If so, Street & Smith's *New York Five Cent Library* went even further in 1893 with five adventures of "Electric Bob." These stories were also directed at Frank Reade, perhaps never more so than in *Electric Bob's Big Black Ostrich; or, Lost on the Desert*, whose cover features a weaponized giant electric ostrich. The caption excitingly reads, "Bang! Bang! Bang! Every report from Electric Bob's machine gun was followed by a yell or splash from the enemy."[43]

—ɯ—

Steam men went on exhibit all over the country shortly after the stories about Dederick appeared. Handbills—large posters pasted onto the sides of any structure that stood still long enough for the glue to set—advertised coming attractions on the fair, circus, and entertainment circuits. They were the finest achievements of the pitchman's art, using the gaudiest of lies to promise infinite pleasure that seldom could be matched by cold reality. Those in St. Louis in December 1868 hyped a machine that had been "exhibited in New York, Boston, Chicago, and other large cities in the largest halls, and run over the principal race courses, and is after the same pattern as the one now creating such a great sensation in Paris. He pulls a carriage a mile in 2:15, and performs other wonderful and almost human feats."[44] Actually more than human: 2 minutes, 15 seconds for a mile would be a respectable time for a racehorse in 1868. The *St. Louis Journal of Agriculture* called the handbills "somewhat exaggerated, as usual," and thought the steam man top heavy and unsteady. "In our opinion the best place for the steam man would be on a tread mill, where it would probably do good work."[45] Designated "the greatest invention of the age," credit is given to "Messrs Z. P. Dederick and I. Grass."

A steam man was also exhibited in New Orleans starting on December 26. "Some persons cannot appreciate great inventions," reported a Pittsburgh newspaper with heavy sarcasm. "The Mayor of New Orleans actually says that the steam-man is a humbug, merely because he can't do the things his owners profess he can."[46]

The St. Louis steam man was certainly underpowered, given a weight of 600 pounds and a 1.5-horsepower engine. The crank that drove the feet could rise 8 inches and move the foot forward 18 inches, figures that do not correlate well with a supposed speed of almost 30 miles per hour. (For comparison, Usain Bolt uses strides near 100 inches to sprint 23 miles per hour.) Nevertheless, smaller cities throughout the Midwest saw advertisements in their papers for "The Wonderful Steam Man!" into the autumn of 1869.[47]

Back in Newark, Thomas J. Winans sought to emulate and even surpass Dederick. Coming from a Binghamton, New York, engineering company, he gained funding from Joseph Eno and other bankers for the $3,000 project.[48] Winans, Eno, and their associates held a grand unveiling on February 22, 1869: "It is six feet nine in its India rubber shoes, and is attached to a handsome carriage, at the back of which is the boiler. In the breast of the iron man are two oscillating engines, working on a crank at right angles, and the joints in the legs are as perfect as they can probably be made. It walks off with the carriage as if it were of no more weight than an feather, and can walk backwards almost as well as forwards."[49] Drawings show a slighter figure than Daniel wearing a crown, giving it the nickname "Steam King," and pulling a light buggy with a top for covering the passengers.[50] A four-horsepower engine gave it oomph, needed to haul the figure (variously billed as 500 or 600 pounds) and the one-ton carriage.[51] Winans and Eno took the same route as Dederick, first exhibiting the mechanical wonder at Crump's Garden and then at 524 Broadway in New York before taking it up to Harrisburg for the Pennsylvania legislators. "He will be no novelty," one newspaper wit claimed, "as all the men steam up there."[52]

The Steam King later spawned a vast number of legends when, in 1909, Eno's son Alfred—by then a rich and powerful New Yorker—displayed him as part of the grand parade featuring 30,000 marchers and 75 floats up Fifth Avenue for the opening of the Queensboro Bridge (better known as the 59th Street Bridge).[53] The Steam King was billed as "the first automobile ever constructed," a claim as disconnected from history as the other stories concocted by publicity men for the ever-credulous papers to print. The "devil car" spewing smoke out of its nostrils, they reported, once lost control and kidnapped General Ulysses S. Grant for a 27-mile ride before it could be stopped.[54] Another time the Steam King was said to have torn through a funeral procession, terrifying the mourners and horses equally, until a Indian threw his tomahawk squarely through the throttle valve.[55] In reality, the Steam King was confined to a small circular track, though it could carry two passengers.[56] Thousands paid their quarters for the spectacle, and it is certainly possible that celebrities like Grant, the Prince of Wales, and General Custer might have been granted the privilege of a ride.[57] However, more robust transportation accomplishments are dubious.

At least the Steam King survived. A forlorn announcement in the *Charleston Daily News* on March 11, 1870, revealed that "the 'steam man' which was built in Newark, N. J., a couple of years ago, has been for some months stowed away at the Union depot, Indianapolis, on account of the non-payment of freight."[58] That merely meant less competition for its rivals. The April 7 *New York Herald* publicized an exhibition at 551 Broadway, ground zero for steam men: "The exhibitors guarantee there is no fraud or trickery of any kind to deceive the public, but that visitors will be shown an iron man that can walk any distance, taking its feet off the ground at each step, without any support whatsoever, so that it can only with difficulty be distinguished from a living being."[59] A chatty article in *Scientific American*, then obsessed with the fertile ground of American technological achievement, assured readers that this steam man was "very much more satisfactory than his predecessor exhibited two or three years since in this city, who could only stand upon fixed crutches, and kick like a spunky child suffering for a spanking." As the new steam man could hobble only "forty feet per minute," or half a mile per hour, "with a rather unsteady gait, and with extremely short steps," it's clear that the only advance was the substitution of "some fluid hydrocarbon" for coal.[60]

The giant Centennial International Exposition in Philadelphia, celebrating the hundred years since the United States had declared independence in 1776, featured (among other marvels) a mechanical horse and some steam men, "simple in construction, and very satisfactory in their movements."[61] (Robots became mandatory adjuncts to expositions in their 1930s heyday. Expositions celebrated invention and technology of all kinds; putting a human face on them, even one of metal, connected with fairgoers.)

Shortly after the Centennial, presumably inspired by what he saw, a Captain Rowe, so the newspaper story goes, started a multiple-year journey from Sandusky, Ohio, to Philadelphia in his boat, the *Experiment*. On it he built a steam man, supposedly exhibiting his creation along the coast from Richmond to Boston.[62] It walked with the support of a wagon, in the Dederick and Grass style, but could not stand alone and staggered in a "feeble and uncertain manner." The story then takes an odd twist, veering far away from the triumphant saga of youthful genius sold by nickel and dime novels. Captain Rowe developed what we now might call a mania and a persecution complex after continued ridicule by the public and failure to perfect his steam man. After his wife died, he docked along the Schuylkill River near Philadelphia with his three children (aged 3, 6, and 12) and made the ship their home, in increasingly squalid conditions.

Captain Rowe came to the attention of authorities in January 1883, as told by a syndicated article written by the Philadelphia correspondent of the *New York World*. The steam man was literally incomplete, with the head and body viewable at Wissahickon (now part of the city but then a separate town) and the lower half in the machine shop on Rowe's boat. When joined, the whole would "resemble a short, thick-set man in ancient armor." The interior mechanisms were in the familiar style but purported to be "strong enough to draw a horse-car and fleet enough to distance the fastest runner."[63]

Yet just two days later the steam man was no more. "His Enraged Inventor Knocks Him into Smithereens," read one headline. The report put the demise "a few weeks ago," meaning that it was already gone before the first story had appeared.[64] These stories cannot now be reconciled. Of interest is the anthropomorphizing of the steam man in the account, making the battle between him and his inventor a tale out of mythology:

> Captain Rowe … struck him a fatal blow in the temple, crushing in his skull, and causing his vapory brains to fill the air for miles around. He fell on the ground, gave out one heart-rending whistle through his iron nose, and then slowly puffed his life away….
>
> It took ten minutes or more for the Steam Man to exhibit signs of life after finishing the meal [of coal, wood, and water]. Then his eyes began to glow; his breath came with a long, powerful respiration; his chest heaved, and he stretched his off leg as if proud of his power…. [A]gain did the curious being start forward, this time waving his arms and snorting loudly. [Captain Rowe stumbled.] The next instant the Man was on top of him and had given him a stunning blow in the chest with his iron arm. The Captain, who is a large, powerful man, and somewhat impulsive, got angry and aimed a blow at his pupil's nose. It countered very cleverly, and the Man struck the Captain on his chin…. [A]lmost livid with rage, [the Captain] jumped nimbly aside, seized a club and struck the Steam Man in the head, knocking him "silly." … Not content with having felled his intractable pupil, Captain Rowe attacked him as he lay prostrate and broke him into small pieces.[65]

Captain Rowe and his steam man could easily be dismissed as a hoax, the fantasy of reporters concocting a good yarn on a slow news day. The original story caught the fancy of editors, certainly, and appeared in smaller newspapers continually through May 1883.[66] The later story, attributed merely to *Philadelphia Press* (a real newspaper of the day) received almost no reprints, perhaps too blatantly fictional even for newspapers. However, one fact argues against a hoax: There are indeed records of Captain Rowe exhibiting the steam man up and down the coast. On July 6, 1878, the *Wilmington News Journal* reported that "Adam Ironsides, the steam man," was on exhibition there.[67] He was moved to Reading, Pennsylvania, in August. "The 'Walking Man' is run by two small engines, and walks over a circular course about seven feet in diameter…. The machine is five feet high and weighs 83 pounds…. Mr. Roe and his family travel with their curiosity, upon a small steamer called the 'Experiment.'"[68] Two years later, "C. C. Roe, of Hamilton, Ont.," again in Reading, was to appear in Harrisburg, Pennsylvania, with the same steam man, said to have been patented in 1874. The newspaper gave the sad news that his wife had died in Washington since their last visit.[69] Roe and Rowe are indisputably the same man.

Those reporters did real reporting. Cyrenius C. Roe of Hamilton, Ontario, identified as a machinist and showman in city directories of the 1870s, received Canadian patent 4175 on December 15, 1874, for a "Steam Man or Walking Machine."[70] The showman took precedence on posters that proclaimed the machine Adam Ironsides (a real name, remarkably, and found in Canadian newspapers of the day, indicating a possible source for the appellation)[71] and falsely told the public that they could boast of having seen "the first steam man that was ever exhibited before the public." The misnamed Adam was five feet tall and "capable of walking four to ten miles an hour."[72] Promotional illustrations show a man-like bearded figure, with smoke pouring out of his hat and steam from a pipe clenched in his mouth, walking by itself past a farm. The patent application clearly supplies the missing wagon that was necessary but subverted the dream.[73]

There was an epilogue to the otherwise phantasmal story of Captain Rowe: The Society for the Prevention of Cruelty to Children investigated the living conditions on the *Experiment*. One newspaper account said the children were found to be "well-cared for," while another claimed that their quarters were "filthy and teeming with animal life" and the girls were never sent to school, although the older boy showed every bit as much inventive imagination as his father.[74] We can neatly excise inventions from history, the way we clip an article from a newspaper, but the real humans who made these devices lived real lives that sometimes did not have storybook endings. Newspapers report the rare and usually fleeting appearances of mechanical men, but they rarely give a glimpse of the emotion-laden humans these creations are meant to supplant.

History, improbable as always, almost repeated itself with the saga of the two George Moores in the 1890s. Professor George Moore, originally of Canada, and George R. Moore of Lowell, Massachusetts, both made national news with their walking mechanical men. This time, perversely, they proved to be separate individuals and not misreported versions of the same person.

George R., a 70-year-old retired mill owner, struck first, as described in a syndicated article deliriously headlined "Electric Frankenstein."[75] As the epithet indicates, George R. made a huge robotic advance by substituting a six-horsepower electric battery for the prohibitively awkward coal-fired boiler as motivating power. The electric man, who wore "a new suit of gray mixed goods ... a soft felt hat [and] a No. 6 shoe, four wide, and sport[ed] a Louis Napoleon mustache," kept upright through being attached behind what writers referred to as a perambulator or an invalid chair—basically a light, wickerwork wheelchair. That gave it a utilitarian purpose, its power "sufficient to enable him to push the fattest dowager up a hill." As always, the articles appear to be based on publicity handouts rather than reporting, given the ratio of superlatives to description: "The action of the feet and legs ... is a remarkable imitation of a human being.... Each foot, as it leaves the ground, rises naturally on the toe, with the same springy motion that is characteristic of the graceful walk among men and women."[76]

Although George R. Moore received patent no. 454,570 on June 23, 1891, his invention never gained traction.[77] He then disappeared until the very end of the century, when the first show of the Automobile Club of America was held in the old Madison Square Garden in November 1900, reviving a huge number of older predecessors to the automobile, including Moore's "walking automaton." The *New York Times* called it "the most ludicrous feature of the lot," while the *New York Tribune* termed it "as striking a novelty as anything in the show."[78]

George R.'s creation appears to have been built with utility in mind. By contrast,

Professor Moore—"Professor" probably being a self-bestowed title to better suit his image—started in show business and never left it, pure but successful humbug the entire way. His steam man, dubbed Hercules, debuted in the surviving records on January 23, 1893, in Scranton, Pennsylvania, traditionally a launching pad for getting the kinks out of performances before taking them to the bigger cities. Hercules received hyperbolic praise in what should be assumed to be a printed press release:

> The marvels of marvels, the triumph of mechanical ingenuity, the steam man, was viewed by large crowds at Wonderland all day yesterday. Truly speaking it is a marvel. It is made of copper and stands about seven feet in heightn [*sic*]. In appearance it looks like a man clad in the armor of ancient times. The boiler is placed in the chest and the machinery of the engine is directly underneath.[79]

Although the above description sounds nearly identical to what Dederick and Grass had produced a quarter-century earlier, the easily impressed *Scientific American* ran an article much reprinted in newspapers. It praised the use of gasoline rather than coal, forgetting that it had similarly praised the use of a liquid hydrocarbon for the same purpose back in 1870, although admittedly the boiler now turned a small "high-speed engine running up to 3,000 revolutions per minute or more." That modification allowed Hercules to be more compact than his predecessors, standing a mere six feet in height (the extra foot in Scranton probably composed of hype). "When in full operation [Hercules] cannot, it is said, be held back by two men pulling against it," though certainly this was said only by Professor Moore. The professor was also said to be working on a larger version that would pull a wagon containing as many as ten musicians, a mobile exhibit guaranteed to draw crowds. As always, the steam man was said to imitate humans flawlessly: "The action is quite natural, and the hip, knee, and ankle motion of the human leg have been very faithfully imitated. The figure moves at a brisk pace and can cover four or five miles an hour."[80] As with other steam men, steam escaped through a pipe clenched in its teeth while smoke rose through the crown of its head.

The wagon is the only hint in these early articles that Hercules still needed support to stay upright, which it obviously did. Hercules suddenly reappeared on the exhibition circuit in 1896, keeping some low company. When he walked around one of the large platforms at Austin & Stone's Museum, a vaudeville theater in Boston, he was on the same bill as Jarrow, the strong boy, along with a three-headed songstress and a Yokohama troupe of lady necromancers.[81] The advertising preceding the "smartest all-around show in the city this week" called Hercules the "Crowning Climax of Modern Mechanism."[82]

In retrospect, Hercules would have been just another in the long line of steam men were it not for the stories told about it coming to life on its own, like yet another modern Frankenstein and his monster (or, as one less-than-literary editor fashioned it, "Frank Enstein's monster"[83]), which was never far beneath the surface of all robot stories. While the steam man was at Austin & Stone's, the *Boston Sunday Globe* reported that Professor Moore had forgotten to bank its fire one night, so that the watchman caught Hercules on the prowl at midnight. The steam man, more in imitation of Samson than Hercules, tore one of the building's large supporting pillars down and then ventured over to the candy and fruit stand. Here "he didn't do a thing, as the small boy says, when he means the opposite." The watchman finally got close enough to pull the steam valve.[84] In 1901, on exhibition in Cleveland's Forest City Park summer resort, the same lack of attention to detail purportedly allowed some of the campers to set Hercules in motion. He "terrorized the park for an hour" but then "marched up to [the bar] just as though he had money," flipped over it, and stood on his head, legs kicking, until his steam ran out.[85]

This tall tale drifted through newspaper legend and was still being printed as recent news as late as 1908.[86]

—⁘—

A subtle evolution marked the steam men over this quarter-century, yet one more fundamental than the mechanics of their boilers. The earliest iron men were designed for a purpose, usually that of cartage or other transportation, before being exhibited. By the next decade, showmen like Rowe and Professor Moore understood that these devices could be designed exclusively for exhibition, with implied purpose giving them greater status than the automatons making merry on vaudeville stages.

Mechanical walking men had two inherent limitations: they could not walk without an attachment that kept them upright, and they could not develop sufficient power to perform useful work from any power source that could fit inside a torso, even an oversized one. Lewis Philip Perew saw in a moment of inspiration that both limitations could be bypassed if the walking man's purpose was changed.

On August 4, 1895, the *New York Times* announced that the long-talked-of "electric man" had finally arrived in a suburb of Buffalo: "All that he is good for at present is to pull a cart around the streets, and this he is doing in Tonawanda, to the delight of the populace, and, incidentally, it is presumed, to the advantage of a certain soap of which the sides of the cart bear exultant signs."[87] People couldn't help but look at an electric man. Perew couldn't charge pedestrians for the privilege of looking, so he shifted the cost elsewhere. This financial model eventually underlay the spectacular growth of television and the internet.

Perew also had the brilliant insight to realize that the public didn't care whether the motor was located inside the mechanical man, and that it would be better elsewhere.

> The figure draws, or appears to draw, a heavy steel carriage, in which is stationed an electric battery which furnishes lights for seven incandescent lamps, including the diamond in the shirt front.
> A gasoline engine of three and a half horse-power is also fixed within the covered carriage. Around this engine winds a network of wires and steel rods connecting with the mechanism in the interior of the man.[88]

Unlike Winans and his Steam King, Perew had actually invented a motorcar. A patent he filed much later (approved in 1910) clearly shows the motor sitting under the carriage, transmitting power to the two axles.[89] The figure—described as six or seven feet tall[90]; made of wood, steel, and brass; clad in a "military" or "Continental" uniform; and said to resemble then-president Grover Cleveland—was nothing more than a larger version of "the Spirit of Ecstasy," the graceful hood ornament on a vintage Rolls Royce.[91] With the figure's feet lightly skimming the ground, there would be no more worries about cobblestone streets, hills, or uneven terrain that had so stymied earlier walking mechanical men.

Even better, there were plans to fit the figure with a phonograph, making its possibilities so unlimited that a Niagara Falls *Gazette* article syndicated nationally couldn't help but end with a reverie about the future:

> The phonograph can say whatever is desired. It can expound the virtues of patent medicine or be used for political campaigns. So, at present, the only form of labor threatened by the invention is that of the sandwich man and the campaign speaker. The men and carts that are used to extol medicines will be very fine pieces of mechanism, and can be geared to go as fast as any one desires. By simply turning on a current, the man, his eyes still fixed on eternity, can hump down the street at a rate far

exceeding any bicycle. The limit has not been reached. In course of time it may be that men can be constructed to do almost anything and the laboring man can sit around and smoke twenty-five cent cigars while a multitude of electric men do all the work. This will not occur for some years yet, but no one can say where it will stop.[92]

Perew was about 33 years old at the time and worked in a merry-go-round factory; not coincidentally, he was the owner of U.S. and Canadian patents for a merry-go-round. He would go on to patent a variety of mechanisms, small and large, before his death in 1946, including a towing system for canal boats, a butter blender, and a dish-washing machine. He seemed to have true inventive genius. What he lacked was money. He had been working on his walking man since 1891, when he created a half-sized version, but, unlike the earlier hypesters, he admitted that his work was still too crude to be fully successful even though he had invested $3,000 in his electric man.[93]

The work was set aside until 1899, when Charles A. Thomas, a moneyman from Cleveland, arrived. With a capitalization of $10,000, the United States Automaton Company (the true predecessor to Isaac Asimov's fictional, yet far more famous, U.S. Robots and Mechanical Men, Inc.) formed in February 1900.[94] Perew threw himself into remaking his automaton, alongside his assistant superintendent of construction, Joseph Albert Dischinger, who was also made a director of the company. Dischinger was a key contributor whose name would appear with Perew's on the 1910 patent.

They built up the figure to a full seven feet, five inches, "well-formed, of heroic stature, and has a dignified military carriage."[95] Nomenclature remained a problem (perhaps explaining why the term *robot* was so immediately embraced when it finally appeared). Various newspaper and magazine stories called Perew's creation an automaton, a mechanical man, a machine man, a wooden and iron man, an automatic man, an electric man, a Frankenstein, a bogy, and even a graven image. ("Great Christopher" appears in a single newspaper from 1902 and never again.)[96] Regardless of the name, the walking man impressed reporters: "He has a quick step of the perfect heel and toe walker. He is dressed in the height of fashion, in a white duck outing suit and cap of the latest shape."[97] The phonograph that was mooted in 1895 was installed to dazzle spectators: "Looking at the party, his lips moved, and I was dumbfounded to hear him announce in a deep bass voice: 'I am going to walk from New York to San Francisco.'"[98] The jump from a sideshow attraction marching in a circle to a cross-country jaunt was a first-class publicity stunt aimed directly at the press, which responded in terms of awe:

A man that walks is a common sight. A dead man that walks is occasionally beheld by sailors on a Saturday Night. But a man that walks long distance that never was alive is something so unheard of that it is hard to believe that such a one could exist.[99]

W. B. Northrup went deeper into philosophical dread when he penned an article on Perew's work for the mighty London magazine, *The Strand*, in the same issue as chapter 1 of H. G. Wells' new work, "The First Men in the Moon":

The Frankenstein of Tonawanda has brought into existence a thing of wood, rubber, and metal, which walks, talks, runs, jumps, rolls its eyes—imitating to a nicety almost every action of the original on which it is founded. All that is lacking is the essential spirit—the Promethean fire, as it were—which would enable one to say to the automatic figure, "Thou art a man." ...

It could be made to carry loads in places inaccessible to ordinary vehicles with wheels; it could ascend heights impossible to men; it could walk distances which would weary the most skilful [*sic*] pedestrian ... it could become a fighting appliance, carrying death and destruction in its machinery....

Would it be allowed on the city streets? Would it not endanger life from causing horses to run away? Would it not prove too great a shock to children and nervous women?[100]

The automaton could supposedly go twenty miles an hour in its size 13½ leather shoes. Either two or four figures would sit in the carriage for the cross-country trip, which had a vital announced purpose: to publicize the Pan-American Exposition to be held in Buffalo in 1901.[101] Officials there no doubt were displeased to see in a newspaper such a not-so-subtle plea for them to finance the trek. They had no interest in spending their money on that kind of craziness. Nobody did. Although a couple of belated mentions of the project appeared over the next two years, Louis Perew and his automaton never set foot outside of Buffalo.

Or did he? Articles appeared in the *Pittsburgh Press* early in May 1913, announcing that a bloodless creature made by Louis Perew would walk the streets of the city as part of a trip from New York to San Francisco.[102] The *Fort Wayne Journal-Gazette* of February 5, 1914, made the same announcement.[103] However, there is no follow-up, no sightings of "Great Christopher," no record of anyone in these cities seeing the automaton walk. These articles read as if they had been sitting in a stack of press releases for more than decade and were suddenly plucked out for use. They are inexplicable. And yet they had their own effect, for they stimulated the thinking of two other men who kept the mechanical walking man alive until after World War I.

Ferdinand "Fern" Pieper was from Alton, Illinois, then a small industrial city located on the Mississippi River just north of St. Louis. In 1913, the local paper began a series of stories about the "ingenious electrical inventor" when he displayed "a miniature mechanical man who can walk, and who does it with remarkable ease and grace."[104] Pieper was about the same age as Perew when he first built his mechanical man, and he copied Perew's logic almost exactly. (The *St. Louis Republic* covered Perew heavily in his 1900 heyday.)[105] Pieper wanted funding from some company that saw the promotional force of "having a mechanical man walking about the country advertising their goods."[106] Still at it the next year, Pieper thought that a nine-foot-tall man could walk the country to publicize the upcoming San Francisco Panama–Pacific International Exposition.[107] Again mimicking Perew, his electric man had the motor in a trailing wagon, but Pieper threw in one twist: he didn't bother to make his model realistic. At this time, it was a headless box on top of a pair of legs in front of a two-wheeled trailing cart that, as the reporter said discreetly, "assists materially in boosting him along."[108] This contraption walked unassisted down Helle Street in Alton.

Two years after that, responding to the technological carnage of the European war, *Illustrated World* magazine splashed on its cover an evil-looking robot giant leading a tank charge: "'King Grey' will be 9 feet tall: his weight will be 750 pounds.... King Grey will be drawing a vehicle, weighing over 1500 pounds, carrying four people, any distance desired." The writer blustered on in full chauvinistic fear-mongering mode about what might happen if a foreign power were to get to this ultimate weapon first. The picture of the real headless model somewhat undercut the achievement, and referring to Pieper as "Vern" rather than "Fern" must have been supremely grating. Yet such publicity seemed to ensure that he would get backing for his full-size man.[109]

Ultimately, Pieper suffered the same fate as Perew. Although authorities in all the warring countries tested scores of small-town inventors with wild ideas, Pieper would not be one of them. In 1918, he passed King Grey over to a Chas. Oehler. Neither Pieper nor Oehler was there when opportunity knocked. "A man from St. Louis" appeared in

Alton in April 1918, wanting "to test out the man prior to closing a contract to have him rigged up to help sell Liberty Bonds." The unnamed visitor couldn't wait for Oehler to reappear, so the workmen in the shop got the mechanical man going. A crank broke. Then a second, newly made crank broke. Six men pushed the automaton back to the shop.[110]

The scruples of the visitor also broke. Almost exactly one year later, a 12-foot-tall walking mechanical man drew admiring crowds in a Victory Loan parade in St. Louis. Its inventor was named as Arthur B. Christopher, who claimed that he had been working for six years, spending a colossal $15,000 to rebuild his figure six times before finally perfecting the "realistic likeness to human locomotion" of the giant's 24-inch-long, 100-pound feet. Under the headline "Uncle Sam to Drive 'Fritz' around City," the *St. Louis Post-Dispatch* described the challenges facing any "walking" robot:

> A mechanic mounted a seat in front of the engine and turned a wheel. The engine began barking. A tremor was seen to pass down one of the great iron legs. A foot advanced itself and plunged down upon the ground. As the earth was soft, it did not get a foothold. With a grotesque simulation of sentience, the iron heel began digging itself into the earth to get a purpose. Firmly planted at last, the leg lurched forward with force enough to kick a man into the air.[111]

Was Christopher the mysterious visitor from St. Louis? If so, the breakdown of Oehler's figure might have been a boon, allowing Christopher to get a good look at the structure of the automaton's legs. Christopher is referred to as a mechanical engineer, so he might have been able to refine the inner workings—at least a bit. His giant moved at a mere two miles per hour, slow even by parade standards. He pulled (or, better, was powered by) a two-wheeled float in the shape of a ship named "Victory." Orators stood on its deck and urged the public to give, give, give. The government, which had run a surplus in FY1916, incurred a $13 billion deficit in FY1918, more than the total deficit that Franklin Roosevelt would run up in his first four years of Depression spending. St. Louis was tasked with raising more than $50 million in this fifth round of Victory Loans, and Missouri was ranked low in meeting its goals. All possible war-related or war-relatable gimmicks were trotted out to spur yet more giving.[112] Victory loan parades were held in all major cities and most smaller ones, allowing slippery operators a chance to gain some publicity for their own efforts.

Christopher was definitely wily: he even stole the name for his giant. The name "Fritz" came from the most famous robot of the early 20th century. *Fritz von Blitz: The Kaiser's Hoodoo* was a short-lived Sunday comic strip from August 18, 1918, through February 23, 1919, from veteran artist Harry Cornell "H. C." Greening. Fritz was a robot, a comic servant that wreaked havoc by faithfully following its orders no matter what a mess they made. Fritz was a reincarnation of Greening's earlier strip *Percy* (a character that Pieper's mechanical man was often compared to). As the *Alton Evening Telegraph* reported in 1913, the inventor "Has Built a New 'Percy.'"[113] (The complete saga of Percy follows in chapter 7.)

For now at least, the era of the walking mechanical man was over. Their appearances had garnered continuous headlines, stunned audiences, drawn enormous crowds, impressed some, and disappointed others. It was the age of wonderful mechanisms. Few doubted that better automatons would be seen on streets regularly in a few years. In fact, they would return in another decade to the same acclaim despite their shortcomings, able to do only another handful of tricks, yet retaining their fascination as humanoid alternatives to living beings because of the wafts of progress they emitted, a continuing reminder of humanity's unique ability to replace itself with something bigger, stronger, and faster.

5

"Quiet, Please—
I'm Talking"
The Westinghouse Family of Robots

The so-called boys' novels came in many subgenres, though all of them concentrated on stalwart and strapping lads overcoming hardships and succeeding through manly application and the sweat of their brows and the brains within. A branch off this flooding river of tales extolled the heroics of the men who built bridges to span rivers and move adventures closer to the wild world outside cities: engineers (probably the one time in history they were celebrated far more than scientists). H. Irving Hancock put forth a Young Engineers Series, chronicling their feats building railroads and digging mines. Hardly documentarian, these books nevertheless reflected a world far closer to reality than science fiction and achievable by any plucky youth.

As proof, consider the background of Roy James Wensley. Born in the Midwest in 1888, Wensley dropped out of school after the eighth grade for the classic reason of needing to support his mother. The dawning age of electricity called him, and he worked days as a linesman and electrician's helper, while he sat up nights at his mother's table studying for a degree in electrical engineering. No fancy college for him—in fact, no college at all. He received a striver's diploma from the International Correspondence Schools (ICS) (still in business a century later, perhaps familiar from a long series of ads employing Sally Struthers as spokesperson). Wensley won jobs with smaller companies but wanted more; unfortunately, the giant Westinghouse Elecric & Manufacturing Company, one of the leaders in electrical equipment, shut him down by loftily informing him that it didn't recognize ICS "degrees." In a move that would seem disingenuous if it appeared in a Horatio Alger novel, Wensley offered to work for Westinghouse for a month without pay to prove his worth. The company accepted.[1]

By 1924, Wensley had established himself as an innovative leader, producing the small but crucial advances in what are now called automatic systems that were needed to lift electronics out of the age of personal services. A newswire spread the story across front pages: "The supervisory control system … is a mechanism which can receive and carry out orders automatically, Mr. Wensley explained." The system worked over existing telephone wires to report information like water levels in the Washington, D.C., reservoir system. "Mr. Wensley dialed hypothetical sub-station No. 3 on an ordinary dial arrangement…. Then Mr. Wensley coded a question demanding how much water was in the pond turning the wheel. Instantly a short series of buzzes came through the receiver.

Each buzz indicated a foot of water.... An inanimate machine in the sub-station had received Mr. Wensley's orders and had carried them out with human direction." The implications were obvious to anyone who had heard of *R.U.R.*: "Such a system would affect an economy by eliminating about 20,000 operators in sub-stations scattered over the country."[2]

In the boom economy of the 1920s, service and safety triumphed over potential job loss. Westinghouse devices saw immediate use in a variety of industries, especially street-car systems, where the ability to control switches down twenty miles of track had required costly and specialized equipment. As an engineer, Wensley saw an engineering problem and a source of profit for his company, and nothing more.

Ordinarily at this point he would have dropped out of history just like the thousands of other engineers who made similar contributions before and since. However, a nameless operative in the Westinghouse publicity department changed his life. Wensley's love of gizmos—he rigged the doors in his lab building to respond to the words "Open Sesame"— was shared by the executive staff at headquarters, engineers themselves, who elevated the remote-control systems into life-size model train sets that they could play with and show off to their buddies. The non-engineers in publicity went wild in their own ways, envisioning the remarkable achievement generating national headlines featuring the word *Westinghouse* in bold type. Therefore, they set up a demonstration for the New York City papers.[3]

Wensley's luck continued its amazing course. The person in journalism best suited to recognize the potential in a box of tubes and wires had just been named editor of science and engineering for the *New York Times*. Moving to the supposedly staid *Times* put no brakes on his sensibilities. Sunday newspapers regularly ran full-page, heavily illustrated articles on the latest technological wonders, competing with the magazines and probably reaching a larger audience. Waldemar Kaempffert gave Wensley's "mechanical slave" a full page in the Sunday *Times* for October 23, 1927. Nicknamed "Televox" because of the vocal control signals sent via telephone, the device was otherwise undistinguished, especially when compared to stage automatons and inventors' robots. An unprepossessing large rectangular box of tubes topped by a smaller cube of relays, the pair stacked one on top of the other appeared at best to be a vaguely anthropomorphic cartoon facsimile of a human head and torso. Fortunately, Kaempffert had as much visual sense as print facility; he discerned the potential before anybody else. He got an artist (thought to be his friend Samuel Johnson Wolff) to add arms and legs and show the mechanical man taking notes on a pad, using a long stick to measure the depth of water, tightening a bolt with a wrench, and, for some mysterious reason, pulling the string on an overhead light to turn it on. A dial on a gauge made for a perfect thought balloon exclamation point, as if Televox were surprised by the readout. If any single person is responsible for the contemporary image of robots, this uncredited artist is that person.

No longer sleek and indistinguishable from humans, robots would now be boxy and cartoonish, friendly and helpful, capable yet clumsy, perpetually smiling through oversized heads, and no more likely to become an evil menace than an automobile would be thought guilty if one ran you over.

Images would conquer words. Kaempffert the careful science writer warred with Kaempffert the showman. Deep inside the article accompanying the illustration, he contradicts every cheerful stroke of Wolff's imagination:

Wensley's invention effectively disposes of the Robot type of automaton dear to writers of fiction and plays, or the artificial man created by a Frankenstein. There will never be a robot—a brainless, tireless, unemotional mechanism fashioned in the image of a man, performing all the functions of a man, moving about stiffly but surely, pulling levers, turning control wheels, wielding broom, pick or shovel. Medieval contrivers frittered away their talents in constructing lifelike automata that could write a name and play a tune or two, and that outwardly resembled Robots. The modern engineer has no patience with such fantastic creations.[4]

Kaempffert was profoundly right and profoundly wrong. Engineers never succeeded in producing true mechanical men, understanding exactly as Wensley and even Čapek had that the human form impedes efficiency and requires simplification for industrial use. The public cared not a whit. They wanted a show. And the very first person to succumb to showmanship was Wensley himself.

Now relegated to a life of demonstrating his invention at the instigation of the Westinghouse publicity department, Wensley served as a mobile billboard. Fellow engineers undoubtedly preferred the open case of electrical equipment so that they could see the wiring for themselves. The general public wouldn't. Assigned to do his song and dance on Washington's Birthday in 1928, Wensley desperately sought something that would impress. Then he remembered Wolff's cartoon.

> [T]his can be credited for giving me the "IDEA." Its importance demands nothing less than capitals.... Why not give my new equipment a body and some limbs? As it was only for symbolism, I made no attempt to have this figure resemble an actual human being. It was but a piece of wallboard crudely painted and loosely dangling arms of the same material.... I arranged a simple mechanism ... so that at my whistle of command the arm of the figure would unveil a large portrait of Washington [hidden behind a flag].... [The judge] blew the note that released the trigger, down came the cardboard arm and up went the flag. The orchestra played the National Anthem which brought every one to their feet for the climax. No more effective ending could have been arranged and I sat down to a thunder of applause.[5]

The *Times* dedicated four articles over the next week to Televox, which now looked like a perfect cartoon robot. In his retelling, Wensley omitted his crucial innovation. He gave the construct a face, drawing rectangular eyes, a rectangular mouth, and a large caret for a nose onto the wallboard covering the small box that served as the "head." No Frankenstein's monster for him; Televox looked about as scary as Casper the Friendly Ghost. Televox's picture appeared in syndication above an artist's rendition of a "lovelorn" woman sitting in his lap, being pleasantly squeezed by the robot's arms.[6]

With more Wensley luck, that very day the Fox Movietone News, then the only filmed newsreel that used the brand-new sound technology, called for a demonstration.[7] Just as the first Cinemascope widescreen cameras would later be placed in a rollercoaster for maximum oomph, sound cameras sought new and novel auditory devices. A machine that obeyed commands generated by a whistle to produce precise sounds seemed like sufficient marvel. Imagine their surprise when, instead of a box of wires, they were given the gift of a performing robot. Televox may have followed Maria, the beautiful robot in Fritz Lang's film *Metropolis*, but few saw foreign art films, as opposed to newsreels that played everywhere. Televox stunned theatergoers, reality that surpassed any fiction, built by a giant corporation rather than a backyard inventor, incorporating technologies that were already being applied in the real world and would soon—as hundreds of forthcoming advertisements promised—find its way into every home.

So likeable a robot needed a friendlier name, and at some point Wensley started referring to it as Herbert Televox.[8] Wensley, himself a Republican, again read the public perfectly. Herbert, the triumph of American engineering, emulated the coming presidential triumph of hugely popular ex-engineer Herbert Hoover in the election of 1928. Of course, metaphors slice both ways. Wensley must have been galled by the inevitable editorial cartoons of a Televox with Hoover's face, a robot controlled by the backroom powers of the Republican Party. Fortunately for Westinghouse, most of the cartoonists were less political. They dragged out their oldest clumsy servant jokes, made new by substituting bumbling robots.[9]

A Televox-like robot starred in a one-act comedy on Broadway in 1929. That same year *The Gold-Diggers of Broadway* (not a play but an actual talking picture and in Technicolor) had the ingénue sing "Mechanical Man," touting the superiority of a controllable beau: "By pressing a button or two, he'll do what I want him to do; if he doesn't love me in a big way, I'll have his batteries charged, and then he'll be O.K."[10] Better known for introducing the song "Tiptoe Through the Tulips," the film was the top-grossing movie of the year.

Westinghouse sent Wensley on a cross-country promotional tour, speaking everywhere from Kiwanis Clubs to engineering conventions. Televox underwent a series of improvements and modifications, the most important being a weight reduction from 600 to 45 pounds, allowing Wensley to store the robot in a case that fit exactly under a Pullman railroad car seat.[11] The wear and tear on the wallboard required new faces at every stop. Every newspaper therefore got a unique robot to tout its community's importance. In addition, Televox added a voice through the use of an endless loop of sound movie filmstrips.

Newspapers ran pictures of Televox even when he wasn't in town, their reporters freed to venture past dull recitations of facts for cheery hyperbole. Hortense Saunders of the NEA syndicate exclaimed, "He's a real live-wire fellow."[12] "The perfect husband!" ran one caption. "It'll do all the housework and never stay out at night, nor smoke, nor drink, nor swear."[13] Women worried that men would snatch away this wonder. Would they "send him off for Sunday papers and cigars, make him clean the car, drive him to town, press his clothes, carry out the ashes, and shovel off snow?"[14] A humorous column had the Widgertons order one "that would put out the cat, feed the goldfish, water the canary, turn the gas on under the coffee and take in the milk." However, the electrician who came to set up the model kept having problems. The Televox "put out the goldfish, watered the milk, turned the gas on under the cat and fed the gas stove!"[15] Columnist Lee Shippey penned a love poem from a supposed female admirer:

> Dear Telly: I have waited long for you.
> Somewhere, I knew, I'd find my perfect mate....
> I know a lot of sheiks who think they're It,
> But, say! I figger in your head of steel
> You've just as much to think with, every bit,
> And quite as ditto in your heart to feel.[16]

Westinghouse loved the publicity, finding new uses for remote automatic systems every year, selling sound-activated units to small airports for automatic control of runway lights and to small cities for traffic light control. An experimental Telelux robot, operated by light beams instead of sound, was introduced in 1929, too early to be practical but a major innovation.[17] Almost all future Westinghouse robots used light to transmit information, as did a number of its everyday commercial control systems. The corporate future appeared bright, a time when robots would be indispensable additions to every household. (Televox got tagged with the word *robot* from his beginning in 1927, although the term *mechanical man* remained in use throughout the 1930s.)[18] Wensley proselytized a future smart home that sounds uncannily like advertisements of today:

> We have already got an order from one gentleman who has a country estate. He wants a robot. Leaving his club, he will telephone to it and say: Turn on the lights! It will obey instantly and when he arrives he will find the driveway brilliantly lighted.... How convenient would it be for the housewife when a storm comes up, just to telephone her robot and say: "Shut the windows." Not only would her robot do that but he would reply: Madame, your command has been obeyed, the windows are all shut. She could then go about her affairs with peace of mind.[19]

Telelux had a bevy of tricks far beyond Televox's abilities. He could sit and stand, talk in a rich baritone using film loops, and spot fires and release a fluid to extinguish them. For crowd-pleasing drama, Westinghouse rigged up a William Tell act in which an executive pointed an arrow at an apple on top of Telelux's head, which was crowned with a cap carrying an explosive charge that sent pieces of apple flying when hit by a light beam emitted by the arrow.

Telelux's appearance was deliberately crafted as a friendly workingman, one that by implication would do your chores. He wore overalls and heavy work shoes on a mannequin body topped by an expressionless face.[20] The slightly creepy figure was replaced in 1930 by a new model made out of molded rubber, by far the most realistically human robot to that time. The new model retained Telelux's stunts as well as the workingman's outfit, but was far more successful as a newsworthy image. Because rubber is black, Westinghouse dubbed him Rastus, a thoroughly unfortunately choice of a pejorative name for Negro servants and characters in ethnic jokes. No articles made it clear that Rastus was Telelux redux. Instead, readers saw a subservient "Mechanical Negro" with a white man pointing a bow and arrow at him.[21] Rastus was soon quietly retired.

—⁓—

Even so, Westinghouse had fully committed to conveying the image of the robot as a friendly servant. Also in 1930, Westinghouse rolled out a "sister" to Televox named Katrina. Probably a leftover Televox system, this robot continued toward humanization, with the wallboard painted with a plain human face and an outfit like that of a Dutch maid's, then a standard character for selling cleanser, cookies, and cigars. She made a short tour of smaller cities, often billed as Katrina Van Televox, "the Westinghouse $22,000 robot servant," a figure ten times a Depression family's yearly income.[22]

Newspaper articles, certainly taken from Westinghouse advance releases, touted Katrina's many household skills, often cross-promoting other Westinghouse appliances. "She will in person demonstrate a cooking range, will talk, run a vacuum cleaner, make coffee, toast and operate a light."[23] One, more flippantly, called her "the kind of servant girl who does exactly as she is told without wanting Thursday and Sunday off, and without saving the choicest pieces of food for the neighborhood cop."[24] While Westinghouse placed many huge ads in local papers prior to Katrina's visits proclaiming that "She Will Amaze You,"[25] the planted publicity articles probably oversold her abilities since she was not a true upgrade over Televox. She may have turned appliances on and off remotely, but no wallboard figure could operate them. An immobile servant gave neither spectacle nor job satisfaction. A world of bustle demanded movement, preferably with sound. The fourth member of the Televox family gave exactly that, if nothing more. It proved to be sufficient.

At first glance, Willie Vocalite must have seemed to be a step backward. Apparently constructed out of old stovepipes and ducting, Willie lacked even the modicum of swank given to rival robots coming out of England (see chapter 6). Though his ungainly body belied the goals of the modernistic movement, Willie held pure function in his form. Those articulated arms and legs were bendable: Willie both stood and sat, and he also moved either arm, duplicating all of Katrina's tiny bag of tricks plus Herbert's old show-stopper of raising the flag.[26] A small bellows enabled him to smoke when a lit cigarette was inserted into a hole at mouth level. (Every robot of the day smoked and would continue doing so into the 1950s. A cheap and easy gimmick, smoking gave the static robots an illusion of motion and a humanizing characteristic. No evil movie monster could ever be imagined smoking, but all the heroes did.)

Even better, as the name "Vocalite" hinted, Willie contained a bank of 78-rpm records, allowing his operator to pick out any of the prerecorded speeches, songs, answers, or stock phrases as appropriate. Vaudeville might have been dying in 1931, but a bit of patter is timeless.

Willie was an instant hit, a promotional tool available to every Westinghouse refrigerator dealership in the country. Willie may have done little, but his very existence signified far more. To engineers, Willie represented the concept of infinitely adaptable remote-control systems, bounded only by their imagination in finding uses. To housewives, Willie promised that Westinghouse appliances contained the latest and greatest technology to simplify their busy lives. To readers of the ads and articles, Willie offered a glimpse into a future world of ever new and improving technology that would lift the United States out of the terrible Depression that had engulfed the country. As a symbol, Willie was a multiplier that turned every penny that Westinghouse spent into dollars of return. Westinghouse's publicity department delivered dozens of stock ads into which appliance dealers, department stores, and electric companies—which then sold appliances as a way of increasing the use of electricity—could drop their names and the times of Willie's appearance, accompanied by an endless rain of exclamation points. Willie—interchangeably called a robot or a mechanical man—did "Everything but Pay Taxes," "Has Many Accomplishments Little Short of Human" and was "The World's Wonder Robot" and "the man who isn't a man."[27] Photographers loved posing "male" robots with pretty women, and Willie was no exception. Exactly what demographic was being pandered to is unclear, but the picture headlined "Robot Gives Girls New Thrill" had a rakishly mustachioed Willie posing with two young lovelies with their arms around him. The caption pondered whether "some psychologist might be able to tell which one caused the reaction that made Willie Vocalite's electric eyes flash a red danger signal."[28]

That same psychologist would come in handy to explain the allure of perhaps the most off-putting description of any putative crowd-drawer in advertising history: "He has No Soul! No Heart! No Brain!"[29] The poster also proclaimed Willie a "Scientific Marvel." Scientific marvels didn't come to western Indiana every day—perhaps that was draw enough. It seemed to be for the 1933–1934 Chicago Century of Progress Exposition, which also housed the faux automatons discussed in chapter 4. Willie had a place of honor in the center of the Westinghouse display on the balcony of the expo's Electrical Pavilion, right next to a refrigerator.[30] Advertising pounded the connection between the two devices into the public consciousness: "Only Westinghouse Could Build Willie Vocalite/and only Westinghouse could build the electric refrigerator that you want for your home."[31] Publicity at the fair made Willie into the wonder of the age.

"WILLIE VOCALITE"—A seven foot, 350 pound steel giant with an almost human brain, who unerringly obeys his master's spoken commands. This robot, on request, will stand up, sing a song, smoke a cigarette, turn on any electrical appliance—even make a speech. Amazing, educational, amusing. Acclaimed by thousands at world's fair. Is sensational; called "scientific marvel" by electrical engineers and leading scientists.[32]

Demonstrations occurred regularly throughout the day, with a handler giving instructions over a telephone. Willie didn't understand the meaning of the words; showrunners e-nun-ci-a-ted rhythmically to send beats down the wire. A story told by Westinghouse engineers has a carpenter tapping nails into a broken shelf in just the right rhythm to make Willie stand up. The workman then ran out of the pavilion and called the police.[33]

Willie made the rounds for the next decade, traveling mostly in his specially designed Williemobile and appearing in all 48 states and Canada, even traveling by ship to Hawaii, with some of the appearances simultaneous with his daily act in Chicago. How did he do it? A twin robot, nicknamed Willie Westinghouse, took over some tour slots. One had a moustache; the other didn't.[34] One final burst of glory came at the San Francisco Golden Gate International Exposition, held during the warm months of 1939 and 1940, where he had his own giant stage in the Electricity and Communication Building. No other robot came close to such ubiquity and visibility. At least 5 million Americans must have seen a Willie in person, with tens of millions more getting a glimpse through newspaper and magazine articles, ads, and newsreels. Westinghouse had seemingly reached the pinnacle of robotdom and somehow managed it during a decade of Depression.

─∞─

Yet, just as Carl Perkins was eclipsed by Elvis within a year of writing "Blue Suede Shoes," Willie Vocalite was just as quickly tossed aside for a bigger star: sleeker, more accomplished, given the better venue, and promoted with reams more publicity. And, long in advance of Data, HAL, and Wall-E, this robot was identifiable by a single unforgettable and futuristic name: Elektro.

"A mechanical man who does just about anything except think for himself today formally took his place among the wonders of the twentieth century," read the lead in the wire service story datelined April 11, 1939.[35] One delirious editor in Munster, Indiana, went with the headline: "Shades of Brick Bradford! 'Elektro' Is Almost Human."[36] (See chapter 7 for the scoop on Brick.) Although the "wonder" had no more capabilities than a boy of four—"Elektro walks, talks, and counts up to ten on his fingers … and he can distinguish two colors—red and green"—its cost (an astounding $110,000) and sheer size made it formidable, as the Westinghouse publicity department's blizzard of imposing figures showed: "Elektro weighs 260 pounds; he is seven feet tall, and has feet 18 inches long and half as wide.… Elektro's 'brain' includes an 'electric eye,' forty-eight electric relays and signal lights and weighs 60 pounds; his 'spinal column' is made of wire— enough to encircle the world at the equator."[37] If readers could swallow that implausibility, the rest went down like honey.

For all that money and technology, only Elektro's left leg bent, and he "walked" via a hydraulic ram sending a foot balanced on Westinghouse vacuum cleaner wheels out a few inches with a loud mechanical whine. The stiff right leg then rolled forward to bring the heavy frame back into balance. All of Elektro's feats were essentially magic tricks using misdirection. The super-heavy "brain" was a separate unit that stayed backstage while the body performed. Power cords, carefully not shown in the publicity photos, connected to outlets and to a microphone so that the robot could be controlled through voice signals. Like Willie, Elektro did not understand the words used to cue his movements; only the number of syllables and their cadence counted. A command was always six syllables, broken into a 3-1-2 rhythm, the verbal equivalent of the whistle sounds to which Willie had responded. "Will you come—down—front please?" started the legs moving. A single syllable, like "stop," canceled the order. The relays were set so that each new command set off another motion in the carefully orchestrated routine. "Tell your age—with—fingers" started a count. "Tell us more—'bout—yourself" triggered a set speech. "Do you want—a—smoke now?" started patter about Elektro not being afraid of stunting his growth. The operator had to memorize the order: one missed command meant the subsequent tricks wouldn't correspond to what was being said and so destroy the illusion. A misconnected relay meant similar disaster, something the publicists left out of the glowing first-day story. When the president of Westinghouse had given the ceremonial first command, the robot's response was "Quiet, please. I'm talking." Elektro then walked backward almost into a wall before a frantic engineer pulled the plug.[38]

The theme of the 1939 New York World's Fair—the first to have World's Fair replace Exposition as a title—was "The World of Tomorrow." American industrial giants gathered a decade's worth of ideas from the screaming Sunday newspaper articles and brought them to such glittering life that all memory of inferior precursors was obliterated. Robots had etched themselves so deeply into popular consciousness that automatons were found all over the fair. No more sharing a building for Westinghouse; it built itself a full pavilion. The futuristic structure consisted of a three-story glass-fronted centerpiece flanked by two giant cubes with glass on three sides. Visitors would be tantalized by the amazing displays inside if they tried to walk past. Few did. Westinghouse took no chances: It drew fairgoers from the entrances with a 120-foot "singing tower of light" that streamed lights, smoke, and fireworks up a huge metal framework accompanied by symphonic music.[39] Full-page ads in magazines like the *Saturday Evening Post* and a 55-minute promotional movie told middle-class Americans—represented by an archetypal all–American family called the Middletons—that they could be entertained and educated by "'The Battle of the Centuries' [old-fashioned hand washing versus a modern Westinghouse dishwasher], the Microvivarium [a sterilizing device that zapped microbes], the Junior Science Laboratories, and the Television Show." And, above all, Elektro, "the amazing Westinghouse Moto-Man."[40]

Elektro towered over the standing audience on a twelve-foot-high stage for 6–10 shows a day.[41] For the fair, Westinghouse devised a copper paint that made the flat gray of earlier robots instantly passé; Elektro gleamed golden under the lights and when photographed with color film. His handlers were always young women, who humanized and tamed any possible threat posed by the mechanical beast, subliminally assuring audiences that the huge metal figure could be ordered about by "a mere slip of a girl." Celebrities, whether pushed by their own handlers or the ubiquitous Westinghouse publicity operation, couldn't get through the fair without posing with Elektro. An eclectic bunch, they included dancer Bill "Bojangles" Robinson, English Channel swimmer Gertrude Ederle, jungle explorer Frank Buck, *Tarzan*'s Johnny Weismuller, Broadway "Dream Girl" Helen Bennett, and the midgets inhabiting the "Little Miracle Town" attraction at the fair.[42] This was an especially shrewd move by Westinghouse. Stunning as it was, the pavilion was dwarfed by other, even more elaborate constructions at the fair. Their vastness worked against them for individual publicity, though. Elektro was a greater-than-life-sized gift to photographers, and the company was more than happy to allow them to go backstage or move the robot around the fair to pose in other settings. His legs were detachable so that he could sit like a metal Santa and have others swarm across his lap, which, of course, willingly or no, included the midgets.

The fair closed 1939 with huge losses, more due to its futuristic level of expenses than to the start of the war in Europe, although that development certainly cut off European vacationers. Stressing amusement over edification, a second season started in May 1940. Westinghouse had sunk too much money into its pavilion to make structural changes, but Electro's act, like any bit of vaudeville, had an easy fix: more and better. The command came from the ever-frenetic publicity department: give Elektro a companion.

Westinghouse engineers raced over the winter months to create something new to draw crowds. Engineer Don Lee Hadley saw inspiration in his Scottish Terrier, Bonnie.[43] The third in the wildly popular series of *Thin Man* movies had appeared in 1939, with Asta the terrier rivaling stars William Powell and Myrna Loy in popularity, which must have weighed heavily on Hadley's thinking. Terriers also had the physiological advantage of being low to the ground—helpful since the final robot weighed a hefty 65 pounds. Encased in an aluminum skin that gave it the look of a friendly cartoon terrier, Hadley's creation, Sparko, had only a few tricks. It "walks forward and backward, barks, wags his tail, and sits up to beg," read the publicity release reprinted by newspapers.[44] As an inside joke, Sparko could cock its head at the exact angle of Nipper, the dog listening to "its master's voice" that rival manufacturing giant RCA had used in advertisements for decades. Newspapers vied to give the publicity shots cute, attention-grabbing headlines or captions. By far the best was the one in the tiny *Brainerd Daily Dispatch*: "Teaching New Dog Old Tricks."[45] Poor, forgotten Willie Vocalite out in San Francisco got a duplicate Sparko for the second season of the Golden Gate Exposition.

After the fair, Elektro and Sparko went back on display in Pittsburgh as well as selected electronics exhibitions. Westinghouse carefully concealed the drab reality behind a robot that did practically nothing (and didn't do that well), except for one jaundiced tell-all by Elektro's handlers.[46] "You put on an act as many times as I have with this tin mummy and see if it doesn't get monotonous," admitted Eileen Detchon. She and Beth Kay were actresses "at liberty" when they answered an ad for fair work to glamorize the robots in 1940, accompanying them to the sticks when no Broadway work was available. The women alternated shifts for 16 twelve-minute performances a day, seven days a week, spending their off time in the basement room that held the huge amount of apparatus needed to make Elektro work—or not work, as he routinely broke down in the middle of his undemanding routines. On top of that, they had to clean the tar left in the tube that controlled his smoking, a disgusting chore that may have been the source of Elektro's greatest service to humanity: most of his operators never wanted to touch a cigarette after seeing what the smoke would leave in one's lungs.

Despite the daily headlines from Europe, Americans remained in a state of innocence through 1941. Pearl Harbor changed everything. After the attack, the Sunday supplements and the popular science magazines featured nothing but war-related stories. Elektro and Sparko, who had flitted around the country fulfilling their *raison d'être* of selling Westinghouse appliances, could no longer waste gas and rubber in their traveling Elektromobile. They survived the war because a Westinghouse engineer hid their parts in his basement and garage.[47] Not as lucky, Willie Vocalite and his version of Sparko were scrapped. No new robots replaced them; wires and metal were precious war commodities—only the most unpatriotic would waste hundreds of pounds of materiel on a robot without a function. Elektro and Sparko later had an inglorious postwar afterlife (see chapter 11).

For just over a decade, Westinghouse had owned the robot world. Then it vanished in a cultural instant. Robots in a nutshell.

6

Iron Monster Turns Traitor
Amateur Robots

The myth of the backyard inventor apotheosized Henry Ford, the Wright Brothers, Nikola Tesla, George Westinghouse, Alexander Graham Bell, and dozens of others. Parts assembled in an old shed could create wonders never before seen on Earth, which (according to the myth) always worked to perfection on the first attempt. Thomas Edison's name became shorthand for genius and achievement. Nothing was impossible. Reporters burnished the mythology by breathlessly announcing each new flying machine, or death ray, or food pill—or robot—as if it were the first, giving readers the thrill of being present at the beginning of the next great thing. Reporters acted as stenographers, passing on verbatim claims, stripping them of any history or context or reminder of past failures (or any investigative reporting to check whether the inventions could do the tricks their promoters promised). For scoop-minded newspapers, the future was always now, the past forgotten.

Occultus led the 20th century down this robot rabbit hole. A German inventor (whose name is given variously as Otto Widman, H. Whitman, and Adolph Whitman)[1] introduced Occultus—the name should have been a giveaway—at the London Coliseum in 1909, a story immediately picked up by American newspapers.[2] A stocky man-sized wax figure studded with pistons, gears, and cogs, powered (or so the inventor claimed) by little electric motors and controlled by radio waves, Occultus' fake eyes always pointed sideways, as if he were annoyed by puny human stupidity. His expressionless face sported an unfashionably gigantic beard, presumably to hide a mouth that didn't move when he "spoke," making use of a stack of phonograph records to supply songs, whistles, laughs, and generic answers. When asked "Who will win the Derby?" Occultus snarkily riposted an undeniably correct "The horse first past the post."[3]

Whitman invented more than a robot. He hit upon the template that successful robot promoters followed for the rest of the century: emphasize declarations that made for sparkling articles over enlarging the mechanical man's diminutive bag of tricks. "I hold the world in the palm of my hand; I could be emperor of the universe; I could make the terrestrial globe my personal property! For I've invented an artificial man who could compel all other men to become my slaves!" he declared in June 1914, just a month before First World War began. He also envisioned "vast factories" turning out these robots in wholesale lots, an image that, regardless of whether it thrilled or terrified, encouraged dozens of fiction writers to adopt it for their work. Soothingly, Whitman was careful to finish on a high note: "But of course I shan't attempt to bring about any such destruction of humanity. Instead, I hope in time to perfect my machine-men to make them a great boon to humanity." That boon promised a life of ultimate leisure.

> In time, his machine-men could be made into a race of slaves, supporting in ease all natural mankind. Each human being would have as many servants as he liked. No one need do a stroke of work other than give orders to his mechanical serfs. Every bit of mining, manufacturing, all the dangerous and disagreeable labors of civilization would be turned over to the unfeeling machine-men. And we humans would live in ideal luxury.[4]

Big words for a machine that literally could not raise a hand to help.

Whitman pushed his mechanical man through a decade of publicity, without a shred of evidence that Occultus did the impossible things claimed for him, like walking.[5] As early as 1911, one reporter apparently grew tired of the charade and put a rare counterclaim into print:

> Writers of fiction, in all countries and ages, have found inspiration in the idea of duplicating the complicated works of nature with mechanical contraptions. To describe such imaginary contraptions is much easier than to actually construct them....
> To construct a mechanical man is merely to triumph over mechanical difficulties—the man being of no use, but merely a curiosity when created—which appears to be the case with Mr. Whitman.[6]

Whitman's hot thrills won the argument over the reporter's cold water. John W. Belcher of Newton, Massachusetts, studied Whitman's plan to perfect effect. His "Miss Automaton," debuted in 1911, was billed as the first female automaton in the world. The figure, dressed in a red silk gown, stood 5 feet, 8 inches tall and weighed 185 pounds. Belcher spent seven years creating her, after a ten-year quest for perpetual motion, or so he claimed.[7] Reporters in turn made sport not of perpetual motion but of women. "The automaton can sing, talk, walk, write and perform every feat natural to a woman, except spend money and scold."[8] Another commented, drily, "Mr. Belcher is said to be a bachelor."[9] "But is there any call for an automatic woman at all?" asked a third.[10]

Dr. Cecil Nixon certainly thought so. As a millionaire dentist, he practiced his profession only in the mornings, leaving his afternoons free to tinker with automatons in his "haunted house," a San Francisco mansion wired like a gigantic cuckoo clock to produce bird-like trills every 20 minutes, flutes every half hour, and a music box concert every hour. The front door opened when spoken to, befitting a man "prominent in the Society of American Magicians."[11] Nixon had two musical female automatons: Isis and Galatea. Isis, a dark-haired beauty in the Egyptian tradition beloved by magicians, could do "anything but think," Nixon claimed, which in real-world terms meant that she could play 64 tunes on a zither.[12] Isis appeared in 1919 after the usual lengthy gestation—in this case, sixteen years (shortened to seven years in later articles that also upped her repertoire to an astounding 3,000 tunes).[13] Four years later, Galatea appeared, a winsome blonde whose preferred instrument was the violin.[14] Isis was obviously Nixon's favorite, though, as she was the one featured in most newspaper articles.

Nixon was as adept at playing the press as his dolls were at playing their instruments. A 1939 Sunday magazine article contained a paragraph of creepy eyeball-retaining perfection:

> Dr. Nixon tells of a young woman who helped to dress Isis, who died soon after the work was completed. A man who had charge of her at an exhibit shot himself for unknown reasons. Three different women who had charge of tuning the zither that Isis holds were engaged to be married. They lost their loved ones and fled to Europe to try and forget. Another young man's interest in the goddess

landed him in jail. Two news cameramen from Hollywood took pictures of Isis and a few days later were drowned when their boat overturned in a lake while taking other news pictures.[15]

—⁂—

Dr. Nixon kept reporters visiting him through the 1930s with his tall tales. Nonetheless, his collected two-decade oeuvre pales next to the headlines that Harry May inadvertently collected in a single day. Certain types of stories were ancestral clickbait. The *Arizona Republic*'s front page on September 19, 1932, said it all in five words: "Robot Turns Frankenstein, Shoots Creator."[16] It was "Man Bites Dog" combined with the monster from the 1931 movie. Newspapers leaped on the analogy: "Owner Finds Robot Is Frankenstein," and "'Frankenstein' Had Nothing on this Robot," topped by "Page Frankenstein: Robot Shoots Man."[17] Sensationalized headlines ran in all capitals: "IRON MONSTER TURNS TRAITOR" and "LIFE OF ROBOT'S MASTER MENACED BY STEEL MONSTER."[18] The irony couldn't have been more delicious if scripted. One day earlier, the *Minneapolis Star Tribune* had warned that "English Inventor Fears Three-Ton Frankenstein He Worked On 14 Years." How did he know? The robot told him! "ROBOT UTTERS A WARNING, THEN SHOOTS CREATOR."[19]

This was America's introduction to Alpha, a supremely ugly pile of tin cans, introduced a month earlier at Britain's tenth annual National Radio Exhibition. Alpha, his ton of metal polished to glistening perfection, had huge saucer eyes and microphone ears that stuck out like today's automobile sideview mirrors. Powered by three Exide car batteries, Alpha could stand up and sit down on command, supposedly activated by sound processed through the ears: "[T]he microphones convey the requests to a series of tuned reeds. When what may be termed the 'keyword' reaches the tuned reeds they release a trip. This brings into use a strip of talkie film containing the answers. The talk comes through a loudspeaker in the base, but as his 'mouth' moves he appears to be talking."[20] Alpha's voice boomed through a quieted exhibition hall, with all other radios turned off for his four 15-minutes performances to prevent interference, a bug turned feature when the thick crowds that surrounded him pushed latecomers far from his booth. Spectators saw Alpha sit on his raised platform and answer questions from the audience or read from newspapers in French and German. And then the *pièce de résistance*: he shot a gun.

Alpha was the product of "Professor" Harry May, an honor he bestowed upon himself in long theatrical tradition, after trying on and quickly discarding the futuristic moniker of "Astra." May said Alpha was the result of fourteen years' toil, a claim as doubtful as his college degree. Behind him scurried a squad of faceless workers, making it likely that May functioned in a role similar to a stage magician, with talented craftspeople building the tricks that his showmanship sold to the crowds. It's not even certain that he knew how Alpha worked. Every explanation he gave differed not only from the next but also from the wiring visible when he removed the robot's breastplate. Did Alpha evolve over time, or did May's spiel change to reference whatever technology was making the news? To this day, nobody knows. Outsiders got as little opportunity to examine Alpha's innards as they did Maskelyne's chess-playing robot.

Expectations for a performing robot were minimal. Alpha impressed with his bulk, the thick cables snaking through his midsection, and his loudspeaker-aided voice. Still photos of the day seldom capture the majesty of his size. Pictures of Alpha sitting holding the newspaper he supposedly "read" demean him, turning his base risibly into a toilet. A publicity shot of a stone-faced cleaning woman giving the nickel plating a polish is less one of humanity serving its new robot overlord and more a glimpse of a dusty and forgotten piece of bric-a-brac.

So, a gun. A robot with a pistol is not our image of futurity. Robots should not need guns when they could crush you with a step. However, the cheap trick worked despite its incongruity. A gunshot in any enclosed space will send an aural shockwave through an audience; it may be all they talk about the next day. No real bullets, of course, just blank cartridges, paper wadding pressed into a casing to hold the gunpowder in place after the bullet has been removed. The danger remained, though. Both the exploding gunpowder, hot and burning, and the wadding, expelled at high speed, can hurt at close range. Actor Jon-Erik Hexum died on set in 1984 from a blank cartridge. A slight pressure on the trigger will fire a gun, an action that fit neatly into the extremely limited range of Alpha's capabilities. Depending on the venue and the nearness of the audience, Alpha would raise his right arm and point or stand up and lift his arm to the ceiling. He always gave a warning, customarily "Look out, or I'll blow off your hand!" a variant on "fire in the hole," meant to alert rather than menace.[21]

Alpha had been moved to Brighton, England, in September when the shooting occurred. May gave the proper command to Alpha, but he pulled the trigger prematurely. The reports that relayed the story to seemingly every paper in America were as skimpy on details as the headlines were huge. Most shared a United Press syndicated story that had a bullet passing through May's hand—an impossibility.[22] May, trouper that he was, returned to finish the act with a huge and easily visible bandage on his hand: "Of course, I shall carry on with the demonstrations, because if I did not it would appear that Alpha had become my master."[23]

May treated Alpha like a magician playing up the potential danger of an underwater trick. Alpha had already lowered his arm onto May's head twice, putting him "three weeks in a hospital," and had given him "several nasty shocks." In an interview given earlier that same week, a reporter portrayed May like a master horror story writer: "Now that he has created a modern Frankenstein, he says he will prevent his robot from falling into evil hands, and vows that he will destroy it before he dies." May planted the image of a robot not with a pistol but the gangster's favorite, a machine gun. "'Think of what an army of well-trained robots could do,' Mr. May exclaimed." Or what robots run amok could just as easily do: "Alpha may be dangerous, and there is always the tense nervous strain watching for tricks like that. Always you've got to be careful and cautious in your treatment of him. He is just a soulless machine, but—"[24]

May reveled in the publicity. "I always had a feeling he would turn on me some day," he enthused.[25] When Alpha appeared at the Canadian National Exposition in Toronto in 1934, the robot knocked May's temporary handler, a Michael Harley, to the floor with its "mailed fist." Harley purportedly was treated for "shock and bruises."[26] A publicity stunt, perhaps, since Alpha is termed "the robot pride of an electrical company" (presumably the Mullard Radio Valve Co. Ltd., in whose booth Alpha performed at the Radio Exposition).

These preludes were like off-Broadway tryouts for Alpha's conquest of America later in 1934. Macy's department store partnered with May for maximum publicity, including a radio interview, with British Pathé filming a segment of a newsreel. The full film interview may have gone something like this, with the robot swinging its head to May after each question to provide time for the proper phonograph record backstage to drop and be played[27]:

Q.—How tall are you?
A.—Six feet.
Q.—How much do you weigh?
A.—One ton.
Q.—How old are you?
A.—Fourteen years.
Q.—How do you feel?
A.—In excellent health.
Q.—What color is grass?
A.—Green.
Q.—What color is the sky?
A.—Blue.
Q.—Do you like little boys?
A.—Nah.
Q.—Do you like little girls?
A.—Nah.
Q.—Do you like the ladies?
A.—Yes.
[In the newsreel, a woman is invited up from the audience and asks the next three questions;
    Alpha swings his head to her instead of May.]
Q.—Are you married?
A.—Not yet.
Q.—Would you like to be married?
A.—Yes.
Q.—What sort of ladies do you like?
A.—Blondes.
[May interrupts on camera: "I say, Alpha, now you just turn your head to the left and look at that
    pretty brunette there."]
Q.—Wouldn't you like to marry this young lady?
A.—She'll do.

Alpha was capable of more when the time was right. A newspaper cynically reported on May rigging him with a special disc for Macy's[28]:

Q.—What do you like to drink?
A.—Gallons of beer.
Q.—What do you like to eat?
A.—"I shall let you into a little secret," Alpha said as the whirring began to increase in volume. "I
    like toast made on the R. H. Macy toaster which sells for…" and this speech continued for
    nearly a minute. It drew a great hand from the employees who had been let into the private
    demonstration.

Bizarrely, this macho dialog, complete with a deep masculine voice, emitted from a now female robot, with a metal wig of curly hair, inset eyes with eyelashes, a metal skirt, and, strangest of all, breasts. (No pictures exist of the latter being polished.) The articles about Harley's mishap in Canada at the 1934 exposition all refer to Alpha as "her,"[29] and a 1938 interview has May explaining that after the Brighton attack Alpha was changed into "Mary Ann" to appear less threatening. Numerous photographs record the changes, but whether the interview contains solid historical evidence is doubtful: the 1934 stunt is given as "last year," May was substituted for Harley in Canada, and his claim about Alpha's voice being changed from bass to soprano is belied by recordings.[30]

Alpha traveled to the 1935–1936 California Pacific International Exposition in San Diego, where he held down a spot in the oxymoronic Palace of Science, with all the same tricks and a new set of phonograph records. When asked whether he loved his wife, he now said mournfully, "I've a heart of steel. I don't love nobody and nobody loves me."[31] Lines like these are the best public evidence for assuming that Alpha continued to be thought of as a male figure. Nobody ever commented on the oddity of these statements issuing from a female body. Pictures of Alpha, "the perfect boyfriend," posing with women are legion, even after he—she?—gained breasts.[32]

Even if their editors hadn't demanded it, photographers loved to invite women, especially petitely pretty younger women, into every possible picture simply because they increased the likelihood of a shot being used. A pretty girl humanized the bulky metal lines of ships and bridges and cars, making them loom even when immobile. A girl in a robot's arms as he carried her off for reasons that audiences could imagine only with shudders gave thrills equal to horror films. Ultimate proof of Alpha's perceived maleness emerged from a stunt perpetrated by some enterprising photographer. Among the most popular exhibits at the San Diego fair was the Zoro Garden Nudist Colony. The "nudists" (g-strings can be glimpsed in many postcards of the attraction) supposedly signified healthy living in San Diego's ideal climate. A surviving postcard shows Alpha standing in the garden, "kidnapping" a wholly naked Zorine, Queen of the Nudists, in his arms. The immobile robot managed this impossible feat by using an actor as a stand-in, whose flimsy costume lacked the adornment of breasts and curly hair.[33]

Alpha's violent tendencies extended into the public realm beyond underground soft-core pornography. Indeed, Alpha is the only display robot known to be a serial attacker. In June 1935, he "went berserk, talked strangely, and knocked out a spectator near him." May again blamed stray short-wave transmissions.[34] The last straw came in February 1936, when May was prepping the robot for the start of the fair's second season. Newspapers again made the claim that Alpha's gun had somehow been loaded with a real bullet, one that creased May's skull. Sent yet again to the hospital, May left his bed, "his head swathed in bandages, and a monkey wrench in hand, to dismember the gadget that makes Alpha shoot."[35] May kept Alpha performing for several more months until the inevitable occurred. A visitor in June bought a ticket for the show, but the robot was a no-show. She said she learned "that in an unguarded moment this mechanical figure had hit a woman in the audience over the head and is now being incarcerated while being sued for damages."[36]

Whatever actually happened, that was the end of Alpha. A machine that inadvertently harmed people in its vicinity was an actual menace. May learned an important lesson, and he announced in all seriousness that his next invention would hurt people with premeditated deliberateness. When he had perfected it, May would demonstrate, that very year and at the fair, a death ray, that "will destroy flies and insect pests, make airplane motors stop in mid-air and[,] with full power on, will be strong enough to kill human beings."[37] It certainly put an end to the high-flying career of Harry May. Unsurprisingly, the death ray never made a public demonstration. Visitors to the Palace of Science saw nothing but a closed door promising the death ray "soon."[38] We are still waiting.

In a just world, Eric the Robot would be mentioned before Alpha: garnering an equal number of excited headlines, he did practically everything that Alpha did, and did them first, *and* without running wild and harming people. That was his mistake. Eric is the most damning evidence against responsible reporting; seemingly no newspaper in America brought up Eric once Alpha came along, even though he toured the country as recently as 1929.

Eric also debuted in London, at a 1928 exhibition run by the Society of Model Engineers, built by real engineers in five months rather than Alpha's truth-stretching fourteen years.[39] Made entirely from aluminum except for a face of steel, he weighed in at 140 pounds and looked like a suit of armor. His eyes sparkled with flashes of light.[40] Motors were hidden in his base and torso, connected by a claimed three miles of wire, allowing Eric to stand and sit, gesture with his arms, shake his head, and answer questions in a baritone voice. He was voice-activated, or at least responded to sound signals. If you told him to "shut up," he shut up without shooting anyone. Captain William Henry Richards, a journalist and Society secretary, was front man, while motor engineer A. H. Reffel connected the springs and pulleys. At some point, they painted R. U. R. on Eric's chest, to make the connection between him and the play that famously introduced the word *robot* into English.[41] Richards swore that "his speech is produced neither by phonograph record nor talking film," which, if true, leaves only a wire connected to a backstage microphone.[42]

Like Alpha, Eric had a stock of patter that wowed the press when he arrived in New York in early 1929. When asked what he thought about Prohibition, he quipped, "The more I think of it, the less I think of it." New York thrilled him: "I am impressed by your tall buildings and compressed by your subways." Though he wanted a blonde female robot companion, the press preferred to write him the equivalent of a Tinder profile.

> "Mr. Robot, do you drink or smoke?"
> "I do not."
> "Gamble?"
> "No."
> "Run around nights?"
> "Certainly not."
> "Married or single?"
> "Single."
> Girls, what a man! Almost perfect.[43]

Almost? No, absolutely perfect. Once snared, Eric would do things that the standard husband of the day would never stand for, à la the hype for Televox. Eric was envisioned as the forerunner to a smart house, assuming he ever learned how to walk:

> This Robot will be stationed near the telephone, and will stand guard while Mrs. Citizen goes shopping.
> She may leave dinner cooking when she goes out, and it may be necessary to turn off the heat before she returns. If so, she simply will call her home on the telephone. "Eric the Robot" will answer, and Mrs. Citizen will know he's on the line when she hears three buzzing sounds.
> Then she will issue her order, "turn off the current in the kitchen." "Eric" will signify he has understood by another buzzing noise, then he will proceed to execute the order. Following this same general principle his uses will be legion. He will be able to close doors, open windows, attend to refrigeration, and handle a multitude of details which might take Mrs. Citizen all day to adjust personally.[44]

Eric toured the country throughout 1929 and continued drawing crowds in England through 1931. He then quietly disappeared, with not even his parts available to historians. When a new Eric appeared at a Robots Exhibition at the London Science Museum in 2017, it had to be rebuilt from scratch, coincidentally also in five months. One possible explanation for the missing Eric is that his innards were repurposed inside "George," Captain Richards' next robot. It's not clear what, if any, advances George had over Eric, except that he spoke in French, German, Hindustani, Chinese, and Danish as Richards shipped him around the world on a three-year tour. Eric's suit of armor appearance made him seem ready to fall apart any moment; George looked to be welded into a single unit except for his joints. He had a thick set of hips similar to those of Houdini's robot in *The Master Mystery* (see chapter 10) and a head variously topped with a flat cap, a crown, and an Egyptian headdress. He came with and without the R. U. R. painted on his chest and later with an "R" in a circle, like a superhero. Or maybe that was a 1952 version, known as Robert, which the press posed with English bombshell Diana Dors with the improbable claim that it had been invented by her husband. Robert finally appeared to make good on all those predictions about housework, as he was shown carting firewood, washing a window, making dinner, and then relaxing with the family over drinks and cards. His feet were of course carefully cut off by the framing of these pictures: showing the immobile robot's stand would have spoiled the illusion.[45] Robots symbolized the future, a pact with audiences as carefully preserved by both sides as the fiction that movie stars were of a higher order than mere humanity.

—⚏—

All subsequent robot-maker wannabes followed the same promotional formula. Take Andrew Bober, a watchmaker from Gary, Indiana, in 1933. He spent the requisite amount of time—seven years—developing a five-foot-tall robot, Little Willie, who looked like a standing Charlie McCarthy, with evening suit, bow tie, glasses, and a bowler hat— a tribute, perhaps, to the ventriloquist's doll that Bober said inspired him. So that no one took him for a real dummy, Willie was always photographed standing, albeit without any trace of the cables and wires connecting him to the control box.[46]

Willie did it all. He could not only dance a waltz but also sing and tap "the cymbal on a bass drum in time with his singing." Even better, "When there's company Willie walks about the room, introduces himself with utmost politeness and shakes hand with each individual." Most amazingly, "His feet shuffle along the floor like any human's, with perfect control in turning left or right. That's where most mechanical men fall down."[47] Exactly right. If any of these claims were true, Willie would have been the most advanced robot of all time. In those photographs, however, Willie's legs look like actual legs, not the contrivances of the modern robots that have licked the problem of walking. How he could move at all is therefore is as much a mystery as why no industrial behemoths rushed to Gary to steal the technology of this marvel.

In 1934, columnist H. Allen Smith reported on the snappy banter of a nine-foot-tall robot double of radio comedian Fred Allen, a creation of K. D. Andres, self-anointed as the "world's foremost builder of robots" (somebody's idea of a great PR gag).[48] The October 1935 *Popular Science* devoted a page to Milton Tannenbaum, whose "lifelike robot speaks, smokes, and drinks." The unnamed robot, built to resemble a casually dressed old man, used compressed air to stiffen or relax its legs.[49] It would have had equal difficulty in maintaining its balance for more than a second.

Big Looie, the creation of Detroit auto mechanic brothers John and Patrick Rizzo, made an appearance on the *Major Bowes Amateur Hour* radio show in 1938, playing the accordion.[50] The husky robot, straining a size 55 suit and size 16 shoes, stood 6 feet, 2 inches tall, though he weighed only 180 pounds. He required six years of labor and also "walks, talks, [and] sings."[51] The wires and cables that controlled him were carefully deleted from photos. More than a decade later, Patrick Rizzo showed reporters a new robot a full foot shorter that operated wirelessly—the first ever, he claimed. "He walks unaccompanied into a room and dances a jig," marveled the reporter. The new robot stood on display in little brother Steve's radio shop in 1950.[52] Could he do more than stand? Rizzo refused to sell or manufacture the robot, so no one got to look at its wiring.

Claimed firsts are always dubious in the world of robots. What would Rizzo have thought of the October 1939 edition of *Radio Craft* magazine? "Clarence is the first mechanical man in the world who can wander around without trailing wires behind him. Furthermore, he uses no records or transcriptions but says anything which happens to come into his mind. The only flaw thus far found seems to be that Clarence is pretty poor at broadjumping, pole vaults and swimming."[53]

Clarence came to the 1939 New York World's Fair in the first battle of show-biz robots. The *New York Times* tried to take him seriously and failed.

> "I am the man of the World of Tomorrow!"
> From the way Clarence (who, by the way, is a robot) stressed the "I" in his declaration, any one could tell it was a none-too-subtle shaft of intimidation, aimed at his rival, Elektro, the mechanical man at the Westinghouse exhibit, who has been performing at the Fair all these weeks.
> There is, however, a wide discrepancy in the appearance, the technique and the employment status of the rival robots. Clarence is a huge, gleaming aluminum, Frankenstein's monster sort of thing, with a barrel chest (8 feet 4 inches in diameter) and a ridiculously tiny head, capped by some sort of decrepit, dented pot, while Elektro is a bulky but smaller and smoother fellow with a nice sun tan effect, effected by a coat of bronze paint.[54]

Clarence was on display much like a school project, as his 22-year-old inventor, Austin O. Huhn, was seeking a job at the fair. A full eight feet tall and 300 pounds, Clarence had eyes, a nose, and a slit for a mouth even though his words emerged from a speakerbox on his chest, just above his ear. The hulk "shuffled" (again, the word is the reporter's) on size 24 feet, had an eye for the ladies, and had trouble getting his canned responses in the proper order: when asked what he thought of the World's Fair, he replied, "Oh, don't mind if I do. Just a little mustard on it, please."

The hugeness of the World's Fair could support any number of robots, including one that gave at least as many performances as Elektro without the accompanying publicity. General Motors stole the spotlight at the fair with its spectacular seven-acre "Highways and Horizons" pavilion. Five million people viewed Norman Bel Geddes' Futurama, which placed visitors on a moving skyway over a metropolis threaded with superhighways. GM carefully designed the rest of the exhibit to force strollers past its showrooms for automobiles and appliances. In those days, GM used its expertise in steel to make refrigerators, washers, and other household goods. To be sure the relationship was pounded home, and to give people a place to sit—enormously welcome to footsore crowds—an auditorium repeated an eight-and-a-half-minute promotional film, *Leave It to Roll-Oh*. The light-hearted presentation was made lighter with the most amateur-looking robot since Televox. In this film, viewers enter a middle-class home of the near future, where the technician (T) from Roy's Robot Repair is mansplaining to the bewildered housewife (H) in a skit dreamed up after a heavy dose of Abbott and Costello.

> (T) "There, Miss, you see the heterodynes were feeding back into the stimulus reaction activators causing non synapse of the motor control resistor units."
> (H) "Oh, that's good."
> (T) "Naw, lady, that's bad. But your re-generative circuits are tuned asynchronously and that causes concatenation in the intermediate amplifiers."
> (H) "Well, that's bad, isn't it?"
> (T) "Naw, that's good. From now on I don't think there'll be the slightest trouble with your robot. Your domestic problems are completely solved."[55]

Just push a button on the control panel, and Roll-Oh could do anything, from answering the door to making the bed to fixing the furnace. (Furnaces must have been balky indeed in those days to require a button for daily service.)

Roll-Oh, who never rolls but moves in what can only be described as a shuffle, is actually a man inside a set of boxes. In the film, his head features a single lightbulb above a mouth whose jaw drops like a ventriloquist's dummy. A horizontal indicator reading from 1 to 15 (but permanently stuck on 3) serves as a chin. Meaningless doodads are pasted across his front. Close-ups show his immobile fingers somehow picking up the repairman's hat, his arm extruding an exacto knife to slice a ribbon off a box of flowers, his foot becoming a vacuum to sweep up dropped petals, and his mouth turning into a nozzle that sends out a foot-long burst of flame to light a candle.

But it's all (including the flamethrower attachment) a housewife's daydream. Her husband (the provider) will get three meals a day cooked by her hands using the numerous gleaming gadgets already in her kitchen, each of them containing a "robot." The toaster has a regulator that turns off the heat before the bread can burn, the coffeepot keeps the coffee hot until poured, and the electric tea kettle unplugs itself if it boils dry. Electric eyes that keep elevator doors from slamming and turn on lights when clouds dim the sun are also robots. Cars were then supposedly as full of robots as they are crammed with computers today. The carburetor, the alternator, and the gearshift each had one or more. "Driving wouldn't be half so much fun if we didn't have that phantom crew of intelligent robots to help us … and leaving us free to live and work and play in greater ease, and comfort, and safety."[56]

The astounding cultural success of *R.U.R.* had made *robot* an instantly familiar term. From then on, writers thought the public knew what a robot was. In practice, the public understood the robot in the subjective, what one *should be* when and if encountered, but the lack of tangible working everyday robots left the concept floating freely in the air, applicable to its whole or its parts, a synecdoche burst out of the dictionary. Roll-Oh was an acknowledgment that, as far as robots were concerned, the fair's World of Tomorrow was far removed from 1940 reality. Embryonic superhighways existed—you could drive one from the fair into the wilds of Long Island—but no obvious evolutionary path led from a gearshift to a household chromium-plated butler.

—〰—

A *New York Times* article from October 23, 1927, titled "Science Produces the 'Electrical Man,'" attempted to stanch the hype by primly setting realistic limits on what the newfangled robots could and couldn't do: "There will never be a robot—a brainless, tireless, unemotional mechanism fashioned in the image of a man, performing all the functions of a man, moving about stiffly but surely, pulling levers, turning control wheels, wielding broom, pick or shovel."[57]

A similar pessimism could be seen in *Automaton, or The Future of the Mechanical Man*, a 1928 publication by H. Stafford Hatfield, who brought a working inventor's perspective to the issue. A 1927 device for "measuring and recording continuously the percentage of a given chemical substance in a liquid" had become known as a "Chemical Robot."[58] *Automaton* was probably the first book to examine the field of working robots in the *Times'* modern sense. In fact, Hatfield defined the robot in almost identical words:

An automaton, by analogy with the human model, should consist of three parts; limbs to work with, senses to perceive what it is working with, or what result it is producing, and a brain to regulate the action of its limbs with the perceptions of its senses. Needless to say, we are striving to create, not a Frankenstein's monster, a Robot, a mechanical servant which can be set to any simple task, but thousands of different automata each specialized for a certain task.... What we still have to develop is the mechanical *brain*, the link between instrument and tool.[59] [italics in original]

Where was the fun in that? Other writers provided a much livelier and provocative look at the robotic future. A 1928 syndicated article provided a forecast under the title "Romantic Old Maids Can Hear the Words of Love They Long For":

In this happy future, no old maid need look under the bed for a man, in vain. He would always be there and such a nice man, a perfect imitation of her favorite matinee idol or film star, with blond or dark hair, moustache or clean shaven, anything her heart desired. These would be stock models, turned out in quantity production and quite reasonable in price.... Or, if the customer is willing to pay a little more and have one made to order, the manufacturer might send artists and photographers to some notorious lounge-lizard and deliver a perfect counterfeit of him. She could order the late Rudolf Valentino's face and John Barrymore's voice or most any other combination.[60]

A humor piece by H. I. Phillips highlighted the difference between an intuitive but independent human and a preprogrammed machine. He noted that an automatic system then in use reported water levels by telephone, "without asking for the football scores and wanting to know if it can have next Tuesday off." But he goes on to forecast a future in which tech support has to deal with irate humans, using the following complaint as an example of what to expect: "Yesterday I was at the church oyster dinner and I called up and tried to get the automatic butler to give my husband's blue suit to the tailor. Do you know what he did? He turned the hot water on in the bathtub and put out the cat!"[61]

Robert E. Martin, in a 1928 *Popular Science Monthly* article, quoted an official of the New York Edison Company assuring the public that "the mechanical man and his ultimate universal practical application will rid humanity of much drudgery and thousands of uncongenial tasks," giving them "the gift of leisure." These robots were merely automatic machinery, not the usurpers of humanity that Čapek had postulated. Humanoid robots might make good servants for humanity, but for Martin the prospect was mostly comical, as he asked (presumably for a readership not bothered by the specter of technological obsolescence or changing gender roles) about a future that is both prescient and absurd.

Will the man of affairs go to his office in an automobile driven by a mechanical chauffeur, who will be directed at busy intersections (and perchance "bawled out," too!) by a mechanical traffic cop? Will that same business man, at lunch time, be waited on by a robot waiter and, in the evening, be guided to his seat by a robot usher? Will his wife have a mechanical ladies' maid to "hook her up in back" and his children a robot nurse to wash their morning faces and take them to school?[62]

Similarly, William Barclay Parsons, writing in the *New York Times*, made an effort to distinguish the hugely useful automatic machine that was then being called a robot from the humanoid creatures of Harry May and his ilk, calling the former "the real Robots, filled with wheels, attending faithfully to their duties and not masquerading in human form with useless legs and arms, and with heads, like those of some people."[63]

The appeal of industrial robots faded rapidly with the coming of the Great Depression. "Hitherto labor in this country has welcomed the machine and abetted the constant introduction of technical improvements which are at the basis of American prosperity," said a writer in a 1930 syndicated article from the *New York Herald-Tribune* titled "Victims of the Machines." Now, "unless the problem of 'technological unemployment' is promptly recognized and its worst effect ameliorated there will come a change of attitude to one of hostility and obstruction."[64]

The *New York Times* editorial board responded indirectly with a cheerful, if economically hollow, endorsement of humanity:

> Having, in short, been duly impressed, intimidated, and almost paralyzed by the thousands of columns of print about the Age of the Robot that is rapidly displacing the Age of Man—
>
> This column hereby declares it has come to see a sudden light, in consequence of which it completely discards and abjures its aforesaid awe and trembling in the presence of the Menace of the Robot, and hereby declares the aforesaid Robot to be largely a bluff and a fraud and so much—even if sliced by electricity and cooled by mechanical refrigeration—baloney.[65]

Not everyone was convinced by the *Times'* bravado. The short-lived Technocracy movement declared that politicians and economists had failed the people; only a government headed by scientists and engineers could solve the country's problems. An article in the *Brooklyn Daily Eagle* titled "Sees Robot Age in Near Future" explained that the movement's founder, Howard Scott, foresaw that "Political maneuvers cannot circumvent the effects on the economic system of the machine with its vast productive capacity, and its diminished capacity for employing labor."[66] Calls for mandatory shortening of the work week, thereby putting more people to work even if the weekly paychecks were lower, appeared simultaneously with predictions that the future would see "Scientific development of industry to the point where no man will work more than 16 to 24 hours a week."[67] Sir Eric Geddes agreed that "robots will perform all the menial tasks of 50 years hence [that is, in 1985] and the great problem will be how to fill in all the extra hours of leisure."[68] Where the money would come from to pay for all this leisure was never explored, nor did anyone ask ordinary workers what they thought of a future in which they would forever remain unemployed.

America's entry into World War II shattered those isolationist dreams of steady peaceful progress into a technological utopia. The word "robot" went to war. Headlines told of robot ships, robot planes, and, worst of all, robot bombs. Industry stopped producing household goods of any kind, needing virtually unlimited supplies of metal with which to build war materiel. Using valuable materials for the frivolity of a sham robot became intensely unpatriotic. Typical of wartime propaganda was a 1943 *Nancy* Sunday comic strip: In it, the titular youngster somehow acquires a household robot. It waters the plants, sweeps the floor, shovels coal and snow, washes the dishes, and even does her homework. Then it hears on the radio, "We should all do our part for the war effort," and immediately turns itself in for the scrap metal drive. "Boy! … That was fun while it lasted!" Nancy says wistfully.[69]

—⁂—

After the war, few looked to robots to solve the world's problems. Knocked down several pegs, they became weekend craft projects. Harvey Chapman, Jr., worked for the AiResearch Manufacturing branch of Garrett Supply Company, one of the leading manufacturers of airplane parts in Los Angeles. In his spare time, working for 45 days and nights, he scrounged through the piles of old parts in his garage to create Garco (sometimes Mr. Garco, sometimes rendered in all-caps as GARCO), a five-and-a-half-foot tall, 250-pound (or 5 feet, 8 inches and 235 pounds) versatile robot.[70] Garco's head was a more angular precursor to *Mystery Science Theater 3000*'s Crow T. Robot, with eyes and ears on a hemispheric structure above a thin stalk with lips. He had a boxy torso; long, thin legs; and arms that revealed the machinery. Not even pretending to walk, Garco stood on a circular base that was useful for photographers, who normally depicted him towering over people. His somewhat skull-like head allowed editors to choose angles from which Garco could look either monstrous or goofy, either expression satisfying some editor's bias about how robots should be perceived.

Garco naturally first met the public at a trade show, the 1953 Western Metal Exposition, and got written up by *Popular Science* magazine, which extolled feats far outreaching any prewar creation. Garco, it gushed, "can saw, hammer, drill, pick up papers, roll his eyes and have the shakes." It actually could. Garco duplicated Chapman's arm movements through a multiple-jointed electromechanical control while Chapman pushed buttons on a remote controller with his left hand. Garco's left hand had pincers to grip, and multiple tools could be substituted for his right, like a real-life Roll-Oh. A transmitter allowed the robot to talk. The sensible Chapman, a true engineer, didn't expect his creation to replace workers. He had a vision of these remote-controlled machines doing work "too dangerous for humans," like handling radioactives, bacteria, or explosives.[71]

Chapman's preteen son, Terry, "the luckiest boy in the world," not only got to play with Garco at home but also operated him on his father's cross-country theater tour. "What bothered me most was the silly questions some people asked," he said. "I'd be telling them all about his electronic brain cells and they'd want to know what he eats. Or could he dance!"[72] Chapman's proximity to Hollywood made Garco a natural for movie and television tie-ins. The Chapmans' tour promoted *Gog*, a 1954 movie featuring robots that also had multi-function arms and were used in an atomic plant. Garco drew far larger audiences when he appeared on *Juke-Box Jury*, *I've Got a Secret*, and *Science Fiction Theater*, and also with the country's friendly uncle Walt Disney on *Disneyland*, when Walt introduced the space-age documentary *Mars and Beyond* on December 4, 1957.

Bill Allen, a science teacher from Wichita, wanted something to excite his classroom. A *Life* magazine article related how "in his garage workshop he gathered together a chemical drum and a paint can, some scraps of wood, wire, assorted pulleys and a couple of electric motors. Drawing on his knowledge of mechanics and electricity, he wired them together, creating an other-worldish, homemade robot he named Magnamo."[73] Cheaper than Garco, Magnamo took a mere 80 hours and $50 to construct.

Even better than a teacher for a catchy story was a teenager. Kids had long scrounged parts to build robots. Bobby Lambert, a 13-year-old from Columbia, South Carolina, built Bugs in 1930. Merely a couple of boxes with legs, Bugs could at least lift an arm and stop when Bobby commanded.[74] Elsewhere, Robert Dupwe started with cardboard and aluminum foil for his first iteration of Cog at age fourteen in 1954. Four years later, at Arkansas Tech, his robot was "fully motorized and made of sheet aluminum, able to move in any direction, carry things about, even write on a blackboard in a childish scrawl."[75] The enormous piles of junked surplus war goods in the 1950s gave even garage inventors more sophisticated parts to work with than most professionals had in the 1930s. Ronald Hezel was a high school senior when he constructed seven-foot-tall Thodar in 1954. Made of pipes and sheathing, with a chimney flue for a head topped by two transcription disks for antenna, Thodar was almost as primitively boxy as Bugs, also lacking elbows or knees. Thodar (the name somehow derived from reversing the initials of the phrase "radio-operated high frequency transmission") was billed (again falsely) as the first wireless robot; Hezel needed a phone line to connect him.[76] Happily, Hezel's copious engineering skills won him a scholarship to New York University. He later taught mechanics and engineering in high school, using Thodar as a teaching tool.[77]

The visual, yet limited, nature of robots was perfect for the small black-and-white televisions and barely mobile TV cameras of the 1950s. *I've Got a Secret*'s "teenage" show on March 26, 1958, featured Donald Rich's Robetron.[78] "It has never learned to say 'yes,' because it's a girl robot," host Garry Moore said, in 1950s-style humor. Rich, then fourteen and a student at the Bronx High School of Science, was already a year-long veteran of the science exhibition circuit. The six-foot robot was unusual in not having legs. Instead, a cylinder dropped to a larger circular base that covered wheels powered by electric motors.[79]

Michael Freeman was a similarly precocious 13-year-old when he won first place in the 1960 Westinghouse Science Fair with Rudy the Robot.[80] Like Hezel, he continued as an adult to build robots as teaching tools:

> Freeman recorded everything that Leachim would teach or say on three thirteen-inch platters called verbal discs. Into this "brain" he poured most of the contents of a children's encyclopedia, the Guinness Book of World Records, a dictionary, a thesaurus, and a series of textbooks. The robot also knew some basic Spanish, what the words in the Pledge of Allegiance meant, some rules of chess, and a few snappy jokes. Stored in a separate memory system were the scholastic records of each of its students, their names, weak and strong subjects, and their hobbies. Once Leachim was plugged into the wall, he was programmed to be able to dip into each memory system and so provide a customized lesson for each child.[81]

In the 1970s, the Mego Corporation, famed for its extensive line of robot toys, mass-produced a tiny Leachim replica, called 2-XL (to excel). The new toy sold in the hundreds of thousands.

Thirteen is the magic age for promotable robots from prodigies, perhaps none more so than Sherwood "Woody" Fuehrer's Gismo, the cutest of all boy-made robots. Just under six feet tall and weighing a mere 92 pounds, Gismo could "hold a tray of cookies and pass them around."[82] Throughout 1954, dozens of newspapers carried either the AP wirephoto or one from UPI, with a similar image appearing in the nationally syndicated Sunday magazine supplement *This Week* in 1955.[83] Gismo cost a bargain-basement $4.94, readers were told, a subliminal hint to any Russians scanning the papers that red-blooded American boys couldn't be topped for their ingenuity.[84] When Gismo won a prize at a Ford-sponsored Industrial Arts Competition in New York City, an alert press agent used the improved talking robot to open the press conference announcing the winners. Gismo not only talked but also walked, or rather rolled better than Roll-Oh when Woody placed him on four steel castors. The press dubbed him "Gismo the Peaceful," and photographs showed Gismo holding the hand of a toddler, the safest robot that ever was.[85]

After a round of radio and television appearances, including two guest spots on the *Today* show, Woody got the ultimate accolade: his own article in *Boy's Life*. "If you hadn't guessed, I've got my career ideas set on being an electronics engineer," he wrote. "My dad has encouraged my experimenting. My mother accepts it with some misgivings." The same gender division is found further down the age cohort: "I became deluged with mail. Many sent clippings. One came from the island of Guam. I received two offers to buy him. Letters came from all over the United States—boys wanting plans and girls just wanting to be friendly."[86]

That was the 1950s in a capsule, the beginning of nerd culture. Girls were kryptonite to nerd boys, who assumed they shared none of the same interests—not merely robots but all mechanics and technology, the space program, electronics and computers, science fiction, comic books, and monster movies. The larger culture aided and abetted female marginalization by applying sitcom standards to daily life. Women became homemakers in increasingly robotic households, a Roll-Oh daydream made real by male inventors creating gadgets to do their work for them. Decades would pass before references to robots invented by girls infiltrated newspaper articles. Today's robotics clubs attract boys and girls together in awesome numbers, with annual competitions around the country. Even so, hardly any of today's humanoid robots would be deemed female. Sometimes the past is still present.

7

Buck, Flash, Tillie
and Mickey
Robots in Comic Strips

Developed from a European model as part of the circulation wars of the 1890s, comic strips quickly became a central selling point for newspapers, the dominant print medium of the era. Within a few years, the funnies (best known as a color supplement to the already fat flagship Sunday papers) had established a permanent place in popular culture. A family lazing away a slow Sunday afternoon, each absorbed in a favorite section of the paper, with Sis and Junior always perusing the comic pages, became an iconic image of middle-class American life. This cliché notwithstanding, comic strips were read by the entire family, with artists using the medium to slip sly and satiric jabs at American life into the text. Although the earliest strips tended to feature mischievous youngsters as leads, like the *Katzenjammer Kids* and the *Yellow Kid*, the strips soon expanded to feature families (*Blondie*, *Gasoline Alley*), adult eccentrics (Popeye, Barney Google), heroic adventurers (Prince Valiant, Flash Gordon), and eventually any possible sets of images (*Ripley's Believe It or Not*, Rube Goldberg's inventions, and so on).

On October 1, 1911, a few lucky readers who opened their huge Sunday papers found a treat in the comics section. There sat, alongside older favorites like *Uncle Mun*, *The Terror of the Tiny Tads*, *Hotoff the Pen*, and *Mr. Twee Deedle*, a new full-color strip simply called *Percy*. Sleek, modernistic, and clearly aimed at adults, the very first panel laid out the premise with marvelous storytelling economy: An eggheaded man wearing glasses and a huge bowtie, the stereotypical German professor familiar from vaudeville, says in broad dialect, "My inventioning is completioned! A mechanism man! No more strikings! No more servant example!" The mute mechanical man stands at attention, the line of his unnecessary mouth raised in a smirk, rows of rivets along his pot-stove–like torso. Percy— we learn his name in the second panel—has been preprogrammed, his back arrayed with buttons that control specific tasks. "Shust push der walking button and walk it is." For more complicated tasks, the buttons are pushed as a code: "Dere, Percy, 2-2-2! Now sweep it der room up, yes?" Percy sweeps. Oh, how Percy sweeps. By panel nine, the inventor, his outraged about-to-be-replaced housekeeper, and every object in the room has been swept into a disaster resembling the aftermath of an explosion. Percy the machine is an utter literalist, mindlessly keeping to his assigned task until he is "runned down." Ignoring who forgot to give Percy an off button, the egghead exclaims, "He is a good worker only he ain't got brains!" Catchphrases, like punchlines, are tricky, always depending on the perfect combination of syllables. Barely a month later, the strip found a variant strip-closer that swept the country: "Brains he has nix!" The machine man did the job of a dozen, but the American worker had common sense, the American value that conquered all.

Percy technically was not the first robot strip. That honor goes to Hans Horina's short-lived robot series, *Professor Dodger and His Automatic Servant Girl*, which appeared in the *Chicago Tribune* Sunday comics section in late 1907. A clockwork automaton with a big wind-up crank in her back, the servant "girl" (who looked like a middle-aged woman) did the washing and wood chopping—or would have if the Professor had ever remembered to push the right button. It lasted a mere three weeks and made not even a ripple in the public consciousness.[1]

By contrast, the *Percy* strip appeared for 67 consecutive, colorful Sundays, filling a full page in newspapers physically much larger than today's and at a time when four-color printing was a rare treat in most homes. (Horina's strip was black and white with a layer of red highlights, a printing technique not unusual then.) Comic strips themselves filled an invaluable niche as a mass art form delivered directly to one's door. Comic strips were pitched at the working classes, especially those who left farms for factories in a gigantic internal migration and immigrants who entered the country at a rate of 500,000 a year from 1890 to 1910. Newspapers had huge advantages for audiences desperate for cheap entertainment. Competition brought prices down to a penny for an item that could be shared by an entire family. (Enterprising kids could get them even more cheaply, using the beg-borrow-or-steal methodology.) Newspapers taught English, showcased pictures of fascinating celebrities and places, advertised goods by the thousands, and served as a repository of job openings. Every city of at least medium size had half a dozen to a dozen English-language papers competing to turn occasional readers into regulars who never missed an issue, plus up to a dozen more in the immigrants' native languages. Moreover, newspapers appeared every weekday, with ever-increasing percentages offering a fat and enticing Sunday edition full of special goodies. Circulation of major papers soared into the hundreds of thousands.

Fans who checked the ninth *Percy* panel saw the signature "H. C. Greening," a star name for anyone paying attention to commercial art.[2] Harry Cornell Greening was a 35-year-old veteran of the business, illustrator of books, magazines, and a potpourri of earlier strips including *Joco and Jack*, *Uncle George Washington Bings*, and *Prince Errant*. Greening's strips played out like cels of an animated movie, jumping with movement and action that drew readers' eyes along to the invariably messy conclusion. *Percy* is as formulaic as a Punch and Judy show; it has one joke, endlessly repeated, as expected as the squirt of seltzer in the baggy-pants comic's vaudeville routine—perfect for an audience whose grasp of English couldn't be taken for granted. Greening nevertheless wrought endless variations within this strict framework, with the robot discombobulating "policers" and "bankists" and using its inhuman competence to destroy opponents at ice skating, sledding, archery, curling, golf, fishing, bowling, baseball, tennis, and even leapfrog. One episode of *Joco and Jack* had used the same plotline with Joco the Monkey as the destroyer. Monkeyshines are simply funny; a too-perfectly-behaving robot said much more about society, turning the gag into science fiction as deep and prescient as that of H. G. Wells.

Cartoonists and humorists are exquisitely sensitive to their times: old jokes are reliable; new jokes create sensations. The lines in that very first panel encapsulated the worries of a generation. Middle-class households had a long-standing "servant problem."

The lower classes who had once worked as maids and farmhands now preferred almost any other type of work. Both houses and farms sought to replace them with mechanical (and, increasingly, electrical) appliances and devices that purportedly did the work of ten for a lower cost and never needed to be fed. *Percy* appeared because the times were ripe for technological satire. The fear of being replaced by a machine lessened when the machine left its world a shambles.

Greening's creation caught on almost instantly. For the next decade, Percy the Mechanical Man was the public face of fictional robots, probably the first such. That he had a name, a rarity among early robots, was a psychological plus. The strip spread from its Boston home base to papers across the country, a potential audience in the tens of millions. Greening's repetition of a catchphrase gave the strip entry into popular slang.

References appeared everywhere in an era when mention of a comic strip was déclassé. Lucy N. Eames, MD, speaking to the Muskegon-Oceana County Medical Society on March 22, 1912, lamented the difficulty of diagnosing illness with the wail that "we sometimes think of ourselves as 'Poor Percy! Brains he has nix!'"[3]

The working classes adopted Percy as one of their own. A workmen's magazine, the *Paper Makers Journal*, published a bit of doggerel about a baseball team in its February 1912 issue:

> The Woodmen are a mighty clan,
> Chock full of slippery tricks,
> Something like "Percy, the mechanical man,"
> But brains they have nix.[4]

The *Leather Workers' Journal* for July 1914 threw in an admonition for worker safety, based on Percy's propensity for leaving the scene in chaos:

Another thing, for God's sake, don't mix up matters here by trying to do things on your own hook. For in some things it seems you are like the mechanical man Percy—"Brains you have nix." Enough said.[5]

Percy's immediate impact is again revealed by an article in the *Pittsburgh Daily Post* for January 14, 1912, which boasted that

John P. Harris has arranged for a great novelty for the Harris Theater for this week. It is called Caanda [*sic*; Gaanda] Humanus, but the Manchester Brothers, who own the novelty, call it "Percy." Percy is a mechanical man, seven feet in height, electrically operated, responds to push buttons in his back which resemble the ordinary doorbell button, walks, runs, rides a bicycle, writes on a blackboard and does numerous other things. He is taken apart in view of the audience and put back together again.[6]

This so-called novelty had been a smash back in 1906, when an identical act had toured the country for a year. Bringing it out of mothballs in 1912 as "Percy" could have no purpose beyond cashing in on the latest fad.

That same year, a totally different act also stole Greening's creation:

"Percy, the mechanical man," was well executed by Roy Bowman, who dressed in a brilliant uniform and painted in gaudy Christmas doll colors, was piloted about the stage in a series of stiff-legged and stiff-armed movements by H. J. Schutje, who acted the part of the "professor."[7]

Percy soon became an all-purpose metaphor. In October 1912, the *Lebanon Daily News* editorialized:

So we go from one triumph to another. Nitrogen direct from the air, made up into edible tablets is the next reform on the table.... Our children may live to see "Percy," the mechanical man doing the work that in less advanced days was done by creatures of flesh and blood.[8]

And the inevitable joke became manifest in 1913, when the *Cherokee Republican* commented, "This new Oklahoma legislature seems to be very much like Percy, the mechanical man of the Sunday colored supplements.—'Brains it has nix.'"[9]

Percy ended on January 12, 1913. The next Sunday, the *Washington Star* ran Winsor McCay's classic *Dreams of a Rarebit Fiend* in its spot. Greening went on to other cartoon work in a long career; yet Percy kept beckoning. The mechanical man had as many lives as a cat and even more names, as in this otherwise inexplicable reference in a 1930 edition of *Time* magazine:

Intermittently from 1915 to 1920 a robot called Mike, then Fritz von Blitz the Kaiser's Hoodoo, then Percy the Mechanical Man, performed prodigies of senseless versatility in the U.S. funny-papers (New York *Herald* et al). Cartoonist Harry Cornel [*sic*] Greening equipped his creature with a row of buttons down the back which, when pushed, set Percy to his tasks. Only trouble—and chief source of comedy—was that, being brainless as well as tireless, Percy would keep on doing whatever he started until someone pushed another of his buttons. Thus, stoking a warship, when he had stoked away all the coal, he shoveled into the powder magazine, blew up everything but his indestructible self. Robert Tyre Jones Jr. likes being called "Robot, the Mechanical Man of Golf," better than a lot of other names to which sportswriters, their superlatives utterly exhausted, have had resort. Before and since his appearance in the golfing firmament in 1916 (one year after Percy), he has had no peer but Percy, and making oneself a mechanically perfect golfer—when one is equipped with temper, indolence, misgivings and other frailties to which robots are heir—is as satisfactory, when accomplished, as it is difficult.[10]

The strip started in 1911, not 1915, but the *Time* writer clearly meant our Percy and our Greening, even if his middle name was missing a final "l."

Greening subsequently hid Percy in plain sight—namely, in another of his strips, *Majah Moovie*, which ran from 1915 to 1916. The Majah (dialect for "Major") was a wealthy eccentric with extremely modern tastes: he wanted to record every moment of his days for his "living moving diary." His faithful servant, 'awkins, after a two-panel setup, ran the camera for the inevitable disaster that would strike over the course of 13 panels, allowing 'awkins to make a sardonic comment in the sixteenth and final panel.

Could Percy be far behind when a disaster was called for? Apparently not. Greening brought him back in the August 15, 1915, strip, instantly recognizable but now named Mike. "Brains 'e 'as nix," 'awkins concludes.

Percy's third life came in 1916 when Greening made a now-lost animated cartoon, *Percy: Brains He Has Nix* (also titled *Percy the Mechanical Man*) for J. R. Bray Studios. John Randolph Bray was himself a pioneer animator. His production company pumped out nearly 100 cartoons in 1916. Sadly, most of his output is gone, and no description of the *Percy* cartoon survives. However, we did finally get a name for the inventor: "Herr Professor Doodlepoodle, N. U. T., famed in both hemispheres—and New Jersey."[11]

Two years later, Percy returned to the *New York Herald*. Germans had become the enemy after the United States joined the Allies in World War I. Sauerkraut was, more or less facetiously, renamed "Liberty Cabbage," and German-accent vaudeville comics hurriedly announced that their unchanged accents were really Dutch or homegrown Jewish. An amiable German professor was a no-go, but pompousness lurked in new guises, always ripe for a takedown. Greening returned with German-dialect caricatures of the enemy, who had their dignity fatally ruptured by the mechanical man *Fritz von Blitz the Kaiser's*

Hoodoo. Fritz is a rounder version of Percy. Nothing else changed, and the one-joke scheme of the strip endured. That version started August 18, 1918, and petered out on February 23, 1919, with peace making the joke even more pointless than it had been before. Greening brought back the strip simply as *Percy* on March 2, 1919. At some point that year, he moved his creation into show business, renaming the strip yet again as *Percy in Stageland*, which lasted until March 28, 1920.

—⚄—

For a generation, "that Buck Rogers stuff" defined science fiction in the minds of the American public. Spaceships, ray guns, planet-hopping adventures, strange alien races, and antigravity belts set the tone in fiction while pundits and columnists used the phrase as shorthand for any current or proposed technological marvel either as praise or as denigration. A 1937 newspaper article described the upcoming regular trans-Atlantic service of Pan American super-clippers as sounding "a little like Buck Rogers stuff, but it is actually taking place under our eyes."[12] Two years earlier, a columnist had heaped calumny on astrophysicist Arthur Eddington for daring to predict scientifically what the end of the universe might look like: "It is Buck Rogers stuff panoplied in jargon that passes for scientific terminology."[13]

Buck was inescapable. A daily comic strip had blasted into newspapers in 1929, and a separate line of color Sunday adventures debuted in 1930. His show could be heard four times a week on the CBS radio network in 1932, with three more programs bearing his name to follow. A ten-minute film played at the 1933–1934 Chicago Century of Progress Exposition, and Buster Crabbe forever associated himself with the part when he starred as Buck in a twelve-part movie serial in 1939. Television was a natural for Buck's visually exciting adventures, and two series appeared, the first on ABC in the 1950–1951 season and a second lasting for two seasons (1979–1981) on NBC. Buck's booming empire caught the eye of *The New Yorker*'s "Talk of the Town" section in both 1934 and 1935, with the latter reporting that "it looks as if Buck Rogers were going to overtake Mickey Mouse in popularity. The Buck Rogers rocket pistol has already set an all-time sales record, we were told, and is now being supplemented by the Buck Rogers disintegrator."[14]

Back in 1928, Hugo Gernsback's pioneering science fiction magazine *Amazing Stories* carried a 30,000-word story, "Armageddon—2419 A.D.," in its August issue, the first published story by Philip Francis Nowlan, a 39-year-old newspaper financial writer from Philadelphia. Readers were introduced to Anthony Rogers, a World War 1 veteran who gets trapped in suspended animation by radioactive gases in a mine and awakens 492 years into his future. Those who go back to read the story might be surprised to find that it stays entirely earthbound and is nothing more than another Yellow Peril racist thriller, a world in which the Han Chinese (called Mongolians for some reason) conquered America hundreds of years earlier. The few Americans who survived retreated to the woods, where they have finally reinvented advanced technology, using ultronic vibrations to create inerton (an antigravity metal) and ultron (invisible but capable of long-distance communications). Rogers helps the American revolutionaries, especially the plucky 20-year-old Wilma Deering, whom he soon marries, to strike back at the Mongolians' super airships and disintegrator rays. The series of battles, close calls, and technological wizardry startlingly resembles an updated version of the Steam Man dime novels, with the Americans replacing the Indians, though here the Americans critically hold the superior technology.

John F. Dille, owner of a comics syndicate, had had an idea for a futuristic strip different from anything appearing in the papers. He thought "Armageddon—2419 A.D." proved that Nowlan could handle the daily grind of spewing out idea after idea. Editorial cartoonist Dick Calkins really wanted to draw a caveman strip, but he, too, fell under Dille's spell.[15] The new team worked incredibly fast. The first daily strip of *Buck Rogers—In the Year 2429* appeared on January 9, 1929 (eschewing that unrounded 492 years nonsense). The strip's title would subsequently advance year by year until it settled into the generic *Buck Rogers in the 25th Century* form used in all later media. In 1930, Wilma was kidnapped by the Tiger Men of Mars, at which point the Mongols were forever abandoned in favor of adventures all over the solar system. Thus the Buck Rogers legend was born.

A 1929 contest in the *Pittsburgh Post-Gazette* asked readers for thoughts on "What Do You See 500 Years Ahead?" Entrants were obviously adults, and the winners included such inspired guesses as "The English Channel tunnels are not being used and will be abandoned"; "Sahara produces half of the world's food. Artificial rain has made a productive country out of a desert"; and "Stored light has eliminated darkness in most places."[16]

Science fiction writers liked to extrapolate from the advances that the public read about in newspaper headlines, and the almost hysterical newspaper response starting in 1927 proclaiming that the Westinghouse robot Televox presaged a future full of robots (see chapter 5) must have influenced Nowlan. Robots were one of the many wonders of AD 2429, with one panel showing robots building skyscrapers, lifting huge steel beams singlehandedly. If construction robots weren't fearsome enough, imagine Buck's archenemy, "Killer" Kane, and his plans for a new "super-robot," introduced in the July 31, 1929, strip. (Newspapers often started strips from the beginning when they added them to their comics section, so that official release dates often do not match the dates when contemporary readers were able to see them or when they are found today in newspaper databases.) The super-robot was only vaguely humanoid, resembling a chess bishop on caterpillar treads. Its head contained television receivers and transmitters, allowing a remote operator to see what it was doing and even talk through it. "Electro magnetic push-pull muscles" gave its arms all the strength a villain could hope for. In the strip, Kane steals a controller and the plans to build more robots. Buck and Kane then pit their robots against one another; anticlimactically, the battle ends after the exhausted Kane falls asleep.

Robots were in the background of several *Buck Rogers* strips that first year, running errands in the Mongolian capital. The Emperor orders nectar from one on January 24, 1930, beckoning it with the command, "Ho, Televox!" Televox, the Westinghouse robot, was still touring the country in 1930. This unexpected shout-out confirms that the Westinghouse family of robots had replaced Percy as the public face of robots in America.

Robots made their first Sunday appearance in the serial titled "Mysterious Saturnian" starting on November 2, 1930. Sunday strips took far longer to produce than dailies, stunning readers with full-color extravaganzas of eye-catching space scenes and vistas of other planets. To meet the extra work, Russell Keaton (inevitably nicknamed "Buster") got hired as a ghost, but his work was painfully static, and Rick Yager took over starting in 1933.[17] Oddly, Sunday strips didn't feature Buck Rogers, although his name was signed to a capsule summary of the action. Feeling that readers might be confused by separate continuities, Dille ordered the otherwise interchangeable adventures to star a pair of doppelgängers: Buddy Deering (Wilma's brother) and Alura, a princess of Mars who fortunately looked exactly like a human—specifically, exactly like Wilma.

In the Saturnian storyline, Buddy and his gang have been battling an evildoer from Saturn when a Hindu lass named Lalla is kidnapped and stashed in the Saturnians' secret Himalayan lair. Buddy snaps into action, pulling out of nowhere a "flying **robot. Radio**-controlled with **attractor**-beam propulsion" (all emphasis and punctuation in quotes as in original). Buddy explains that "with this control box I can make it do anything and talk through it." "Oh! I think it's **just** too **clicky**," says Mary, the president's daughter.[18] The robot flies in and grabs Lalla, who believes that she's merely being kidnapped again by a different alien. (Keaton's anatomical ineptitude is embarrassingly proven by the image of the robot lifting Lalla by her breasts.) After a short round of back-and-forth action, the robot again swoops in and captures all the Saturnians. Why such an incredibly handy gadget was immediately written out of the strip is unfathomable, unless it made escapes *too* easy.

Robots reappeared in "Mekkanos of Planet Vulcan," which started on October 24, 1934. Breaking out of the fourth dimension that fall Sunday, Buddy and Alura find themselves about to crash into Vulcan, a once-theorized planet even closer to the Sun than Mercury. Fortuitously, their ship gets pulled into a gigantic abyss at Vulcan's north pole and lands safely at the planet's center, which is improbably but necessarily cool, well lit, and a source of breathable air. (Such hollow planets had been a trope in fantasy fiction for almost 200 years, most recently popularized by Edgar Rice Burroughs' *Pellucidar* series, two books of which were published in 1929.) The adventurers are greeted by a radio-controlled robot with "television eyes" and "microphone ears" (much like those of the super-robot), which leads them up, down, and across a buried city. The robots are individualized around a set pattern of arms that reach down past nonexistent knees, legs that are pinned to the sides of their chests, and heads that resemble European clowns, all in shades of blue and purple. The few remaining Vulcanians hide inside their luxurious homes in fear of their fellows, doing all business and communication through their robots' eyes and ears.

This story is an important midpoint in science fiction history. Keaton vividly copies William Wallace Cook's remote-controlled robots battling for supremacy in wildly cartoonish style, the robots hydraulically extending their necks, arms and legs maneuvered by the giant thought helmets worn by the Vulcanians. The agoraphobic and sybaritic Vulcanians anticipate the Solarians in Isaac Asimov's 1957 novel, *The Naked Sun*, who have retreated from active life in favor of a world run by robots. Both Nowlan and Asimov were young men fascinated by science fiction yarns, making the connection provokingly plausible. The short stay on Vulcan ends with a canny future prediction. Lost in the gigantic city, pursued by an army of Mekkanos, Buddy and Alura enter a "fact room," "a kind of reference library" with "automatic talking and thinking machines," that gives them directions back to their ship.[19]

Four years later, robots formed the backbone of both the daily and the Sunday strips. In "The Secret City of Mechanical Men," Killer Kane and his moll Ardala Valmar are employed by the Fiend of Space to help him conquer Earth. He sends Kane and Valmar to a city hidden in the mountains of Peru, run by robots—in this incarnation, rude cylinders with legs. "Slaves of Steel!" Kane exclaims, "With mechanical brains—sensitive to our every thought command—and with these, we shall conquer the world!"[20] Alura saves the captured Buddy by setting one "mech man" against another—they, like most robots, can think for themselves when convenient but require orders before moving when the plot requires idiocy—only to be captured in a second secret city filled with ancient Incans.

Kane destroys their city with a squadron of flying aerial torpedoes, horribly prescient of the robot bombs of World War II (except for their art deco design and fins). Buddy and Alura are ultimately saved by another *deus ex machina* escape.

Separate continuities enable the Fiend of Space to imperil Buck in the overlapping daily strip. The Fiend is kidnapping the most brilliant scientific minds on all the planets, so naturally he comes after Dr. Huer, Buck's friend and mentor. A series of increasingly gigantic mechanical men are employed in the attempted captures, the Fiend being an immaterial cloud of malignant evil so terrifying that when Dr. Huer captures an image of it via his "penetra-reflecto-television using his next radio beam as a carrier wave," he faints at the very sight. A rocket bomb to the Fiend's spaceship headquarters finally finishes it off, but not before the entire Earth reels from its assault.

—◦◦◦—

Piles of newspapers were hard to store; cut-out strips often got lost. One solution appeared as Big Little Books (BLB) (originally a line started by the Whitman Publishing Company in 1932, and today a generic term for the many series of that type). The typical BLB opened to hundreds of double-page spreads, with the left-hand page containing about 50 words of large-type prose across from a single-panel cartoon illustrating the action. Whitman's books were 3⅝ inches wide and 4½ inches high, easy to hold in small hands but plenty of value for a dime (later 15¢). The first book, in 1932, starred Dick Tracy. More than a thousand volumes followed, both from Whitman and in a variety of sizes and names from competitors.[21]

Buck Rogers: 25th Century A.D. vs. The Fiend of Space appeared in 1940, after the series was renamed Better Little Books and ballooned to 425 pages. Both text and art were extremely faithful to the strip, some of the panels directly copied and others skillfully adapted for the purpose. The unshowable Fiend didn't make the cover: instead, a gigantic robot overshadowed a swarm of human gnats in spacesuits futilely shooting a ray gun at its bulk, the quintessence of a cover selling a book.

Robots sit comfortably in science fiction, and they were also used in fantasy, following the example of L. Frank Baum's *Tik-Tok of Oz*. The first BLB with such an adventure was adapted from the comic strip *Tiny Tim*. Written and drawn by Stanley Link, a 40-year-old veteran cartoonist, the Sundays-only strip was a humorous look at eight-inch-tall Tim and his slightly smaller sister Dotty in fantasy lands taken from storybooks and modernized. During a 1935 stay in the kingdom of Erewhon (reprised as the 1937 BLB *Tiny Tim and the Mechanical Men*), Tim deliberately ignores a gigantic (to him) warning sign and falls into a chasm, landing in a raging river and escaping up a narrow path on the side of a cliff. All this is harrowing enough, but he is nearly trampled on the path by "a man made of iron!" Many adventures later, he follows the robot deep into the earth, into an artificial city filled with thousands of mechanical men. "Zorax the First, ruler of all Boogaboo Land," tells the robots what they should already know: they are "the only invincible army in the world. Soldiers who never eat, who never sleep, who can fight twenty-four hours a day—I have created you. With you I will CONQUER THE UNIVERSE!"[22] Zorax is the sole human inhabitant of his kingdom. Before he conquers the universe, he plans on kidnapping Princess Philomena of Erewhon to be his queen. (Women had a dismaying tendency in comic strips to exist solely to be kidnapped, leaving them in weeks of peril until heroic men could effect a rescue.)

Zorax is sufficiently brilliant to create iron men who obey his spoken orders and yet "have not ears to hear" (their heads have binoculars for eyes and a slit for a mouth without other decoration), as well as a mechanical horse, a fire-spitting flying mechanical stork, a sub-sea dreadnaught crab, and a terror turtle tank. Zorax's one failing is his inability to create a robot that can talk. When Tim crawls into the head of a robot, he inadvertently responds to one such lament. Zorax is so pleased by this miracle that he makes that robot his generalissimo. Even more miraculously, the robot's head contains an otherwise useless set of control buttons that allow Tim to move him around and sabotage the army by destroying all the factories. The seeming triumph is upended when a newly captured Philomena is brought before Zorax. In one of the greatest *deus ex machina* rabbits ever pulled from a writer's hat, a Jinnie suddenly appears out of thin air and hands Tim a magic wishing ring. Zorax can't win. Tim even reduces the villain to his own eight-inch size so that he can deliver a sound thrashing, fair and square, in archetypal American style.

Link added some Easter eggs for the worldly adults reading the story with their children. The Kingdom of Boogaboo is a knockoff of Baum's Kingdom of Oogaboo, visited by his robot figure in *Tik-Tok of Oz*. The larger Kingdom of Erewhon is named after the country in Samuel Butler's 1872 satirical utopia, *Erewhon: or, Over the Range*. (Tim gets to Boogaboo by climbing a cliff.) *Erewhon*'s last section is titled "The Book of the Machines," suggesting that machines, like animals, will necessarily evolve until they can reproduce themselves, mirrored in Zorax's huge robot factories run by robots. Never underestimate the erudition of cartoonists. Later cartoonists in strips and comic books would duplicate many more elements from *Tiny Tim* than from the Mekkanos of Planet Vulcan.

Maximo the Amazing Superman is that extremely rare object, an original BLB character. Russell "R. R." Winterbotham is thought to have written 60 BLBs in the six years before he started full-time work as a reporter in 1943, along with dozens of other children's books and adult science fiction stories.[23] *Maximo the Amazing Superman and the Supermachine* (1941), the third Maximo BLB, sits at the intersection of these works. Maximo Miller is yet another stalwart young American at the peak of physical perfection and gifted with a superbrain. In the movie *Lucy* (2014), Scarlett Johansson ludicrously gained the ability to work wonders when a drug opened up what the script called the 90 percent of her brain humans normally don't use, but Maximo's powers are explained in the same way, only more so. Instead of being "ten times as brilliant," his brain is 1,000 times more powerful than a normal brain. Breaking down walls and hypnotizing people with a glance are minor powers. Maximo also potentiates positive and negative charges and so can fly by magnetic repulsion, save a city from a broken dam by making the water flow uphill, and protect his body from a lethal death ray. Foreshadowing the way computers would shrink from room sized to pocket sized in the future, Steinmark, the villain, muses about the difference in scale between the supermachine he builds to duplicate Maximo's powers (to "conquer the world") and the size of the human brain and wonders whether, by examining the real thing, he could shrink his apparatus. Like most evil inventors, Steinmark can whip up any mechanism needed to propel the plot. That includes a robot watchman who "walked on two legs like a man, but there the human resemblance stopped. Its body was a square of metal on which was mounted a large brass ball.... Whenever a figure moved between the lamps and the receiving electric eye, the robot watchman moved toward the fence to intercept the intruder."[24] In the end, Maximo uses his brainpower, both in clever stratagems and in pure force, to make the squad of robots destroy one another with their own lightning bolts.

Weird adventures and nonhuman villains made the line between science fiction and fantasy blurry, with a middle ground of "science fantasy" often a better descriptor for adventures involving swords and sorcery. Edgar Rice Burroughs set his plots in jungles, on other planets, and at the center of the Earth, with millions of readers caring only about the spectacle. His *Tarzan* comic strip debuted the same day as *Buck Rogers* and soared to the same immediate and continued success. Numerous attempts to launch such a strip for his other major hero, John Carter of Mars, failed over the years, with ironic unanticipated consequences. The King Features syndicate decided that an original comic strip (i.e., one that didn't have to pay royalties to an outside creator) was a better way to proceed. The result was Alex Raymond's *Flash Gordon* strip, which quickly rose past its blatant imitation of John Carter to rival Buck Rogers as the king of science fiction. In another irony, Buck Rogers' origin itself surely contains memories of Burroughs' books, as John Carter also is overcome by fumes in a cave; his spirit, encased in a new body, finds itself on Mars, fighting the differently skinned enemy and falling in love with the local, human-in-appearance Princess of Helium, Dejah Thoris.

Not until 1941 did Burroughs find the perfect solution: his 28-year-old son, John Coleman Burroughs, would write and draw a Sunday-only strip. Unfortunately, it had the exceptionally bad luck of starting on December 7, 1941. The strip lasted less than two years, out of step with the wartime atmosphere, the lush rendering of fantastic landscapes, medieval castles, and weird monsters presented in dreamy pastel colors apparently not resonating even as pure escapism. Nineteen of the episodes in 1942 pitted Carter against robots, creations of Vovo, the green-skinned, four-armed Martian Wizard of Eo. Vovo is an inventive genius, a technological wizard, exiled by other Martians out of fear of his amazing technology. However, when Dejah is turned to stone by the "strange vapors of Go-la-ra," Vovo is the only one who can restore her to life. He, with his mechanical servant Oman, shows up on his degravitated (i.e., flying) mechanical horse to bring her to Eo, where his machines can cure her. Vovo, exactly as evil as the Martians thought, then announces a plan to hold her for ransom, blackmailing the people of Helium, leading to the spectacle of John Carter battling Vovo's robot horde, all armed solely with swords. Carter can't win such a fight, but he finds an unlikely savior in Oman. Oman looks perpetually surprised with his circular eyes and bulging earcones, perhaps because he is revealed to be the only robot with free will. "This is the end of your cruel rule," Oman declared, "for now I shall rule my own people."[25]

Two other strips often lumped in with *Buck Rogers* (and rightly so) kept their science fictional adventures on Earth. Our planet had plenty of room for lost worlds in the 1930s, so when *Don Dixon and the Hidden Empire* debuted in 1935, readers might easily believe in Pharia, a place not found on any map but containing a beautiful princess named Wanda with hair as blonde as all–American Don's. Don and Wanda endure deprivation, suffering, and evil foes as they flit from one dangerous place to another. One such locale is Robot Island, where Wanda is taken by the villainous Strunski early in 1939. The robots were invented by a kindly professor who wanted only to use them for service to humanity but made the mistake of hiring Strunski as his assistant. The professor has been a prisoner for ten years; Don comes along just at the moment when he has perfected the device that will allow the copper robots to be controlled by pure thought. Don is no match for the ten-foot-tall metal beings whose half-dome heads look a bit like sleek period radio sets. Substituting brains for brawn, Don tells the professor to build a control signal blocker. The uncontrolled and uncontrollable robots revolt, giving Strunski his just rewards.

Imitators often tend to be more like each other than the creative originals, and Brick Bradford's robot encounter has remarkable similarities to Don Dixon's, just as their backgrounds and exploits were often nearly interchangeable. An expert pilot and deep-sea diver, fast and deadly with his fists (not to mention as handsome as a movie star), Brick battled villains wherever the plots led him—into an atom, under the sea, in the far future, or in contemporary America against the usual run of villains working with foreign powers to steal our technological might. Writer William Ritt epitomized the "of all the gin joints in all the towns in the world, she walks into mine" style of plotting, with absurd coincidences and connections serving to force disparate elements together. Clarence Gray's figures were barely more than outlines, with many panels simply marking time with drawings of cars, ships, and airplanes. Regardless, something clicked. The strip grew in popularity from its 1933 debut on the small-town circuit, leading eventually to major newspaper appearances, a 15-part movie serial taking Brick to the moon in 1947, and a short-lived comic book around the same time.

"Brick Bradford and the Metal Monster" ran for a full thirteen months from February 13, 1939, to March 16, 1940. The first strip announced the plotline ("THE METAL MONSTER—A titanic robot menace to all mankind! Can Brick defeat this treacherous adversary?") but readers had to wait until April 7 for the robot to appear, an ungainly collection of mismatched parts, from its crinkled ductwork arms to its skeletal legs to the bullet-shaped head with binocular eyes that closely match those of Tiny Tim's robots. Kindly scientists who have their creations stolen by evildoers would comfortably fill many pages of both comic strips and comic books. The Metal Monster storyline ran nearly simultaneously with Don Dixon's Robot Island adventure, with writers Ritt and Moore likely drawing from a common well of inspiration rather than each other. Ritt's kindly scientist, Kalla Kopak, wants to "manufacture these [robots] by the million—to do humanity's labors." Brick sensibly objects that "you'll need a vast crew of workers to build these!" Kopak ripostes, "Under my direction I'll have robots build additional robots."[26] Such utopian justification for robots, repeatedly mouthed by well-intentioned fictional scientists convinced of the benefit of robotic technology, is hard to fathom in an America barely recovered from the Depression. Ritt himself evidently didn't believe in this future, since he had evil Avil Blue build the ultimate robot from stolen Kopak plans in an abandoned ship factory that dwarfs the two workers depicted. This amazing robot can literally crush airplanes and submarines within one squeezing hand and also "sweeps up hundreds" of the army of robots that Kalla throws against him in that one supersize paw. Just before the robot starts marching undetectably underwater toward the city of Metropola to crush it with his spiked feet, Brick sneaks into the robot's head. All the robots are radio-controlled from the outside, but Brick conveniently finds a manual titled "Instructions for Operating the Robot from within the Mechanism Itself" that Blue had thoughtfully put in the head for interlopers. Perhaps Blue should have said "yes" one of the many times Brick was captured and a henchman asked to kill him.

Newspaper syndicates ranged from the gigantic to the barely there, with the strips' quality proportional to circulation. Perhaps the lowest of the low was *Dash Dixon*, created in 1935 and credited to "Dean Carr." Larry Antonette was the artist behind the pseudonym.[27] Though Antonette was in his mid-twenties in 1935 and later taught art in several schools, the strip looks like a portfolio of a flunking high school student. The sheer awfulness of the linework is matched by the dialogue, though the blame cannot be fixed on a known writer. In the strip, Dash Dixon and Dot Smith are motorboating around the Pacific, like ordinary people do, when they get stranded on a tropical isle and kidnapped by a Martian spaceship. The Martians wear helmets inside their own ship, but the humans need none. Any science in this science fiction strip was as imperceptible as the invisible planet with no gravity that Dot and Dash land on, only to get kidnapped again, with Antonette handling the odd format by dangling a cliffhanger every week. A ray pulls the hapless heroes, along with a good guy scientist, into the spaceship of the evil Mogo, commander of "Mecho" men. Dash knocks out a Mecho with one punch, but Mogo imprisons them until the scientist changes the Mechos' speech receptors to respond only to his voice. Mogo then disappears in a puff of smoke. The brief adventure constitutes the eight most bizarre strips in robot history.

—⚍—

Little can be gleaned from these otherworldly strips regarding an understanding of the general public's view of robots between the two world wars. Robots mostly were mindless menaces under the control of a nefarious villain, just another tool in their deadly arsenals. To keep pace with the impossible grind of 365 gags or narrow escapes a year, cartoonists floated on a sea on cultural clichés, exaggerating tropes and stereotypes for comic effect, with familiarity scoring higher than originality. *Percy* had succeeded by providing endless repetition of one joke based on the innovative notion of deceptive machinery, technological perfection always only a step away from running amok. Strips set in the real world rarely had characters encounter robots, unfortunately, but the cultural moment can more easily be read when they do.

Hairbreadth Harry's name says it all. The hero, formally Harold Hollingsworth, is the prototypical red-blooded all–American boy, forever getting into jams that he slithers out of by a hairsbreadth. His foil is the top-hatted and mustachioed villain-of-all-trades Rudolph Ruddigore Rassendale. (If you've ever wondered what Dudley Do-Right and Snidely Whiplash were parodying, wonder no longer.) Starting in 1906, *Harry*, by far the most famous of the thirty or so strips created by Charles Kahles (pronounced Kaw-less), begat a rage for serial cliffhangers in both comic strips and silent movies. Kahles had a fondness for robots, first using them briefly twice in 1927—a boxer described as an "electric mannikin" and a miner called an automaton—and then blatantly parodying Televox in 1929 when Harry introduces his girlfriend Belinda to the push-button-controlled "most perfect radio-electrical machine yet invented," Mr. Televolts. The machine is "almost human" (at least in comic strip terms)—in fact, it's an "ape-man" resembling a gorilla. Just two panels later, a real gorilla escapes from the zoo and Harry the hero dashes off, warning Belinda not to touch that middle button! Kahles overstuffs the action the next day when Rassendale dresses up a minion in a gorilla suit and turns him in for the reward. Just when things couldn't get worse, Belinda presses the middle button. It's the love button, and the ape-man-machine literally sweeps her off her feet. The zoo must be located in Gotham City, for there's a cave within easy walking distance in which all three stages of gorilla evolution wind up, with Mr. Televolts triumphant over all, an inch away from killing Harry. In mid-clubbing, Harry is saved by … a hairbreadth: the robot's clockwork runs down. Dangerous devices these automatons, fit only for war.

So war it shall be, during a 1931 sequence introducing a robot named only "robot." The nation of Emacia has attacked the plucky underdog Hepatica, which calls upon Harry to be its general because of his knowledge of the inventions of the late Professor DuBall.

Rassendale immediately offers his services to the Emacian "local big shot," and the hijinks begin, serious war being a fortunate decade off in the real world.

Harry controls the robot on the battlefield by walking behind it, issuing orders into a microphone. Nobody gets hurt: the robot throws grenades containing sleeping gas— luckily for Harry, as he becomes a gas victim after an exploding shell turns the robot around. Any semblance of realism is abandoned when Rassendale kidnaps Harry's sweetheart in America and stows her in an abandoned building that a film crew has wired with dynamite for a scene. Belinda is thrown sky high—literally so, because she lands in a passing open-cockpit airplane. Where is it going? Hepatica, of course. That might have been too much even for Kahles, for he winds up the war in a whirlwind week of dailies. Shades of *R.U.R.*, Harry's robot factory turns out thousands of soldiers an hour, certain doom for Emacia, until Rassendale calls in an air strike. Harry's ahead of him, as always. The factory is protected from bombs by ... a net, an idea still ahead of its time. Then a robot struck by lightning shoots Harry, and he is left for dead, when who should wander by but Belinda. The armistice is signed two panels later, and another adventure begins the next Monday.

Westinghouse's robots left another comic strip legacy in 1933. Soon after Katrina Von Televox, the company's housemaid female robot, toured the country, artist Russ Westover parodied the female robot worker in his strip *Tillie the Toiler*, which began as a cruel caricature of the Jazz Era working girl and didn't soften later. Westover introduced Tillie in 1921 as a stenographer in the office of magnate J. Simpkins. Tillie wore a different outfit to work every day, each an extravaganza of supersized bows, ruffles, and winglike shoulders that would have seemed stupendously over the top in a Hollywood musical. Sunday strips printed outfits that little girls could cut out to dress Tillie paper dolls. As tall, thin, and beautiful as her forerunner, the Gibson Girl, Tillie towers over the officemate who pines for her, Clarence "Mac" MacDougall, who by the time this sequence ran daily for four months was literally only waist high to her magnificence.

Tillie wants Mac to be an ever-faithful puppy dog but has her head turned by every pretty male face in her sightlines (the only humans in the strip allowed to be taller than she is). They all fall instantly in love with her, of course. Even so, Tillie, like virtually every popular culture female of the era, is jealous to the point of dementedness. When Mac sneaks off for a secret project, Tillie is positive he is seeing another woman and finds proof in his every move, even when he buys hacksaw blades and rivets. In fact, Mac is building the perfect stenographer, named Rosie the Robot. Rosie is a frump, the second banana comic friend of the movies, as tall as Tillie but with electrical wiring as curly hair, a long plain metal dress, and oversized shoes. Mac explains that "these hair-like wires draw electricity from the air that furnishes the power."[28] Self-sufficient Rosie never turns off.

Poor Simpkins needs a good worker, as Tillie is anything but a reliable employee. When the strip was introduced, promotional literature told readers, "She moves slower than a crippled worm and works less than an I.W.W. in prison." (The International Workers of the World were an early militant union, often jailed for violence.) She's always late, frequently takes vacations, and thinks nothing of calling in sick when she wants to go dancing. Stenography is beyond her. In a contest, Rosie finishes typing the dictation as soon as Simpkins stops speaking while Tillie is lost by the third word, which is "it." Rosie not only breaks the glass ceiling but also chews it as a delicious dessert.

Rosie needs to have a flaw, so when Mac makes some adjustments, he inadvertently reassembles her without her governor. Superstrong Rosie wreaks havoc throughout the city while Tillie dallies with the world's handsomest and dumbest detective (he found the lost chord). After recapturing Rosie, Mac improves on her perfection by giving her the power to read minds (an uncomfortable development for those around her, as robots have no social skills and no sense of discretion). Such perfection gains the attention of "electrical wizard" Professor Stymedoo (like the earlier Steinmark, a reference to General Electric's wizard Charles Proteus Steinmetz). Stymedoo, somehow even handsomer than the detective, wants both Tillie and Rosie, easily conquering the former and offering Mac $50,000 for the latter. Although Mac flatly refuses this fortune, Rosie's lack of discretion allows the professor to inveigle her into handing over her plans. Naturally, Stymedoo builds a duplicate—leading to the first female robot catfight in literature. Rosie may be the easy winner of a one-on-one fight, but Stymedoo wants to build thousands of robots to release office girls from their daily drudgery. Tillie, sensible for the first time, points out that robots would put those thousands of girls out of work. Rosie, her plans now destroyed, must disappear for the good of humanity. Comic strip logic comes to the fore when the representative for Ali-Undi-Pundi of Babologna marches in and offers to make Rosie their country's goddess of the moon, with all the tin cans, razor blades, and nuts and bolts she can eat. And with a "Hey, Nonnie, Nonnie, and a Hot Cha Cha," she's gone.

Rosie generated many letters to the editor in favor of robot workers, not all of them serious. One, signed "Just a Human Robot," presented a sardonic list of advantages of a robot maid over human help:

1. A robot would not be sensitive to personal and catty remarks.
2. Insolent and ill brought up youngsters could not ruffle a robot.
3. A robot could get along nicely on a salary of two to five dollars a week.
5. Robots would belong to the category with vacuum cleaners and refrigerators, etc.—in which most maids are placed.
6. Robots would be in no danger of overfriendly husbands.
7. When robots would appear run down a little more current would revive them—they don't need rest.
8. Robots would not offend the Madame by having friends as intelligent and prominent as the Madame herself. [#4 was skipped in the original][29]

—⁂—

Some comics run unchanged seemingly forever—after a radical reversal of its first few years, *Blondie* has run the same gags for seven decades—but both individual strips and the field as a whole tend to turn over about once a decade, as the zeitgeist zigs and zigs. The dawn of the 1940s brought forth a very different culture from the early 1930s. Robots moved from a marker of a fantasy future to a familiar guest subject in a broad variety of contemporary strips, from realistic to very much not.

Superman changed popular culture in an incredibly short time after his introduction in *Action Comics* #1 in June 1938. *Superman* #1, the first comic to name itself after a super-hero, debuted in 1939. The first iteration of a *Superman* radio show began on February 12, 1940. Seventeen Technicolor cartoons hit theaters in 1941 and 1942 (see chapter 10). The first of thousands of tie-in products appeared in ads that the comics ran. As early as 1939, Superman had both a daily and a Sunday comic strip, estimated to run in news-papers with 20 million in circulation. The earliest issues of *Action Comics* saw Superman face off against horrendously outmatched human criminals, so the numerous Superman writers realized that they needed villains to match Superman's mettle. The Man of Steel—something he'd been called since the first Sunday comic strip sequence—required a Villain of Steel, and a literal incarnation in the form of a robot was a natural jump. "The Bandit Robots of Metropolis" (as the sequence was named by reprinters) ran from October 27 to December 15, 1940, the unusually short stretch typical of *Superman* Sundays. Not much can be done in only eight strips; the plot is an extremely compressed version of "Brick Bradford and the Metal Monster." The huge robots, several times the size of a man and built along the lines of stovepipes connected to a boiler, smash up Metropolis while their leader demands ransom from the city. Superman easily quashes these "oversized toys" but has to capitulate when Lois Lane gets kidnapped. Fortunately, Lois manages to send a signal from where she is being held—apparently the world's biggest submarine since robots the size of telephone poles easily walk around inside it. Superman defeats the robots (which are controlled by men hiding in their heads) so quickly that we never learn who made them or who their leader might be. It all seems too easy for him.

Anthropomorphic animals were the forerunners of superheroes, starring in adventures that would kill any normal human. Felix the Cat was, as comic historian Don Markham notes, "animation's first superstar," appearing almost a decade before Mickey Mouse was born.[30] Mickey demonstrated that mice would always win out over cats, but Felix hung on to a parallel career, showcased in comic strips, comic books, and television. A Felix comic strip could easily be swapped out for a Mickey comic strip, each character being a fearless adventurer who travels the world and beyond, winning with pluck and grit and ceaseless optimism.

In March 1941, Merlin the Magician sends Felix on a trek to hunt down sorceress Morgan Le Fey, finding her in her hidden castle in the clouds. She's the only living inhabitant, letting her squadron of robots wait upon her hand and foot. The superstrong robots are impervious to all sorcery except her own. Le Fey has a nefarious plot: she is going to destroy the Earth and everything on it, down to the last creature, because if she leaves even one alive it might be Merlin in disguise. When he does arrive, he sweeps away her magical attacks with a wave of his hand. Yet he's helpless against the robots. They have only one weakness: they must be "re-sorcerized" (i.e., recharged) every twelve hours using a magic word only Le Fey knows.

A secret magic word is exactly like a modern computer password: the longer it is, the harder it is to guess. OZYMBAGLOFTYPETYLCZENCKROWORZEPAXYHONKETY-LIPOCH is a dandy that way; yet impossible-to-remember passwords have a weakness of their own: they need to be written down. The intrepid Felix miraculously finds the magic word written on a wall in invisible ink that shows only when a certain flashlight is trained on it. Just as the robots are about to take him back to Le Fey in pieces, he speaks the word and gains command.

If Felix the Cat fought robots, could Mickey Mouse be far behind? He more than caught up in a daily series named "The World of Tomorrow," running from July 31, 1944, through November 11, 1944. We no longer have a coherent image of Mickey Mouse as a character rather than as a corporate image, but in his early cartoons, comic books, and newspaper strips, he was an all-around adventurer, fighting foes on land, on the sea, and in the air, and in any time period a writer might want to send him. Mickey was the ultimate Everyman, a mild-manned suburban house owner, beset by conniving relatives, wacky friends, a self-obsessed girlfriend, and the world in general. Not until the end of a given adventure did he bear down and wind up smashing the criminal gang and thrashing their leader, usually the bear-sized cat Peg-Leg Pete (still called that at times even after he somehow grew back his leg in the 1930s).

Heavily influenced by the 1939–1940 New York World's Fair, itself promoted as "Unlocking the World of Tomorrow," the strip sent Mickey and his dog Pluto into the future by the simple device of having Mickey hit his head. This undated future, much like the later *Jetsons* cartoons, is culturally identical to contemporary America except that everyday activities take place in the sky. A dairyman (his horse riding shotgun), a street cleaner, a dog retrieving a newspaper, buses, and police cars all take wing with overhead propellers. Even a panhandler bums a dime from a floating hammock. Other classic future tropes like food pills, gigantic vegetables, and two-way picture phones are piled on until a troop of mechanical men suddenly appears and kidnap Mickey's girlfriend Minnie. These "mekkas" look like cartoon elves with hinges and are Mickey's size but far more numerous. Pete, the only criminal in a world where police have nothing to do except traffic duty, has a factory with an assembly line of mekka men making millions more mekka men. He dares say it: "Frankly, old pal…. I decided to take over the univoise" (ellipses in original).[31]

Pete also introduces the robot femme fatal Mimi, created for no apparent reason other than to make Minnie jealous. Just for fun, he makes a robot duplicate of Mickey as well. More insidiously, Pete has ready-to-go robot duplicates of important men, including the president. "We removes important people … and dese dummies take dere place!" Pete says in his Brooklyn-esque accent. "Den … when my mechanical army moves in … d'woild falls like a rotten apple."[32] (Pete was years ahead of his time. Using robot duplicates of world leaders eventually became a cliché in television and movies, but this may have been the earliest use of the device.) The world is doomed until Mimi, who naturally has fallen in love with Mickey, shows him how to sabotage the robots and sacrifices herself by taking a ray gun blast meant for him. Even Minnie is touched. One final gag twist is left: Pluto needs to be saved. Too late. Police Inspector Gluesome has already vanquished the "horrible robots." "How did you tame them?" Mickey asks in amazement. "Robots fear nothing!" The inspector holds up his ultimate weapon: a can opener.[33]

Invisible Scarlet O'Neil debuted in June 1940, her inventor father bathing her with a mysterious ray that turned her into one of the earliest female superheroes, assuming that invisibility is a superpower. (Universal released *The Invisible Man Returns* as a horror movie in January 1940 and *The Invisible Woman* as a comedy in December of that year. Invisibility was seen everywhere.) Derring-do was not Scarlet's style; she was more like an invisible Lois Lane, always caught up in other people's business. She looked as real as Lois, but her plots, courtesy of creator Russell Stamm, were far sillier. In 1942, Stamm had the milquetoastish Wilbur Wilt, tired of being pushed around, build a robot in his basement. Wilt wonders why it doesn't work; perhaps because it's so hollow that Scarlet can crawl inside. In a Stamm strip, a Wilbur Wilt can not only accept that his truly wireless robot has come to life and speaks in a woman's voice but also names it *Mr. Clinkety Clank*. Wilt naturally lives with a gigantic, overbearing wife who sells the robot—with Scarlet unconscious inside—to the junk man as part of the war's metal scrap drives. Here's where worlds collide: Mr. Clinkety Clank is moved in a horse-drawn wagon to a metal dealer using the latest electromagnetic crane—the old, the new, and the future co-existing in a single week's worth of strips. Much science fiction depends on new inventions immediately and thoroughly replacing our world's devices; *Scarlet* reminds us of the enormous inertia that human societies retain. Changes dot civilization pointillistically rather than with the all-encompassing brush of watercolors.

The robot, sans Scarlett, eventually winds up at the bottom of the river, seemingly defunct. A year later, a crook running from police jumps into that same river and finds Mr. Clinkety Clank. With perfect Stamm logic, he instantly realizes that it's hollow, swims inside, and walks the water-filled, rusted, and airless robot along the river bottom until he reaches shore, and then he uses the robot for a string of hold-ups. Even Scarlet thinks, "I still can't understand how that robot could operate after being in the river for a year and a half." No time to ponder implications. Wouldn't you know that at just that moment Wilt has finally created a perfect working robot—except that he goes berserk whenever he sees a woman. At this point, every comic reader old enough to turn a newspaper page knows that a battle of robots is in the offing, with Scarlet in the middle. Except Stamm, that is, who junks the new robot in a speeding truck doing ninety and schemes to put Scarlet back inside Mr. Clinkety Clank. Doing the unexpected allowed the strip to last more than a decade.

The supposedly non-funny *Tim Tyler's Luck* followed the adventures of an orphan boy battling fifth columnists on the home front. The Octopus sends a mechanical monster to destroy Professor Ames' flying tank; this muscleman has no problem hoisting the tank over its head. Tim and his buddies stumble onto the lair of the radio-controlled robot, but it captures them even after they smash the control board. To their surprise, the robot is hollow, and the Octopus can wear it like Iron Man's suit—a perfect disguise. He's in another speeding truck doing ninety when a crash ends both menaces forever.

Far loopier gags brought readers to *Dinky Dinkerton, Secret Agent 6⅞* by Art Huhta. Dinky looked like a road company Sherlock Holmes, complete with deerstalker cap and calabash pipe, but sleuthed in the modern world—so modern that he gets kidnapped by a robot that resembles a deep-sea diver with a propeller attached to its head. The robot is an advertising device, sporting the message "EAT AWNUTS/THAT CRISPY CRUNCHY BREAKFAST CEREAL" across its chest. It's controlled by a gang of crooks who want to impersonate Dinky and steal "The Headlight of Salami" diamond from the Shah of Bah. A squadron of dinkier half-size Dinks appear out of nowhere to save the day. Of course, nobody believes any of this when Dinky tells the tale. Flying robots! Good one, Dinky!

Even Lee Drake's Mandrake the Magician (from the strip of the same name), who after a decade had seen more of the uncanny than a mere Brick or Dash, couldn't believe that the "handsome and fascinating" Prince Zombaga could be a robot. There were clues, to be sure: Zombaga demanded to eat in private, be seen only in dim candlelight, and sit on a marble bench. Royal eccentricity gets stretched to the breaking point when he starts taking walks at the bottom of a river. Zombaga—radio- and television-controlled to steal both Mandrake's girlfriend, Princess Narda, and the Van Stoff jewels—is the perfect foil, for no matter how often Mandrake gestures hypnotically, the robot cannot be fooled. (Hypnotism therefore is not transmittable by television, a scientific breakthrough surely worthy of comment.) In the end, however, the robot destroys its maker.

—⚊—

After the war, robots fell out of favor in comic strips, which reflected the more timely menaces dominating the headlines. One rarity that combined both threats was *Barnaby*, the strip most beloved by intellectuals in the years immediately preceding the appearance of *Pogo*. From its 1943 start, *Barnaby* wowed everyone but the masses. Dorothy Parker gave a hardback reprint of the first year's strips—before it reached its peak—a "valentine": "I think, and I am trying to talk calmly, that Barnaby and his friends and oppressors are the most important additions to American arts and letters in Lord knows how many years."[34] Barnaby was the name of a five-year-old who gains a fairy godfather, Jackeen J. O'Malley, a boy-sized Irish pixie formally dressed in a coat and hat, with an omnipresent cigar and butterfly wings. Crockett Johnson made the strip the spiritual forerunner of *Calvin and Hobbes* and *Doonesbury* with wit that was the dictionary definition of bone-dry and running commentary on the political scene. In 1945, Johnson turned the strip over to the writer/artist team of Ted Ferro and Jack Morely so he could concentrate on children's books, including the *Harold and the Purple Crayon* series. His replacements took the strip much deeper into the rich vein of adult hysteria that *Pogo* would soon mine. Ferro seems to have invented the notion of MAD (Mutual Assured Destruction) in 1947, two years before the first Soviet atomic bomb. "No nation would think of attacking a stronger nation," O'Malley says. "So if every nation agrees to increase its armed might beyond that of any other—well, a perpetual armistice results! Simple, isn't it?" As Generalissimo-in-Chief of the Pixie Nation, O'Malley has "a moral obligation to get hold of the most terrifying weapon of all.... Peace demands it!"[35] The Cold War, summed up in a sentence.

One slight hitch intrudes: Atlas, the builder of the secret weapon, can't imagine "anybody inhumane enough to set it off." Therefore he invents a button pusher, Blotto the robot, whose inhumanness makes it the ideal soldier for the era of "Robot War." Blotto is the robot as pixie, a pot-bellied stove with a periscope for neck, head topped with a tiny antenna, and a skyrocket attached to its back. Blotto plays with a plutonium atom for a yo-yo, useful as a disintegrator beam for clearing away rubbish. Like everything else O'Malley does, Blotto's mischief gets Barnaby in trouble with his parents, so Blotto must go. The answer has been wonderfully in plain sight for a month: O'Malley's cigar lights the skyrocket, and Blotto zooms off to Mars.

Using the robot as a symbol instead of comic relief was to be a rarity at any time, let alone the repressive 1950s, when political commentary became the equivalent of Soviet propaganda. Even *Pogo* skirted the substance, although during the 1952 presidential campaign Howland Owl, using an old alarm clock, invents a "mechaniwockle man" that's better than a machine candidate—it's a machine voter. "We'll make **thousands** of them li'l fellers an' put them all out to **votin'** round the clock … a twenty-**four hour day**."[36] Too bad that was the only alarm clock in Walt Kelly's Okefenokee Swamp.

The heroine of *Ella Cinders*, like those of so many other strips of the era, wandered around having adventures pop up under her nose, vicariously thrilling her followers. In 1949, she came into possession of half a blueprint from the late inventor and practical joker Cromwell Shnook. When she tracks down the other half and the Shnook family, she discovers that Cromwell drew a plan for a robot (operated, of course, by push buttons) bequeathed to his nephew Horace. What is the robot designed to do? No, you're not thinking like a cartoonist. It was made to give Horace the kick in the pants Uncle Cromwell had always dreamed of. Robots had hit their low point.

Scraping along at the bottom of the popular cultural food chain, robots continued to pop up here and there in the postwar era, especially in the science fiction and fantasy strips. Buck Rogers rescued the impossibly beautiful Flame when she was imprisoned by the Black Patch's robot troops in 1946. It was easy. The robots wore full-face radio-control hoods that Buck and Flame could don. By making "click-clack-clickety-clack" noises, they could fool, well, no one, as it turned out. In 1952, Brick Bradford encountered an alien robot—looking unaccountably like the Earthian one from 1939—reassuring long-term readers with a knowing wink that he's "had some experience with robots in the past." Veteran science fiction writer Jack Williamson wrote a short-lived Sunday strip called *Beyond Mars* that in 1953 included the Robot Master invading the base of Victoria Snow, "spaceologist."

Robots sneaked into other postwar strips trying to update their prewar plots with a slathering of modern décor. *Oaky Doaks* was a Li'l Abner figure, a muscle-bound farm-boy with a homemade suit of armor who cavorted through a series of comical kingdoms. In 1955, he roamed the land of Uncertainia, ruled by King Corny and his exceedingly fickle daughter Princess Pomona, who treated Oaky like a farmboy except for the several dozen occasions when she needed to be rescued by him. Her kidnapper this time is Sir Robot, the "biggest, meanest, strongest" knight the king ever saw, who carries Pomona off under one massive arm as a hostage for the "hundred thousand bucks" he demands as ransom. "I've got muscles like **steel**!" he says, altogether truthfully. Much like Baum's Tik-Tok, Sir Robot is tinker made—literally so, by Jason Tinker. Nobody knows this until, lances drawn, he and Oaky tilt and Sir Robot's head goes flying. Pomona is outraged, first at Oaky for killing the man she loves, and then at the robot for not being human. "It had an **iron deficiency**!" Tinker exclaims when he hears the news.

Also in 1955, it seemed for a moment that the *Mandrake the Magician* strip had introduced another android. An inventor has been demonstrating a female stenographer just like Rosie—comic strips are not the place to expect originality—except almost perfectly human, asking businessmen to invest in his surefire company. He then disappears along with their money, all in cash. Mysterious disappearances might be supposed to be Mandrake's forte, but the reader figures out long before Mandrake does that the android is a real woman, the inventor's moll. He is tripped up in the classic fashion: rather than skipping out with their millions, he lingers for one more sting. A robot wouldn't have made that mistake.

The venerable boxing strip *Joe Palooka* pitted man against robot early in 1956. Professor Draykopp, a German caricature almost as badly accented as Percy's Professor, brings Joe his robot, Apax, and proposes a "scientific boxing exhibition." It's a trap. The nutty professor turns up the juice on his radio controller during the bout and orders the robot to crush Joe in a clinch. Joe manages to tear off Apax's antenna and then literally knocks his block off. Draykopp is dragged away muttering, "'**Apax**' .. '**Apax**'... He has killed '**Apax**'..."

Freckles and His Friends, a strip that had been running since 1915 under indefatigable originator Merrill Blosser, had Nutty Cook, its resident intellectual (he wore glasses), invent one of the rare mechanical women for a Sunday strip in 1958. "Remote Control Katie" is a woman for scientific reasons: Nutty couldn't make legs work, so he gave her rollers hidden under her skirt. Swathed in a veil, Katie is sneaked into a party and, Percy-like, wreaks havoc when Nutty's control box shorts out.

Robots returned to make Superman's life miserable in late 1958 when the writers introduced Metallo—again. The first Metalo (with one "l") appeared in *World's Finest Comics* #6, dated Summer 1942. Originally mistaken for a robot, Metalo instead was a man wearing a supersuit of armor, complete with cape. The story was mostly an excuse for the title: **MAN OF STEEL VERSUS MAN OF METAL!** At the end, Superman knocked him into a pool of molten lava and seemingly certain death. Metalo actually survived, although Superman didn't know it. To avoid depicting Superman killing a villain, the caption blatantly ignored Superman's roundhouse punch, having Metalo fall into a convenient "crevice." As if that weren't enough, in the June 1956 issue of *Superboy*, the Boy of Steel found himself trapped on an asteroid hidden within a blanket of Kryptonite dust. Of all the asteroids in all the solar systems in all the universe, this one was inhabited by a Kryptonian robot named Metallo, who sacrificed both "l"'s to save Superboy's life.

The 1958 Metallo appeared as a cyborg in the daily *Superman* comic strip. John Corben is a murderer who crashes his car while fleeing the scene on a rainy night. Professor Vale rescues the mangled body and tries a desperate cure: he puts Corben's brain into a superstrong robot body, powered by a type of uranium with a bizarrely short half life, forcing Corben to steal uranium pellets to replenish his strength daily. A former reporter, he gets himself hired by the *Daily Planet*, figuring that a science reporter would have access to uranium stores. Lois Lane fends off his cheap advances until bullets bounce off him and she leaps tall assumptions to declare that he must be Superman. That sends both John Corben/Metallo and Clark Kent/Superman into identity-confusing mishaps stretching the sequence out for another three months, finally climaxing when Corben discovers that kryptonite will give him permanent life. Superman tricks him with fake kryptonite, again effectively killing the villain. Reader reaction must have been positive, since the same story was told in condensed form as the padding to the May 1959 *Action Comics*, far better known as the issue that introduced Supergirl to the world. Yet no villain ever truly dies in comics. Metallo has been brought back time after time in a huge variety of forms and guises in comics, cartoons, and video games throughout the DC Universe.

Popeye versus a robot: that's such a great idea it's surprising that it took until 1963 for readers to see it. Popeye, Olive Oyl, and Wimpy go off to a deserted island for a vacation. Surprise again, the island is full of robots who have declared their independence so they can "protect mankind against itself" and pay them back for inventing robots. Humans will never need to work again: "The world will be a **utopia**! There will be a machine or robot to do **everything**!" Not an easy task: Wimpy eats so many hamburgers that he breaks the hamburger-cooking robot, a rolling grill on wheels with a head and two arms with spatulas. Elsewhere, Olive gets hat-, shoe-, jewelry-, and perfume-making robots to wait on her. Only Popeye, with his sailor's work ethic, is immune. "Ya crazy tin cans will turn us into **vegetables**," he rants, forty-five years before *WALL-E*.[37] He asks to meet their creator, Robot #1—actually a computer with eyes and ears everywhere and an IBMer's dream of an endless bank of tape drives filling the inside of the island. The computer, however, has no interest in being turned off. No matter. Popeye, the original superhero, of course wins—after an infusion of spinach, he beats the computer's army of one, a caveman robot wielding a giant club.

Flash Gordon was the last of the classic comic strips to introduce a full robot plot, and it didn't do so until its second incarnation as a daily strip (although robots had been mentioned a few times earlier). Starting on Christmas Day in 1961 and lasting for almost three months, the sequence—it can't properly be called an adventure—took place in a conception of the future entirely different from that Buck Rogers stuff. Real-life astronauts dominated the news: midway through the sequence, John Glenn would become the first American to orbit the Earth. Flash's near-future depicted with detailed verisimilitude a space construction crew striving to build a solar-power satellite to beam electricity down to the planet. The crew is as ethnically accented as the archetypal GI platoon in a World

War II movie and similarly working class. Their conflict is only partially with the rigors of working in space. A robot is the strip's villain, not for any evil deeds—he is morally much purer than any of the humans, save possibly Flash—but because he is a quality-control officer brought in to ensure safety and efficiency for the rush job. The prissy bureaucrat versus the regular guy workers is an old theme, but this appears to have been the first time that a robot was cast in the bureaucrat's position. Workers had spent most of the past century worrying about having their jobs replaced by automation, sometimes literally represented as robot workers. The rise of the computer, the mechanical brain, meant that white-collar workers could also be threatened, with the class conflict raised by a form of racial conflict. In the strip, the workers eventually destroy the robot, not realizing that one of their own was responsible for sabotaging the satellite. The strip simply ends there, with the workers not suffering any visible consequences. Any similarity to the ease with which Southern bigots went scot-free for the concurrent murders of civil rights activists may just be an artifact of hindsight, but the freedom that science fiction writers have had to employ metaphors to comment on contemporary issues without spelling them out has always been one of the strengths of the field, one easily overlooked as timely commentary is forgotten in favor of timeless adventures.

Newspaper adventure comic strips began a long, slow demise in the television era as cartoonists switched over mainly to gag strips. The strips mentioned in this chapter died off one by one: *Tiny Tim* in 1958, *Superman* in 1966, *Mickey Mouse* in 1975, *Brick Bradford* in 1983, *Buck Rogers* in 1987, and *Flash Gordon* in 2003. The appearance of the first robot strip showing a human thrashing a robot due to its usurping her value as a worker is mirrored by the robot in the last surviving classic science fiction strip being thrashed by workers upset by its lack of humanity. Little changes for the image of robots. They are not us, and apparently they never can be.

8

A Tribe of Living Mechanical Men!
Robots in Comic Books

Originally a newspaper bought the exclusive right to run a comic strip in its city. That was shrewd marketing; comics were huge draws and therefore circulation builders. Major cities could have half a dozen competing papers, and few people subscribed to more than one, let alone all, guaranteeing that most strips were never seen by potential fans. Comics were paradoxically ubiquitous and out of most people's reach. Both the creators and outside entrepreneurs therefore saw a tremendous market for the wider distribution of strips in more permanent form. Literally cutting and pasting strips (usually the full-page Sunday strips but sometimes dailies) into cheap booklets was an idea independently invented multiple times during the first decades of the 20th century. No series lasted very long, but so many individual titles were given away as promotional items that millions of Americans were already familiar with the format when the modern comic book finally was born.

That occurred during the Depression, fulfilling a need for the cheapest possible entertainment at a time when even the 25¢ pulp magazines lost circulation. Comic historians tout numerous titles as the first comic depending on their criteria, but the 1934 appearance of *Famous Funnies* is an acknowledged milestone. The first to be sold through newsstand distribution, it contained a mixture of reprinted Sunday strips and original work. Seeing market gold, other publishers rushed in with lookalike products, and the format became firmly ensconced in about a year.

Two different comics dated October 1936 vie for the honor of introducing the first comic book robot. One, however, is barely a footnote. Dell Publishing had been the first to launch a semi-successful comic book. *The Funnies*, only a third of whose 24 pages were in color, staggered along for 36 issues starting in 1929, losing money all the way.[1] Dell tried again in February 1936 with *Popular Comics*, which did well enough to add a second title, again named *The Funnies*, in October. That *Funnies* issue managed to pack 60 strips into 64 pages, many of them two to a page, just as they would have appeared in the papers. Although all were purportedly reprints, evidence for an earlier incarnation of *The Adventures of 'Sparagus and Chubby* is lacking. (As of the second issue the strip's title changes to *The Adventures of 'Spargus and Chubby*, and so does the character's name in the text.) Each of the first three issues contained two pages of a continuing six-page story. The titular boys, Depression victims, are trying to hustle money in odd jobs when they find a classified ad from Professor I. Canem looking for assistants. The Professor is a standard comic scientist, complete with a bulging bald dome and a white labcoat that brushes his shoetops. He wants the boys to take care of his six cats (Iggy, Piggy, Wiggy, Figgy, Jiggy, and Biggy) and warns them "nevair, nevair enter deez door," replete with warning signs. Naturally, as soon as he leaves, the boys free the cats they believe are going to be used for experiments and barge past the "don't even look!!!" signs. Behind the door are robots: a robot man, a robot dog, and a robot horse. They ride the horse right through the side of the building, busting the robot into smithereens. The Professor comes home to the empty carnage and calls the cops, who round up the ruffians and march them off to jail. Presumably they never got out, for that is the last anyone ever saw of 'Spargus and Chubby.[2] One item of interest is how calmly the boys take the discovery. They know exactly what robots are and don't show any surprise when one talks back to them. Robots were thus sufficiently familiar, often from comic strips, to portray in a comic aimed at kids without need of further explanation.

The other October 1936 comic book is also a footnote of sorts, a forgotten harbinger of a long and celebrated career: Major Malcolm Wheeler-Nicholson's National Allied Publications countered *Famous Funnies* by offering nothing but original material in *New Fun* and *New Comics*, both started in 1935. Needing vast amounts of product on a shoe-string budget—his first office contained nothing but a chair and a card table—Wheeler-Nicholson pulled heavily from over-the-transom submissions by hungry amateurs.[3] That he even bothered to look at one package from Cleveland showed his desperation: one strip was drawn on butcher's paper, the other on the back of wallpaper. Nevertheless, the crude art had energy and the scripts briskly moved the action. "Henri Duval, famed soldier of fortune" and "Dr. Occult, the ghost detective" both debuted in *New Fun* #6 (October 1935).

The two men who had produced these offerings weren't yet old enough to vote. Jerome "Jerry" Siegel (the writer) and Julius Joe Shuster, Jr. (the artist), had been trying to break into the comic business, strip or book, for a quarter of their lives. Just as the robot writer Luis Senarens was shaped by the revolutionary technology that was the talk of his day, Siegel and Shuster lived and breathed mass commercial entertainment: radio, movies, pulp magazines, novels, popular science magazines, and the wide-ranging contents of newspapers all formed a bottomless pool of inspiration. Their fourth strip came directly from a crimebusting movie: "Federal Men," *New Comics* #2 (January 1936) borrowed from the 1935 film *G-Men*.[4]

Siegel was the print fanatic, subscribing to the few science fiction pulp magazines, reading every story, sending in letters of praise, and trying to sell his own work, including the prototype for the Superman strip that later made him and Shuster famous. His skillset tended toward the gigantic idea rather than the mechanics of paragraphs and dialogue. He needed Shuster's bold graphics, full of spaceships, mile-high skyscrapers, and rockets—and robots. The pair diverted "Federal Men" into science fiction. *New Comics* #9 introduced a Crime Empire whose futuristic technology terrorizes the population. Stalwart G-Man Steve Carson (and fearless girl reporter Jean Dennis) take them on. In that issue the bad guys unleash a giant robot from their undersea base. In the next, Carson steals a rocket flyer to battle the "triumphant metal monster," a robot so huge that its claw alone dwarfs a large ship. Shuster outdoes himself with a panel showing the robot looming over the tallest skyscrapers of the metropolis, swatting away attacking planes and a blimp (King Kong foreshadowing Godzilla and a thousand future superhero comics). Climbing into the radio-controlled head, Carson seizes control and sends the robot back to sea, where the Empire produces two more equally huge robots for a final battle.

"The robot is like some fiend of darkness: terrible, yet majestic, in its fury!" Siegel captioned a panel.[5] His skill lay in encapsulating a notion. Robots would always bear this conflicting nature: capable of destruction when controlled by evil, yet an unrivaled tool for good when wielded by the forces of righteousness, and all too easy to swing from one side to the other at a moment's plot whim. Siegel and Shuster's pioneering robot would be endlessly copied, adapted, and repurposed by succeeding comics.

The first strip to feature a robot as a continuing character drew heavily on "Federal Men." The creator of the new strip, George Brenner, undoubtedly knew of the earlier story. *Smash Comics* #1 (August 1939) stuck a robot right on the cover, where it fights a gorilla brandishing a club. Gorillas always sold, according to comic legend. Selling the comic was more important than fidelity to the contents; no gorillas were to be found inside. Readers were expected to forget that detail when they thrilled to the opening panel of "Hugh Hazzard and His Iron Man," which featured a newspaper with headlines blaring, "IRON MONSTER AT LARGE AGAIN/SPREADS HAVOC AND FEAR THROUGH-OUT CITY/ROBS, PLUNDERS, AND KILLS." And, in smaller type, "KIDNAPS BABY FROM IN FRONT OF HOME," as if murder and pillage weren't heinous enough. The man-sized robot is the invention of the genius crook, the "great Von Thorp." The police, at their wits' end, summon millionaire playboy Hugh Hazzard by firing a flare gun into the sky (two years before the Bat-Signal). After Hazzard conveniently spots the robot breaking into a jewelry store, he climbs inside the hollow body and waits in surprise when Von Thorp brings the radio-controlled robot back to his lair.[6]

After Von Thorp's defeat, Hazzard adopts the robot as his own, now controlled by his voice. How the robot could work without interior parts is never explained, but its ability to contain Hazzard is perpetually useful. So is its ability to fly, installed by Hazzard in issue #2, anticipating Superman by more than a year. Brenner has the robot function both as a bodysuit protecting Hazzard and as a separate steel figure responding creatively to fulfill Hazzard's instructions, ingeniously depicting the superhero and his secret identity in the same panel as well as providing two lead characters for more supple plotlines. In comics the super identity inexorably takes precedence, so as of #13 the name of the strip changed permanently, to "Bozo the Robot." While that name is today fixed in cultural memory by the later Bozo the Clown, in 1939 a bozo was merely a beefy but not very bright guy—appropriate for a robot that looked like a bunch of stovepipes fastened to a barrel topped with a cartoonish face slashed with a horizontal white grille resembling a fixed grin. Although Bozo scares bad guys with his mere presence, the young readership saw a friendly hero performing superhuman deeds. Appearance is character, even for robots.

If Bozo doesn't qualify as the first robot with a series, perhaps Iron Skull does. He debuted in *Amazing-Man Comics* #5 (September 1939) from Centaur Publishing, a successor to CMC. As if to confuse readers, #5 is the first issue of the magazine and Iron Skull is introduced without an origin as if he were a continuing character. All we are told is that Iron Skull has steel fists and a skull that is presumably also metal, as a gangster's hand breaks from socking him on the jaw. Is he a robot? In later issues he needs to breathe, so the better descriptor might be *cyborg*, although that term had not yet been coined. Those few lucky kids who got their hands on all early comics found a familiar plotline: the villains in that story used radio-controlled controlled robots for robbery just as Von Thorp did.

Two issues later, in *Amazing-Man Comics* #7 (November 1939), Iron Skull's origin is given as the strip reboots:

> In the year 1950, during the Second World War, this time in the United States, a soldier battered and smashed beyond recognition was brought into the base hospital in Chicago where under the capable hands of Dr. Watson,—steel and iron plates replaced flesh and bone—the soldier now looked like a living image of a skeleton.
>
> We never learned who or what he was,—but ten years later, a strange being arose during the period of reconstruction and became known as the enemy of crime!—The underworld call him—"the **Iron Skull**"! [all emphasis and punctuation as in original for all quotations]

Iron Skull, however, resembles a living skeleton no more than today's similarly experimented-upon Wolverine. Looking exactly as human as he had in his first two appearances, he battles mad scientists standing in for Nazis. Using transparent aliases for Germans was common in the two years that Europe was at war while America remained technically neutral, but an alternative world was a far cleverer ploy than just namechecking Batzis. In issue #9, the setting suddenly changes to 1970. Robots are back in issue #11, "grotesque" robots that look a great deal like Bozo. *Amazing-Man Comics* #15 (August 1940) settles the questions of Iron Skull's robothood: A nameless but brilliant turbaned scientist—he has a flatscreen "video machine" that can spy on any scene he wants to look at—creates a special giant magnet to pull Iron Skull to his hidden lab. His fiendish plot is to inject the Skull with a serum that will turn him completely into iron, and then he "shall transfer his blood to the lifeless bodies of my robots and bring them to life…. Think of it! … A tribe of living mechanical men!!" How a being made completely of iron still has blood is not explained (nor is what becomes of his human brain), but Iron Skull is outwardly unchanged after his transformation, just more invulnerable than before. Hero robots cannot be too powerful to suit justice.

If the stricter definition of a non-human-brained creation able to think independently of radio or human control is applied, then the first comic book android is Carl Burgos' Manowar, the White Streak (neither name making any sense in context). The "super-robot of a new terrifying type" appeared in *Target Comics* #1 (February 1940), an early title from Novelty Press, a line that would become obsessed with robots. Manowar is described as "the last of the servants of a dead civilization, dedicated to mete out justice to those murderers who prey on the weak—**the warmongers.**" The White Streak resembled a human wearing a colored mask, very similar to Marvel's 1968 android the Vision, down to the pointed green skull covering. His weapons were electronic eyes that blasted lightning. Created when Americans loudly insisted upon isolationism, Manowar disappeared when America entered the war and peacemongers became anathema, memorable mainly for his South American Indian origin.

"Marvex, the Super Robot," from *Daring Mystery Comics* #3 (April 1940), qualifies for the nitpickiest definition of all firsts: he's everything Manowar is and wholly mechanical, a robot rather than an android. After watching humans, evil aliens in the 5th dimension mix up chemicals and a strong steel body to be their slave. However, he rebels against their plans and lands on Earth after he destroys the aliens in a super-explosion. Modeled after humans, Marvex has a normal human face above a body made distinctive by overlapping plates covering his shoulders and upper breast. Comic book artists wavered between making androids able to blend into human society and creating distinctly metallic robots; the next issue retrofits Marvex with a robotic head bald as an egg and Clark Kent's blue suit. "You're the most wonderful man I know," says his admirer Clara Crandell. "I am not a man—only a machine!"[7] He has most of Superman's powers and a few to spare, but his kryptonite is nets: strong ropes hold Marvex even though he can tear through steel like paper. Captured after battling a squad of evil robots and about to be taken apart, he sends out an S.O.S. radio signal with his mind. Clara hears it while in bed on a special radio tuned to Marvex's brain. In a grand display of pure comic book plotting, not only does she know where to go, but no time at all has passed when she arrives and saves him. Marvex returned to being nude and weirdly robotic before the strip abruptly ended after 13 appearances.

In between the Iron Skull and Manowar, Burgos devised a brilliant visual strategy that allowed an android to look human *and* visibly freakish as needed. "The Human Torch" debuted in *Marvel Comics* #1 (October 1939), a huge success. That was the *Marvel Comics* around which the modern company built its corporate identity. (To endless confusion, the publishing company was then called Timely and the name of the book changed to *Marvel Mystery Comics* with issue #2.) In that memorable first issue, Professor Horton calls the press to his laboratory to witness a remarkable unveiling: "As you all know, I've been working on a synthetic man—an exact replica of a human being!" But "something went wrong with my figurings somewhere. Every time this robot, the Human Torch, contacts oxygen in the air, he bursts into flame! **Now watch!**" The flame being is possibly the most visually exciting creation of the Golden Age of comics, perhaps even more so on the cover, drawn by pulp science fiction magazine maestro Frank R. Paul, who gave the Torch a face that Burgos' pictures lack.

That first page and the first page of issue #2, recapping the origin, are the only times that the words *android* or *robot* are used to describe the Human Torch. The next three issues refer to him as a scientific creation or scientific marvel. By issue #6, he is introduced as "that most amazing person." *Marvel Mystery Comics* #7 (May 1940) shows the Torch adopting the human identity of Jim Hamond and joining the police force. The Fall 1940 *Human Torch* #2 (actually the first comic with that title) gifts him with a completely human flaming kid sidekick, Toro, and they soon turn into a less wealthy Batman and Robin. By the end of the Torch's run, a panel now infamous among comics fans appeared in *All-Winners Comics* #1 (vol. 2, August 1948), showing the Torch shaving. No good explanation was ever given for ignoring the Torch's origin, although, as Iron Skull proved, no one at the time gave much thought to continuity issues. The same *Marvel Comics* #1 also introduced the Sub-Mariner, master of the sea, and their fire-versus-water battles lifted Timely (now Marvel) into the majors in a way that CMC and Centaur never achieved. The Human Torch appeared in seven different titles in the Golden Age. If only his robotness had been acknowledged, he would be the premiere robot/android of the era.

—⁓—

Elektro, the Westinghouse robot at the 1939–1940 New York World's Fair (see chapter 5), undoubtedly influenced the spate of robots introduced in comic books over the next few years. Two stand out for their more direct connection. The first appeared in Centaur's oxymoronically titled *Amazing Mystery Funnies* vol. 2, #7 (July 1939). The Fantom of the Fair literally lived in an underground laboratory at the World's Fair, emerging to fight the constant menaces that one expects would have a deleterious effect on attendance.

The opening panel of the November issue says it all: "Catastrophe strikes at the electric utilities area of the World's Fair. As the electro-robot is turned on a rampage by some mysterious power, with fury in its electrical mind and a million volts of electricity in his hands, death for everyone is certain until **The Fantom** appears."[8] The robot carefully fails to resemble Elektro, bearing a pot-bellied stove torso over extremely long, spindly legs and sporting a skull with a sagittal crest in a style that numerous other early robots would adopt.

The other robot, forthrightly named Electro, started appearing as a backup strip in *Marvel Mystery Comics* with issue #4 (February 1940). Electro sports one eye in the center of his face but otherwise seems inspired by a Roman legionnaire's outfit, complete with cassis-style helmet, tunic, and shoulder plates (although he later got a sleeker look that incorporated a televisor into his face). "Prof. Philo Zog [a nod to television pioneer Philo Farnsworth], mechanical wizard and electrical genius, has contrived a marvelous robot, or mechanical man…. All powerful and intelligent. With his wonder robot, Prof. Zog hopes to work for the welfare of humanity, towards the ends that justice and universal happiness may take the place of wars, crime, and suffering," reads the first panel, words that could almost serve as a mission statement for the real-life World's Fair.[9] Of course, lofty goals rarely pan out. Electro first is put to standard use fighting gangsters; then, in issue #8, he is kidnapped by Jago, the King of Dragon-Men, and taken to his planet to fight the peaceful Empress Nara. Robots seem to irresistibly drive writers toward science fiction, and Electro battled various aliens before he disappeared a year later, the fair's end stopping its subliminal promotion for the strip.

Jerry Siegel kept trying over and over to replicate his smash success with Superman. One obvious combination of his core interests was Robotman, first seen in *Star Spangled Comics* #7 (April 1942). Crooks attack scientist Bob Crane, leaving him near death. His lab assistant Chuck Grayson sees only one hope: he transplants Crane's brain into the robot body they've been developing. Robotman has one of the most human-looking bodies of his era, making it easier for him to don a face mask and clothing to appear in his secret identity of Paul Dennis. Siegel went into the army the next year, but "Robotman" lived on, escaping its grim beginnings to become a kid-friendly strip about a man and his dog, Robotdog, nicknamed Robbie. Robbie could talk and was delightfully self-centered, breaking the fourth wall to tell kids how much help he was. That aspect helped the strip survive after the war when DC repositioned its lineup to counter the crime and horror comics that swept through the medium as superheroes faded.

An unclassifiable oddity appeared in Eastern Color Printing Company's *Reg'lar Fellers Heroic Comics* in July 1941. Reg'lar Fellers was the name of the comic club that its readers could join for free (unlike those of the publishers that kids actually cared about). "Man O'Metal" was created by H. G. Peters with the least likely origin story in an era that frequently strained credulity. A vat of molten unnamed metal drenches Pat Dempsey. Afterward, "through some strange chemical reaction of the skin, whenever he contacts heat or electricity, Pat becomes a being of super-**heat**, capable of burning his way through the most resistant steel barriers." Like the Hulk's, however, his pants are made of the strongest material of all: the edicts of censors.

Charles Biro was just starting his decades-long career in comics when he as artist and Abner Sundell as scripter created Steel Sterling for *Zip Comics* #1 (February 1940). To avenge his murdered father, John Sterling swears to fight crime. Rather than just put on a bat costume, however, he jumps naked into a vat of molten steel. Unlike Pat Dempsey, he thought ahead and first coated his body with a special chemical. Though he looks normal-skinned, he now has an all-steel body with magnetic powers, giving him at least as much right to be called a cyborg as the Iron Skull.

Three years later, Biro (this time as writer) and Bob Wood put an original twist on the notion of a robot sidekick when they co-created "The Boy King and the Giant" for *Clue Comics* #1 (February 1943). The land-locked Ruritanian nation of Swisslakia has just been invaded by Hitler's forces when the Quisling Grousse orders the murder of the royal family. Shot and taken to be buried in secret, the king and his son cling to life so that the king can reveal the country's secret: Nostradamus "created a mechanical giant who would have but one *master* … the person who screws the *bolt* in the giant's *head*." Prince David, now the Boy King, digs up enough of the giant so that he can put in the bolt, and the giant rises … and rises … and rises, probably the largest character in comic history up to that point.

Biro had taken a trend to its logical end. Before the era of miniaturization, Americans liked their machines big and powerful. Robots almost always were larger-than-life sized. Comic book artists, restrained by nothing in their bids to awe potential readers, vied to outdo one another with ever-larger robot menaces. "Federal Men" had led the way. Three years later, in 1939, Burgos created a bizarre black and white science fiction strip: "Air-Sub, DX." The rocket, plane, and submarine hybrid is captured by a "mechanical monster" so huge that it simply tucks the Air-Sub under its arm and walks off in *Amazing Mystery Funnies* vol. 2, #8 (August 1939). Elsewhere, Dr. Diamond must fight a gigantic robot (perhaps the only one to wear boots) in *Cat-Man Comics* #8 (July 1941). Few outdid Alex Blum, the artist of "The Giant Robots of Kilgor," *Fantastic Comics* #4 (March 1940). In a story thought to have been written by Will Eisner, Kilgor, a mad scientist, convinces the Hitler stand-in Rigo to build a factory to produce 5,000 giant robots that simply trample puny humans under their feet. Only the mighty Samson, a descendent of the biblical strongman, can defeat them, which he does by toppling the factory's pillars (in their modern form of steel girders) and crushing the robots—though not himself this time— under the debris. These robots, incidentally, have eyes but no mouths—and look all the more terrifying for that lack. (One exception to gigantophila is seen in *Planet Comics* #8 (August 1940). Captain Nelson Cole of the Solar Force is investigating a robbery on Mars, which proves to be perpetrated by Tobor, the Evil, who is not himself a robot, hard as that is to believe considering the plethora of future robots with that name. However, Tobor is building an army of pint-size robots for the simple reason that they are cheaper to manufacture. Finally, a mad scientist on a budget.)

In "The Boy King and the Giant," Biro's nameless, silent giant risen from the earth clearly reflects the golem legend out of Jewish folk history. According to one version of the story, Judah Loew, the chief rabbi of Prague from 1597 to 1609, grew increasingly concerned about anti–Semitic attacks on the Jewish ghetto there. In response, he created a superhero, a massive figure of clay, brought to life with a sacred amulet bearing the name of G-d, mimicking the creation act in Genesis. Modernized and secularized, the purpose of the Boy King's giant is the same. The writers take its enormous size and power, now a reflection of American might during an active war, to almost hallucinatory exaggeration. The giant walks to America, wading across the Atlantic, pausing to pick up a Nazi battleship and shake out the sailors into the sea. Inevitably, his first sight of America is the Statue of Liberty. He doesn't just figuratively fall in love with America—he literally does so, covering Lady Liberty's face in smooches while flying hearts circle his head.

Though they vary the giant's height as needed for a panel's purposes, Biro and Wood seem to realize that they've written themselves into a corner. A giant taller than New York skyscrapers cannot move through the city without wreaking as much carnage as his foes. Therefore, in issues #4 and #5, they move the action back into the ocean, as the giant fights to the death with … a Nazi robot Tyrannosaurus Rex of similar size and wearing a swastika leg band to ensure that it gets no sympathy from dinosaur-loving youth. (Amazing as it might seem, this was not an original idea. In *Hangman Comics* #4 [Fall 1942], the title character also battles a Nazi robot Tyrannosaur.) After this "Battle of the Ages," though, the robot's appearances are limited, as the Boy King becomes a routine Americanized battler of evil. One thoroughly odd exception was presented in issue #9: A crook coats a robot named Paris' insides with diamonds so precisely assembled that it can do any stunt perfectly, as it proves when it invades a circus and outdoes all the performers. For unknown reasons, the tall, lanky figure is dressed in a top hat and tails. It also bears a striking resemblance to Lucille Ball. This robot has more than merely feminine features—it has eyelashes and lipsticked lips, an infallible telltale in 1944. Yet everyone refers to the robot as "he." Only these unique touches made the strip worth reading, and without them it died after a few more issues.

Robots had quickly become as routine in comics as bad German accents. Acknowledging this fact was an episode of "The Wizard" in *Shield-Wizard Comics* #13 (Spring 1944). The usual elderly scientist is presenting his latest creation to the press. They are unimpressed. "What? Another automaton?" one asks. "**Wait!** I know robots are not new, but this is different!" the professor pleads. He might be referring to the fact that his robot is the rare one with eyebrows, but he understates his case. Robots appeared from 1941 to 1944, in addition to those mentioned above, in strips as diverse as "The Amazing Man," "Dick Cole, Wonder Boy," "Silver Streak," "The Black Terror," "Sergeant Spook," "Captain Midnight," and "Mekano." Elektro's all-encompassing media presence surely accounts for robots in the strips "Landor, Maker of Monsters," "Major Mars," "Dr. Mortal," "Inspector Dayton," "Captain Future," "Three Comrades," "The Blue Blaze," "Power Nelson, Futureman," "Captain Venture and the Planet Princess," "The Green Claw," and "Tom Kerry, District Attorney," all of which appeared in 1940 alone, along with Flexo the Rubber Man and Dynamic Man, the first of two totally independent androids of that name (one in *Mystic Comics* #1 [March 1940], the other beginning in *Dynamic Comics* #1 [July 1941]). Some 1940 covers—*Pep Comics* #1 (January 1940) and *Fantastic Comics* #3 (February 1940)—showed heroes fighting robots despite their complete absence from the inside pages.

—⁓—

Comic writers had no intention of giving up a plot device as useful and versatile as robots. They would pop up at random in strips by the dozen from then on. Perhaps more than other media, though, comic books are particularly susceptible to changes in the zeitgeist and have the ability to swerve almost instantly to exploit them. The industry boomed during World War II, thriving on endless stories of good guys battling Nazis and Nazi saboteurs (or both: the robot in "The Wizard" is appropriated by an actual group of Nazis drilling on U.S. soil), with the occasional Japanese or Italian stereotype providing some change of venue. Soldiers were said to be especially fond of comics as reading material, and those too young to join up were treated to vicarious killing, torture, and mayhem at a point when few were willing to decry their use against the hated enemy.

After the war, comics suffered. Obvious villains were in short supply, and the soldiers went on to more mature forms of sadism. Readers voted with their wallets. Tough guy literature à la Mickey Spillane—a former comic writer—surged while sales of superhero comic books plummeted. Comic publishers responded like Stephen Leacock's Lord Ronald, who "flung himself upon his horse and rode madly off in all directions." The major new trends in comics were leggy superheroines, romance, horror, and science fiction, with the old staples of funny and kiddie strips (overlapping but distinguishable subgenres) now playing a larger role. Robots would be found in each subgenre, with the possible exception of romance.

Early on—in that seminal year of 1940, in fact—"Dickie Dean, the Boy Inventor" first appeared in *Silver Streak Comics* #3 (March 1940). Created by Jack Cole (better known for the classic Golden Age hero Plastic Man) Dickie was a fourteen-year-old "lad of exceptional ability—a mental wizard." Two issues later, however, he was downgraded: "**Dickie Dean** is no superhuman.—He is just an average American boy who has developed within himself the ability to figure out problems by logical deduction."[10] Just as any boy could grow up to be president, any boy could apply himself to become an Edison or a Batman.

Dickie first encounters robots in March 1941, made by his archenemy, Professor Skinn (who incidentally has a henchman who becomes stronger as he gets angrier, foreshadowing the Hulk). Skinn unleashes an army of robots with odd spherical bodies and claws for hands upon Dickie's city of Castleton. (Cole had grown up in New Castle, Pennsylvania.) It wouldn't take long for Dickie to start making robots of his own. Befitting a boy inventor, they are toy-sized, barely a foot high. Their purpose is also suitably boyish: he plans to send them searching for sunken treasure. Naturally, their controller is seized by crooks who use the army of robots to rob the city before Dickie foils them. We don't see the robots again, but they apparently work spectacularly offpanel. In the next issue, the now fabulously wealthy Dickie altruistically builds a gigantic, ultramodern skyscraper laboratory where "hundreds of skilled scientists will match their minds against the mysteries of nature and seek to throw light upon the hidden-evils of disease and all things harmful to humanity."[11]

"Edison Bell, Young Inventor" soon followed Dickie, appearing in *Blue Bolt* #1 (June 1940). (Comic books frequently mixed adult superheroes with kid strips in those early days.) Sent to the store by his mother to buy her a pound of lard, the preteen Eddie and his pal Nicky seek a way out of such boring chores. Using some copper piping for arms and legs, along with a box for a torso, and making a cookie jar head into a Leyden jar—glass lined inside and out with tinfoil to store static electricity—all that is necessary is for lightning to strike, just as in the *Frankenstein* movie. Quicker than you can say "it's alive!" it's alive; Eddie and his mother wake up the next morning to find the robot making breakfast. Nicky gives it the name of Frankie—Frankie Stein. Even better, most of the third page of the three-page strip is devoted to plans for making a robot doll at home, using a large matchbox, cardboard tubes, and an empty spool for a head. Frankie is the first of many comic robots to have a head shaped like a spool or a cork. Eddie, Nicky (renamed Jerry in issue #2), and Frankie race off to have adventures, literally so after Eddie learns how to "smash the atom" in issue #3 and builds an atomic-powered rocket car. The strip would last until 1950, but Eddie's inventions were soon toned down to the ordinary and Frankie disappeared after a year of wild fun.

The difference between kid strips, often light and fun with silly villains, and comedy strips featuring kids is exemplified by "Tommy Tinkle," a strip by "Sandy Klauz" in *Hit Comics* #6 (December 1940). Tommy visits the Professor's lab and sees Stoop-pendous, who is "thirty feet tall and has the strength of an army." Naturally, crooks steal him to rob Toyberg. Tommy defeats them by stealing the robot—that is, stealing the robot's heart by building a lady robot. The end in two swift pages.

Robert Robot, created in *Ding Dong* #1 (1946), is aimed at even younger children. A cross between Pinocchio and a teddy bear, the robot is created by "Dr. Xandu, the mad scientist," out of "plastic metal" but soon finds himself with a middle-class American family. Robert's naïve enthusiasm and enormous strength create havoc in adventures that parallel those from the Katzenjammer Kids and Dennis the Menace. In fact, in issue #5 he meets his match in mischievous twins.

Most kids writers simply adapted the new vocabulary of the modern era to their strips. Atomictot also debuted in 1946, in *All Humor Comics*. The "world's mightiest mortal," atomic energy flows through his veins even though he is as young as Robert Robot. In issue #8, he defeats the robots of evil Professor Zounds but really would rather be fishing. Cosmo Cat, the Cat of Tomorrow (a pure Superman takeoff), followed in late 1946. Though his setting is a children's comic, the writing carries far more sophisticated notions. He battles "The Revolt of the Robots" on a distant planet. "We Titanians," explains one victim Cosmo Cat saves from a robot attacker, "created robot machines to do all our work for us—then one day the machines revolted and took over the city! Now we must either flee or be slaves to the robots!"[12] Cosmo Cat defeats Klang, the robot leader, and pulls the plug on the central control device, recapitulating the entire history of robot stories in one seven-page adventure.

"Axle and Cam on the Planet Meco," starting in 1952, is the rare (perhaps unique) strip about a robot family, although we only see the father and son. It ran, improbably, in *Popeye* comics because its creator was Bud Sagendorf, one of the many artists who ran the Popeye empire when its originator, Elzie Segar, died in 1938. Cam is a young robot boy with the usual problems: no money to have fun and a parent who tells him not chew batteries between meals and ruin his appetite. Oh, and a "terrible uranium eating dragon" that Cam lures out to sea. He flies there via the propeller beanie he copied by watching Earth boys on his people-vision machine (known to humans as a TV set).

Charleton Comics, exclusively a kids comic house, got into the superhero animal business with Atomic Mouse, the title character in his own comic book starting in 1953. He gained his powers from U-235 pills provided by Professor Invento. (Not to be confused with Charleton's Atomic Rabbit, who gained his powers from U-235 carrots.) Among other adventures, Atomic Mouse faces an evil robot duplicate only he can defeat. Charleton couldn't help itself and continued the theme with Atom the Cat, whose enemy was the sinister Dr. Mole and his robot Kule. (They missed an opportunity by not naming the evil robot henchman Krule.) Supermouse, from competitor Pines Comics, has to battle Terrible Tom's huge and nasty-looking robot, TV Stevie, and only wins after he, Popeye-style, manages to sneak down a hunk of supercheese. It's Terrible Tom's evil plot from 1957 that has resonance today: while doing television repair, he sneaks tiny transistorized TV cameras into the sets and uses them to spy on every move made by their owners, robbing their houses when he's sure they are gone.

Humor tends to exploit prominent themes in contemporary culture, with technology and its unpredictable changes a constant source of inspiration for the past century. *Mad* magazine provided a lucrative model in the early 1950s as a comic book that parodied popular culture in general as well as individual comic characters. More than a dozen imitators sprang into existence, with blatantly similar names like *Nuts*, *Wild*, *Bughouse*, *Unsane*, and *Crazy*. *Crazy* was written largely by Stan Lee, some of the thousands of stories he cranked out to keep the company alive in the pre–Marvel days. For *Crazy* #7 (June 1954), he came up with "Robert the Robot," one example of robots in the year AD 2000 that "walk like humans, talk and think like humans, and even act like humans!" Acting like humans means chasing all the pretty girls in the office, just like all the scientists do. The mistake was using the recently released movie about Toulouse Lautrec, *Moulin Rouge*, for the robot's vision test. That was the worst picture to show to a robot: it featured can-can girls.

Another vintage name, Jerry Siegel, needed the work and took odd jobs in comics. *Amazing Adventures* wasn't intended to be a humor comic, but Siegel's "Invasion of the Love Robots" made the cover for issue #4 (July 1951). Blep and Shlep from the planet Blyntzyn need to conquer Earth for its precious wood and sand. They scheme to fill their robots, Adonis and Venus, with personal magnetism to make them irresistible. It works: "I'll give you anything … everything I possess … for just one kiss!" cries rich Genevieve McCoin. The robots, however, fall in love with each other and overthrow their masters. The Earth is saved by true love.

Disney also brought in a Robert the Robot in 1953. Donald Duck's three nephews want to fish, but he makes them accompany him while housesitting for Cousin Marmaduke, a retired magician. His house, Bad Manors, is tricked up with gags, one of them being Robert. Wanting to scare Donald out of the house so they can get back to fishing, the boys dress the robot up like a ghost. It works too well. Robert sees his reflection and is equally scared. He and Donald flee … to the beach for relaxing fishing.

Otto Binder sold his first science fiction story in 1930 at the age of nineteen. Among the dozens of stories he wrote during the 1930s are the "Adam Link, Robot," series (see chapter 9), the seminal portrayal of a good robot trying to find its place in human society. (The stories were closely adapted, with most of the text taken straight from the original stories, in the comic books *Weird Science-Fantasy* #27–29.) Binder then moved over to comics and, in 1941, took over primary writing chores for Fawcett's Captain Marvel, the most successful Superman imitator, eventually writing about one thousand stories about the Marvel family.[13] (The characters are now owned by DC, causing interesting copyright problems.) Captain Marvel was a youngster named Billy Batson who became the adult hero by saying "Shazam," the name of the wizard who granted him his powers. Though just as heroic as Superman, Captain Marvel inhabited a goofy, self-contained universe with partners like Mister Tawky Tawny (an intelligent humanoid tiger) and foes like Mister Mind (a brain-controlling worm). The ever-expanding Marvel family included Mary Batson, who became a teenage hero by saying "Shazam," and kid sidekick Freddy Freeman, who became Captain Marvel, Jr., by saying "Captain Marvel," leading to endless complications.

The similarities between a Captain Marvel story and a Donald Duck story are surprisingly large in the postwar era. Both feature extended families of eccentric characters engaged in light adventures with bloodfree consequences. The humor appealed to adults as well as kids, and the plots, which emphasized good behavior, made them more acceptable than standard superheroes. Numerous robots found their way into stories with strong moralistic endings, even those Binder didn't write. A 1947 story introduced Mr. Atom, an atomic-powered robot with "the invincible power of the universe." As strong as Captain

Marvel himself, the robot is barely subdued and locked away in a lead-lined prison. Mr. Atom warns humanity with heavy metaphor, "You, who have made me your prisoner, **beware** lest I return to destroy my keepers!"[14] The family builds the Shazam Robot to take a job too dangerous for humans: driving truckfuls of nitroglycerin. "Captain Marvel Trains Another Captain Marvel" in 1949, a silicon man that an evil scientist wants to make Marvel's equal. Dr. Sivana. Jr., the "world's wickedest boy scientist," builds a robot to destroy Captain Marvel, Jr., but the robot is so impressed by Freddy's goodness that it refuses to do so. In addition, the entire Marvel family is replaced by robots in an evil scheme concocted by the overlooked janitor at the lab.

—⚏—

Other postwar trends produced comics wildly different from these kid-friendly strips, appealing to different niches in the buying public (or perhaps the same niche at different ages). Leading ladies were one such—sadly not for egalitarian reasons. The sex appeal of a cartoon may seem juvenile, but juveniles of all ages were the audience. Women had been title characters in comic books for the entire decade but now beckoned readership with skimpier costumes, barer midriffs, and longer legs, gratuitous to the plot. Action panels alternated with cheesecake poses, the character drawn head-to-toe, arching her body back in a sultry come-on, even if talking to a district attorney.

Phantom Lady is the epitome of these cheesecake heroines. Though she first appeared in 1941, she was back-of-the-issue filler material until the Fox Feature Syndicate took her away from Quality Comics and starred her in her own title after the war. Covers displayed Sandra Knight in provocative poses, a costume now slit to the navel and essentially backless, showing off a body that had grown several cup sizes. The cover for *Phantom Lady* #17 (April 1948) made the hall of infamy in Fredric Wertham's attack on comic books, *Seduction of the Innocent*, for its depiction of her massive "headlights" and the ropes symbolizing "the sadist's dream of tying up a woman."[15] A few months earlier, she had battled a robot duplicate of herself, which entailed much stripping and switching of clothing as she took the robot's place to infiltrate the gang. That story was titled "The Beauty and the Brain" but left no doubt that Ms. Knight was both.

The glamour craze had one positive aspect—namely, showcasing heroines with outstanding qualities: they were beautiful and brainy, as well as brave, resourceful, omnicapable, and always ladylike, even when punching out bad guys. Most gave little doubt that they were avatars of Athena, the Greek goddess of wisdom—few more than Mysta of the Moon. In *Planet Comics* #35 (March 1945), the splash panel has her skimpy bathing suit eclipsing the god Mars, even though he's the title character. He's been busy inhabiting earthlings' bodies to get them to burn all the universities because "without knowledge, the fools are no better than pigs." This scheme for universal enslavement is countered by Dr. Kort, who raised a boy and a girl on the moon to inculcate them with literally "all there is to know." He plans to send them off, with protective robots, to "plant anew the seeds of learning in the universe," but Mars gets there first and the boy dies. Mysta swears vengeance, traveling with around the universe with her robot (named Robot and later replaced by Robot II) to fight evil. She eventually finds her way back to Earth, protecting it from her Citadel of Science, although Robot does most of the heavy work. The robot's appearance varies greatly from issue to issue—originally a sleek, silvery homunculus, later a gray bullethead with a rectangular grille for a mouth, and then a cork-headed goof with a circular grille that allows him to talk. Mysta likewise goes through changes of her own. By the end of the 1940s, the anti-comics crusaders had forced cover-ups of the leg-baring cartoon bodies. Mysta acquired a regulation-length dress, and Phantom Lady's torso got fully covered.

In the interim, *Planet Comics* used its futuristic and offworld settings for exploitation of some of skimpiest (though often most poorly drawn and least erotic) outfits in contemporary comics. Its science fiction stories often turned to robots and artificial beings as touchstones of an increasingly technological era, along with spaceships, televisors, and ray guns. Issue #46 (January 1947) was the high point for both flesh and robots, with three separate stories incorporating them—perhaps the record. The cover, by Joe Doolin, dared readers to look away from the magnificently toned woman in a metallic strapless brassiere swinging from a rope ladder and rescuing a mere man, chained at wrist and ankle (never mind that the unlucky slob is Auro, Lord of Jupiter). As usual, the scene is found nowhere inside but is a twisted metaphor. The woman is the evil Naga, who drugs Auro and forces him to land her and her ship full of Mecho-Men, made from steel bodies and petrified human brains. Naga wears a strapless harem costume but fortuitously fits in: all women's outfits on Jupiter are strapless. Earlier in the book, once the reader has slowly paged through Queen Lyssa of "The Lost World" battling the Voltamen in her strapless two-piece outfit, is "Futura." A few issues earlier, Marcia Reynolds, a "second-grade technical secretary," had been whisked into the future on the planet Cymradia, "pearl-city of the universe," by Lord Mentor and his squad of synthopoids, "more robot than human," as Mentor needs healthy bodies in which to implant his race's immortal brains. Futura's skimpy outfit rides up often, showing her long legs, and she has a magnificent six-pack. Mysta, who changes outfits in almost every issue (in issue #43 she is comparatively overdressed in standard superhero tights below her strapless bra—until page 8, when the colorist, perhaps out of sheer force of habit, forgets and leaves her legs bare), is here in issue #46 wearing a *green* two-piece as she flies to Uranus, where her testimony could convict Vitor on the charge of using "damp light." These design decisions were quite deliberate. Editors knew their readers. In "The Visigraph" (the letters-to-the-editor page in that issue), Billy from East Cleveland wrote, "The stories are good science fiction, and the girls are hubba-hubba material."

—⟋⟋⟋—

It's not clear whether Wertham and his ilk had a spectrum of depravity—the denouncements were equally vitriolic whatever the subject matter at hand—but surely the horror comics of the early 1950s enraged nearly everyone except their eager readers. Depicting evil that the good guys must fight was the *raison d'être* of superhero comics; in the early, supposedly innocuous years, the gangsters and war fiends freely inflicted torture in graphically delimned scenes. Removing the superhero and allowing torture, cannibalism, vivisection, and general carnage for its own sake gave the remaining pool of comic workers a welcome respite from beat-'em-ups and, for the best, a way of working subtle social commentary into their pages.

Horror being a deeply personal terror, the robot tales delve deeply into identity and the upending of issues of personality and surface appearance. "Beauty and the Beast" from *Mister Mystery* #11 (May–June 1953) is the most elementary of these examples. The exceedingly ugly dress designer Van Schnapier repulses his models, so he designs a robot woman to wear his clothing. Unfortunately, he accidently pushes her into the steam presser, melting her wax face and turning her into a thing uglier than even he. Herbert

Phillips has no such excuse for the robot woman he builds (*Mysterious Adventures* #16, October 1953). He detests how obese his once beautiful wife Hilda has grown since they married, so he kills her and uses her skin to create a new, thinner Hilda. "The Monster Doll" in *Forbidden Worlds* #1 (July–August 1951) is another creation of an ugly man, one a hundred years ahead of his time. She kills him and every one of her succeeding husbands, until her last husband destroys her and is sentenced to die since no one knows the truth about her. As it turns out, women have similar urges: In 1954's "Break-Up!" Marcia Lomax hates the time her scientist husband spends building a perfect humanoid replica of himself, especially since he only married her to use her wealth as funding. (Apparently no female robot builders existed in this time period. No matter how wise women might be in the halcyon 1950s, they are never technologically so.) Marcia smashes the duplicate, telling herself, "Now Zane will have to give me more of his precious time or I'll cut him off without a cent!"[16] But an automobile accident reveals that she killed Zane and not the duplicate.

Robots are fatal, no matter how innocuous or human they look. When in "The Silent One," *Men's Adventures* #21 (May 1953), Hugh Maxted goes to work for robotics expert Dr. Kronkeit, he can't keep his hands off the beautiful Margo, even though the doctor warns him. Unable to conquer his urges, he pulls Margo in for a kiss and gets electrocuted. His coat was wet and she was a robot. "All for Love," in *Space Adventures* #8 (September 1953), combines all the variations into one tale: When Jason Casan crash-lands his spaceship on the asteroid "Vulca," the beautiful Clio not only nurses him back to health but also falls for him over the "scrawny, book-reading old man" she is with. In fact, she sinks a hatchet into the old man's head. She and Jason then escape in a derelict ship. It can't fly to Earth, but they can be together forever. And they shall be even though the heat of the ship melts her face and reveals the robot beneath. "G-Good God! Doomed to float around in space until I die … with … with **that!**"[17]

—⁂—

Science fiction is a large and loosely defined genre. For some, the mere presence of robots in a story automatically turns it into science fiction. The Adam Link stories take place in a wholly contemporary America, with the single exception of the inclusion of an intelligent robot, yet were published as science fiction and have been considered so ever since. Others would demand the presence of basic tropes like space travel or a futuristic timeline before considering a story true science fiction. Comic book publishers tried to have it both ways. Most of the comic titles and strips classified as science fiction take place elsewhere than contemporary Earth. For example, Jet Powers, the Space Ace, wears an outfit indistinguishable from that of a superhero and battles robots with superior strength, similar to his dozens of predecessors, but happens to do so on Mars starting in 1951. *Robotmen of the Lost Planet* (1952) is a one-shot comic written by Walter Gibson, famed as creator of the Shadow (who battled the Robot Master in his 1945 comic book, *The Shadow Battles the Robot Master*). Despite the title, the setting is Earth in the far future, long after humans gave over all work to robots, who could perform it better. The robots revolt, of course, and the humans must spend years regaining their technical know-how to be able to defeat them.

In another example of an Earth-bound hero, physicist Gordon Dane is given a mechanical brain by a crash-landed alien to help Earth against the invading aliens from RAK in a scene eerily foreshadowing the way that Green Lantern would later be given his power ring. As Captain Science, he uses his scientific knowhow to defeat the giant robots that RAK sends to Earth. Captain Video, "famous star of the Dumont television network," also does his video rangering on Earth, traveling not in a rocket ship but in his Whirlojet, whose outer shell "makes ten thousand revolutions per second." When Makino, the super-intelligent robot, kills all scientists who lost to him in a knowledge contest, Captain Video bests him via an unanswerable question for a robot: "Describe the feeling of sorrow in man."[18]

"Revolt of the Robots," a title surprising not in how many times it was reused, but rather how few, appeared in *Space Detective* (February 1952), depicting mine workers put out of a job by robot workers, egged on by the mysterious Targan, who insists that "the future of the world is not with humans! It is with robots!"[19] He's right, being a super-robot from 300 years in the future. Another "Revolt of the Robots," in *Space Adventures* (September 1952), also dealt with robot mine workers, these on "Decima, the tenth moon of Jupiter." There a man is stirring up the robots, but not for political reasons: they've struck gold, and he wants it. *Weird Tales of the Future* #1 (May 1952), an attempt to blend horror and science fiction, foretells the war between robots and humans in 2032, but, in a clever twist, it's not clear which side the reader should be rooting for.

Lost Worlds #6 (December 1952) told of "The First Man to Reach the Moon," in far-off 2021, an intelligent, self-sacrificing robot, who takes the astronaut's place when he finds a flaw in the rocket. He wasn't the first robot on the moon, though. Four years earlier, in 1948, Robotman (still going strong a decade after his invention) had volunteered to be the astronaut on a planned moon mission, not knowing that his enemies have taken the place of the real scientists and sabotaged the rocket. Supposedly stranded on the moon, he reaches into his vast store of knowledge to remember that escape velocity is lower there, "only 1½ miles a second," or 5,400 miles per hour. Pumping his motor-driven legs, he reaches that speed and leaps up so that he will simply coast back to Earth; apparently he sprang for the optional navigation system. That same year, *Wonder Comics* featured the hero robot Roboroy, the kid sidekick of Wonderman: "An artificial man! Built by Brad [Wonderman] ... Roboroy has a real brain powered by a tiny uranium atomic-pile!"[20] Roboroy's egg-shaped head manages to look like a young boy's, down to the eyebrows that the artist includes when he remembers them. Robots were still doing favors for humans in *Space Adventures* #36 (October 1960), when Tondo, "The Omnipotent Robot," sends his Venusian masters a false report on the vulnerability of Earth, for he sees that Earthpeople want only peace and happiness.

Perhaps the best-remembered and best-regarded science fiction story of that era of comic books appeared somewhat unexpectedly in EC Comics' *Weird Fantasy* #18 (March–April 1953). (The firm also had a *Weird Science* title, but the stories were essentially interchangeable.) Written by Al Feldstein, probably from an idea by publisher/editor William F. Gaines, and illustrated by Joe Orlando, "Judgment Day" is an exemplar of the lushly illustrated, yet text-heavy, style that EC brought to its line of horror and science fiction comics. Tarlton, of Earth Colonization, has come to Cyberia, seeded with robots thousands of years before in the hope that "*in time* [they] would develop a society worthy of **inclusion** in Earth's great **galactic republic**." The population of orange robots has advanced to 20th-century-level technology and democracy. Tarlton approves of the great advances made; yet he keeps asking about the blue robots lurking in the background. "You ... differentiate between **blue** robots and **orange** robots?" he asks. "Of course! **Otherwise** there'd be **trouble**! Have to keep them in their **place**, you know." Their place is in blue town, held there through cultural forces and their lack of education. But nothing is different about the blue robots except the color of their sheathing. Tarlton is forced to tell the robots that they are not yet ready for the republic. The end panel is the twist: "And inside the ship, the man removed his space helmet and shook his head, and instrument lights made the beads of perspiration on his dark skin twinkle like distant stars." Thuddingly obvious as the parable might be, it was heady stuff for a comic book in 1953. It was equally powerful in 1955, when Gaines intended to reprint the story but first needed to get approval from the newly organized Comic Magazine Association of America's code administrator Charles F. Murphy. As retold in David Hajdu's *The Ten-Cent Plague: The Great Comic Book Scare and How It Changed America*, Murphy denied permission. Feldstein quoted him as saying, "You can't have a Negro."[21] Murphy relented only after Gaines threatened to call a press conference and blast the association as racist. *Incredible Science Fiction* #33 carried the story, but that was the last comic book EC ever released. Gaines subsequently reinvented the *Mad* comic as a magazine, not covered by the code, launching an empire. Comics would remain stifled and sanitized for another two decades.

In 1956, National Periodical Publications (better known as DC) revived an old gimmick to explore the increasingly knotty question of what the remaining comic-buying public wanted. Using a title specially as a testing ground for new strips dated back to the beginnings of the field. *Showcase* #1 (March–April 1956) not only put the premise into the title but also claimed in so many words that it would build its strips from suggestions that readers sent in. The state of comics in 1956 can be read precisely from the fact that the covers of the first three issues featured "Fire Fighters," "Kings of the Wild," and "The Frogmen." A second, non-superpowered foursome called the "Challengers of the Unknown" appeared in *Showcase* #7, along with a giant robot, Ultivac. Created by one of the numerous ex–Nazi scientists who continued making convenient foils in the 1950s, Ultivac relies on readers' recognition of the trope of the computer as a giant brain that also marked 1950s culture. He and his creator vie for supremacy until the comic turns unusually thoughtful. The government steps in, sending its "greatest authority on robots and calculating machines," who is perhaps the first female represented that way. (Tiny steps: she may be the experts' expert, but she's never created a robot of her own.) She wants the robot for the great good it can provide the government, its creator denies that the government has the right to take away his private property, and Ultivac is "willing to apply my powers to the cause of helping mankind—if mankind meets me halfway."[22]

The exceedingly odd "Doom Patrol" started life in DC's *My Greatest Adventure* #80 (June 1963). A superbrain in a wheelchair leading a pack of outsiders feared and hounded by normal humans also describes the X-Men, which would debut three months later, but these characters were self-described freaks rather than mutants and included a rebooted Robotman. Like his 1940s counterpart, he had a human brain inside a robot's body. When given their own title in 1964, the Doom Patrol also received an evil countergroup, the Brotherhood of Evil, one of whose members was a super-giant robot controlled by a man sitting inside the head.

DC is not credited with much experimentation in this era, but the Doom Patrol appeared just a year after *Showcase* spent four straight issues on the most innovative robot concept in comics history: the Metal Men. Iconic comic characters almost always depend on exaggerating childhood daydreams about speed, strength, competence, or wisdom. The Metal Men might easily have emerged from a talented youngster doodling in a notebook during boring lectures in chemistry class, Platonic ideals of elements made into people. Iron was powerful; Lead stopped radiation; Mercury turned liquid; Gold was ductile; weak and timid Tin provided comic relief; Platinum, otherwise redundant to Gold, embodied femininity. Yet they were robots, constructs of the brilliant Dr. Will Magnus. The text of the inaugural strip in *Showcase #37* (March–April 1962) is peppered with melting and boiling points (in Centigrade, not even Fahrenheit) and other handy facts about elements that might have helped readers ace their next exam. That the group of highest-technology robots battles a prehistoric radioactive mutant creature freed from an iceberg seems like a bigger oddity today, but virtually every DC comic in the early Silver Age featured similar monsters. The height of strangeness occurs at the end of the battle, when every one of the Metal Men perishes in victory.

They are back in the next issue, of course. Robots can be rebuilt. In modern comic books mourners sometimes skip funerals because superheroes always come back from the dead. However, back then the ability of robots to take fearsome punishment and simply be hammered out at the body shop was a rare asset for plot writers. The Doom Patrol's Robotman seldom made it whole through an entire issue. That the Metal Men could be melted, scattered, and endlessly reshaped was as essential to their appeal as it was to the writers' and artists' imaginations. Readers responded with fanatic enthusiasm. "I 'demand' that the METAL MEN have their own magazine," wrote Edward from Brooklyn. "Why, the excellent educational material contained in it makes it worth the price." (Charles from Uniondale, New York, more penetratingly wondered how Mercury stays a solid at room temperature, without receiving a good answer.)[23] Comic writers had long used scientific facts to drive plots and provide seemingly magical solutions to problems, making schoolwork come alive and practical. Even a creative bombshell like the Metal Men proved to have a one-shot precursor, *All Star Comics #26* (Fall 1945), "The Mystery of the Metal Menace." Here the Justice Society of America fights a spaceship full of "metal men" from Jupiter who literally eat metals and take on their characteristics. Silver, Iron, Magnesium, Copper, and Gold groups of metal men are defeated by using their metallic properties against them. Fortunately, 1960s villains seemed to have no memory of Golden Age comic books any more than their young readers.

The Metal Men (despite Platinum's presence and enduring love for Dr. Magnus, nobody at the time thought twice about their name) probably had more robot opponents than almost any other comic. Their first foe after being reformed is a robot taller than skyscrapers. Titles in the first twenty issues of their own comic include "Rain of the Missile Men!," "Robots of Terror!," "Menace of the Mammoth Robots!," "The Robot Juggernaut!," "Raid of the Skyscraper Robot!," "The Headless Robots!," "The Revenge of the Rebel Robot," "Robots for Sale," and "Birthday Cake for a Cannibal Robot!" Each comic requires an archfoe; theirs was Chemo, an artificial being made of radioactive, toxic chemicals who can no more be eradicated from the Earth than pollution.

Like the Doom Patrol, the Metal Men were at best cult favorites. Their comics lived and died in the 1960s before the endless modern nostalgia for forgotten characters (at least on the writers' part) brought them back again and again. A third critical robot title followed the same pattern, though issued by Gold Key rather than DC: *Magnus, Robot Fighter*, set in the endlessly perilous world of AD 4000. "From the sea comes Magnus to fight the evil robots who are the masters of men!" claimed the cover for issue #1, cover dated February 1963.

The comic became the repository for every robot trope developed over the past quarter-century, much like *The Jetsons* simultaneously trotted out every cliché about the future. Issue #1 (Gold Key had no anthology titles to test with) starts with a paraphrase of Isaac Asimov's Laws of Robotics: "No robot may harm a human, or allow a human to come to harm." Yet renegade robots somehow exist, and the ancient and wise robot, symbolically numbered 1A, takes an orphaned baby and raises him to the peak of human physical perfection, able to smash robot steel with his bare hands. (Magnus' first foe is H8, who takes the "hate" implied by his designation literally.) The need is great. Some robots, like 1A and H8, have become self-aware, leading to friction between robots and the outnumbered humans. All of North America is covered by a single city, with literally billions of robots doing all the chores. Of course, Magnus fights a robot duplicate of himself in issue #2 and a giant alien robot in issue #3.

Russ Manning, *Magnus'* creator, who produced the art (and often the stories) for its classic first 21 issues, made interesting design choices to create a more complete future world than was the norm. He inverted Tarzan's concept to make Magnus the hero of a futuristic jungle and battler of its wild beasts. Manning alternated between simple close-ups that sped the narrative along and intricate vistas that let the reader get lost in the levels of his future city, awe-inspiring alien landscapes, or unending banks of electronic machinery. One panel that sold the concept from the first issue was, in H8's odd robot speech, "A—computer—using—human—brains ... just—as—you—humans—once—made—computers—of—mechanical—brains!" (composed of a thousand humans with thought helmets arrayed *Matrix* style to power one super-brain). Magnus' costume is also unusual, a barely-to-the-bottom-of-the-buttock tunic and boots showcasing his mighty thews the way that artists made Phantom Lady's thighs iconic. Girlfriend Leeja wears an almost equally short black minidress whose bottom few inches are transparent. Nobody else in the future dresses this way, making them stand out in crowds.

Issue #7 is a trove of quotes mirroring comics' robotic history. Robots emerge from a ship at the spaceport berserk with rage from a "metallic plague." All nearby robots catch the disease, and soon a mob of diverse robots attacks the civic center, only to be stopped single-handedly by Magnus. Fortunately, the disease destroys the robots after five minutes, but the thought of it spreading terrifies the government. Magnus is unfazed: "As the council is already aware, I believe man has grown **weak** by depending too much on robots.... Perhaps if man had to do with a few **less** metal servants, he could once again become **strong**!" "No!" a councilman exclaims. "We could not **live** without our robots." (Substituting "internet" for "robots" shows how prescient this notion was.) All robots are ordered evacuated, but the literal and figurative lower classes, "petty criminals and antisocials ... gopher-like people who never leave the lowest levels except to cause trouble," refuse to give up their robots. Idealistic Magnus goes down to explain to them the reasonableness of the government order—not a good idea in any year. He is saved from a mob by the timely intervention of the archvillain Xyrkol, he of the giant alien robot, who teleports Magnus to his home planet of Malev-6, "The robot world! An electronic brain that fills and uses the power of an **entire planet!** A brain capable of controlling **the universe!**" Xyrkol is himself a robot and the instigator of the metallic plague. He and the robot planet are one; all that stands in his path of galactic conquest is Magnus. Magnus, though, is not as alone as he seems. A thousand minds he saved from H8 on Earth beam Leeja's love to Malev-6, and he uses their combined power to short-circuit the planet (probably the first time that phrase could ever be used).

Robots may manifest by the thousands, but they are eclipsed by the might of the singular computer brain. Robots are useful tools; computers can rule empires.

9

Utterly Alien
and Nonhuman
The Robot in Golden Age Science Fiction

The word *romance* goes back to around 1300, when it was first used to describe inventive tall tales about knights conquering all in fabulous lands. A specific subgenre, the scientific romance, congealed into a category in the 19th century. The editors of *The Galaxy: A Magazine of Entertaining Reading* prefigured the next century's worth of science fiction in a September 1873 review of Jules Verne's *Around the World in Eighty Days.*

> We hardly know whether the thoroughgoing moralist can approve of the publication of Verne's books or not. We are pretty sure, however, that the inexperienced boy, or the weary adult reader, will find much entertainment in them. They belong to the order of "scientific romance"; that is to say, what science they contain is eked out by romance, and an air of truth is given to the romance by the introduction of a great deal of science.[1]

Readers wanting more stories of scientific romance found them in the newfangled pulp magazines. Pulp magazines were adult versions of the weekly boys' papers found in chapter 3. Instead of flimsy newsprint, however, publishers chose the thickness of pulp stock for a greater heft at even cheaper prices. Neither paper was designed to last through more than a handful of readings; the fact that so many have testifies to the attachment readers had for these stories. *The Argosy* is usually granted the honorific of "first all-fiction pulp magazine"; its December 1896 issue switched to what became the standard 7x10" pulp format, offering 192 pages of pure fiction for one dime. Its title evolved from *The Golden Argosy*, a boys' weekly that Frank A. Munsey started publishing in 1882. Munsey eschewed literature for pure formula and made fortunes by doing so. Under an unfathomable variety of titles (including *Argosy and Railroad Man's Magazine* for several weeks in 1919), the Munsey magazines remained among the privileged heights of the pulps until 1943. His writers included the earliest major American science fiction authors: Garrett P. Serviss, William Wallace Cook, Howard R. Garis (who would ghostwrite the first Tom Swift series), Edgar Rice Burroughs (for all three of his major series), and fantasist A. Merritt. Literature never sold like these storytellers with enormous popular appeal.

Abraham Grace Merritt was an anomaly among pulp writers: he had a real job. As assistant editor of the magazine *American Weekly*, he pulled down an annual salary of $25,000 per year, increasing to $100,000 when he later became editor, allowing him the luxury of working slowly to craft distinctive prose.[2] His style is today considered florid, but by comparison to the bare-bones English of most pulpsters, Merritt's best work immerses the reader in a golden flow of atmosphere and description, standing out from the pack as Ray Bradbury would a generation later.

All eight of Merritt's novels would be serialized in Munsey magazines, most of them lost-world romances in exotic, far-away settings with gods and goddesses from ancient times. *The Metal Monster*, almost the definition of the lonely subgenre of science fantasy, ran in *Argosy All-Story Weekly* in eight parts from August 7 through September 25, 1920. In the story, a Persian goddess controls geometrical shapes made of metal, all-purpose Legos of every size that can join seamlessly to create bridges or bunkers or terrifying robot warriors.

> Sphere and block and pyramid ran together, seemed to seethe. I had again that sense of a quicksilver melting. Up from them thrust a thick rectangular column.
>
> Eight feet in width and twenty feet high, it shaped itself. Out from its left side, from right side, sprang arms—fearful arms that grew and grew as globe and cube and angle raced up the column's side and clicked into place each upon, each after, the other. With magical quickness the arms lengthened.
>
> Before us stood a monstrous shape; a geometric prodigy. A shining angled pillar that, though rigid, immobile, seemed to crouch, be instinct with living force striving to be unleashed.
>
> Two great globes surmounted it—like the heads of some two-faced Janus of an alien world.
>
> At the left and right the knobbed arms, now fully fifty feet in length, writhed, twisted, straightened; flexing themselves in grotesque imitation of a boxer. And at the end of each of the six arms the spheres were clustered thick, studded with the pyramids—again in gigantic, awful, parody of the spiked gloves of those ancient gladiators who fought for imperial Nero.
>
> For an instant it stood here, preening, testing itself like an athlete—a chimera, amorphous yet weirdly symmetric—under the darkening sky, in the green of the hollow, the armored hosts frozen before it—
>
> And then—it struck![3]

The famed horror writer H. P. Lovecraft, a dedicated *Argosy* fan, later would say that *The Metal Monster* "contains the most remarkable presentation of the *utterly alien and non-human* that I have ever seen" [italics in original].[4]

Munsey also pioneered the single-genre pulp, with *Railroad Man's Magazine* in 1906 and *The Ocean* in 1907. A decade passed before a competitor, Street & Smith, narrowed in on mysteries with *Detective Story Magazine*. Fantasy took even longer and got off to a faltering start with *The Thrill Book*, by Street & Smith, which lasted for only six months in 1919, possibly because it contained mostly standard adventure tales rather than the fantasy it promised. *Weird Tales*, subtitled "The Unique Magazine" when it debuted in 1923, tried enticing readers with horror; the first issue's cover story was illustrated by a tentacle creature rising from the muck, a limb coiled around a handily placed flapper. It hedged its bets when editor Farnsworth Wright took over, giving Lovecraft a near-monthly home side-by-side with scientific romance, including robot stories.

Edmond Hamilton graduated high school at fourteen but dropped out of college. Like other bright youngsters who couldn't hold a steady job, he turned to writing. At twenty-one he started selling Wright a stream of stories. "The Metal Giants" was his third, in the December 1926 issue. It reads like a prototypical boys' fantasy: build big things and make them go smash. Professor Lanier of tiny Juston College is fired for making unprovable claims about an artificial electronic brain. Four years of intensive work at a rural location are sufficient to perfect his work, adding more connections that could learn proportionally more. The resulting gigantic brain became "a mind that was far greater than man's, aided as it was by cold, ruthless reasoning power, precise, perfect memory, and quite unswayed by the thousand and one emotions that affect human intelligence, untroubled by love and hate and fear and joy and sorrow."[5] When Lanier is hospitalized, the brain filters aluminum out of the very earth beneath it and builds itself a body hundreds of feet tall, along with an troop of equally huge subservient robots. They stomp off, utterly destroying a small town before heading toward larger cities. Lanier returns at the crucial moment. With no time, money, equipment, or supplies, he builds an even more gigantic wheel, studded with spikes, that he rolls over the central brain, dying in the process. There is no moral lesson learned. Robot menace stories seldom have a good ending. The spectacular must be its own reward, a barrier to the genre's literary growth for a half-century.

The next try at a specialty pulp seemed equally unpromising. The first issue of *Amazing Stories* in 1926 contained nothing but reprints by writers like H. G. Wells, Jules Verne, and even Edgar Allan Poe. The second issue had one story apiece from Wells and Poe and two by Verne. Wells would be in every issue for the next two years. Robots would not appear once. Yet *Amazing* had the one element that almost always ensures magazine success: an editor with a vision.

Radio (and soon television) obsessed Hugo Gernsback, a born tinkerer in the best backyard tradition—he would ultimately amass 80 patents. In 1905, he marketed the first home radio kit, the success of which he parlayed into a catalog of radio equipment and then, in 1908, a propaganda magazine for all things radio, *Modern Electrics*. Gernsback wrote fewer than a dozen pieces of fiction over a sixty-year career—a snarky critic might say that he never wrote any—but his first and longest attempt skewed the entire field of scientific romance the way that the sun skews light from distant stars. *Ralph 124C 41+: A Romance of the Year 2660* ran in twelve issues of *Modern Electrics* starting in April 1911. Ralph's name translates into "one to foresee for one," with the "+" designating that he is one of the ten top geniuses on Earth—an inventor, naturally: he is basically Reed Richards. *Ralph* is a lecture describing wonders that humanity can look forward to in a mere 749 years, a list long and often uncannily prescient: videophones (*telephot*), fax machines (*telautograph*), three-dimensional television (*tele-theater*), radar, tape recording, synthetic milk, space stations, heat pumps, and motorized roller skates (*tele-motor-coasters*: Gernsback succeeded because of his broad stroke of monomania).

The only trope missing from *Ralph* is the robot. Any charismatic visionary picks up disciples as a matter of course; Gernsback's would fill in the blanks. Most notable among them was Clement Fezandié, who for five years in the 1920s contributed a work of "fiction" to Gernsback's *Science and Invention* magazine on a near-monthly basis. One was "The Secret of the Tel-Automaton," in which Fezandié's genius inventor Dr. Hackensaw creates a radio-controlled female robot. Superstrong and nearly invulnerable, she is stolen by crooks to rob banks. As we've seen, about a thousand comic book writers stole this plot.

Response to the Dr. Hackensaw yarns was so strong that Gernsback made the August 1923 issue of *Science and Invention* its "Scientific Fiction Number" (as in a musician's next number). The uncredited cover artist graced it with a spectacular metaphorical scene of a bubble-helmeted astronaut floating in deep space between Saturn and a solar eclipse. The table of contents listed the stories under the heading as "Popular Scientific Articles"—exactly how Gernsback would always think of them. When he asked his readers whether they wanted an entire magazine of such stories, they overwhelmingly voted no; his nonfiction readers thought of stories as sprinkles on a cake, not the cake itself.

Visionaries listen only to themselves. *Amazing Stories*, subtitled "A Magazine of Scientifiction," debuted in April 1926. (Earlier stabs at naming the genre included "different," "weird scientific," and "pseudo-scientific.") Gernsback's core insight was that a cadre of fiction fans existed; all they needed was to be targeted, cosseted, and fed bountiful quantities of newness. He was magnificently right. Fans leaped upon the new magazine, quickly driving circulation up over 100,000. They wrote letters to the magazine and to each other, setting up fan clubs, correspondence circles, and fan magazines. Fezandié was dropped after two stories. New writers debuted to huge acclaim. E. E. "Doc" Smith's saga of super-science, "The Skylark of Space"—the first of his endless tales of heroic figures romping through galaxies—appeared in the same August 1928 issue as the first Buck Rogers story, "Armageddon—2419 A.D.," by Philip Francis Nowlan. Gradually the term *science fiction* (also coined by Gernsback) pushed out the Victorian-sounding *scientifiction*. A new genre coalesced around *Amazing Stories*, and from there it spread as fast and far as a Doc Smith space battle.

The value of Gernsback's conception of what became known as science fiction and his influence over the field has been hotly debated. Fiction without style, characterization, and deeper meaning was derided even in the 1920s as subliterary, fit only for the troglodytes that were occasionally the subject of a caveman yarn. Gernsback's one and only concern was to proselytize for the improved world that could be had today, or at least the day after tomorrow, if only people would make it so: propaganda today, utopia tomorrow. To his chagrin, writers he inspired lacked his singular vision and monomania. Their wild, nominally scientific futures merely showcased the clever way they substituted space battles for Wild West ones, creating a subgenre that would retroactively become known as *space opera* in homage to the phenomenally popular radio soap opera. Ripping the human out of the future became a taint that science fiction has yet to fully recover from.

Contemporary robot depictions veered between mindless machines and imitation humans, which made them difficult to fit into Gernsback's cosmology. The first robot story in *Amazing Stories*, "To the Moon by Proxy" (1928) by J. Schlossel, also generated the first robot cover, which showed a bug-eyed robot wrestling a lion. Lions on the moon? Never in a Gernsback publication. The robot's inventor, a paralyzed man confined to a wheelchair, has sent his robot out on a test mission, and it just happens upon a lion escaping from a circus. As the inventor explains, "Radio television provides it with sight; that is, it enables me, sitting here on the chair, to see through its artificial eyes. Radio telemetrics, or wireless control at a distance, guides its legs, arms—in fact, every movement of the body." All this is prologue to the short moon voyage, where the robot finds intelligent life in underground caverns, so intelligent that in a few moments it crushes the robot under a falling stalagmite. The inventor berates himself for "not putting the controlling apparatus in a less vulnerable place than the head." To pound home the purportedly short distance between this future and the present, a brief insert article touts the Westinghouse Televox robot, which was controllable at a distance (albeit only by wire). Schlossel was a pure expounder of Gernsbackism, and it's surprising that, with a single exception, this was the end of his short career. His vision meshed perfectly with the technological optimism that Gernsback and most of the nascent science fiction crowd radiated. "In this age of ours everything is possible," his inventor proclaims. "Ideas and beliefs are changing constantly to conform to the present day standard. What yesterday was accepted as an unchanging truth, is today looked down upon with a feeling something akin to contempt."[6]

David H. Keller was twice Schlossel's age and well established in his profession when his first story appeared in *Amazing Stories*. Writing was his after-hours hobby, a supposed pleasure that irritated him when he didn't sell. One day his wife brought him a copy of *Amazing Stories* and suggested that a market had finally emerged. He would have ten stories in *Amazing Stories* and its companion, *Amazing Stories Quarterly*, in 1928. "The Psychophonic Nurse" in *Amazing Stories* (November 1928) is prime Keller. (The nurse is "psycho" for psychological and "phonic" because of voice control.)

Susanna Teeple is a proud modern working woman greatly inconvenienced by giving birth. As a work-from-home freelancer for the *Business Woman's Advisor*, she finds herself interrupted "exactly one hundred and ten times every twenty-four hours." A machine could care for a baby; in fact, it would be better. Hadn't modern science proved that "untimely love and unnecessary caresses" would result in complexes when the child grew up? (Keller namechecks Hermine Hug-Hellmuth, the first child psychologist, and John Broadus Watson, the first behavioral psychologist, as proof of this assertion.) That gives her husband an idea. He commissions a machine nurse in the shape of a black mammy, which he calls Black Mammy. "Naturally, there is no intelligence," he says, "but none is needed in the early days of child care." Soon a substitute father, Jim Henry, is introduced to do the heavy work of pushing the child's stroller. (It says everything about Keller that the mechanisms have names, but neither the father nor the baby do.) Men are just old sentimental fools, though. The father bonds with his baby and concocts ways of secretly snuggling with it, including swapping himself for Jim Henry on stroller duty. One day a blizzard appears out of nowhere, trapping them in a snowbank overnight, Susanna assuming they have died. In response, Susanna junks the robot, closes her typewriter, and starts singing lullabyes while cooking. "And that was the end of the Psychophonic Nurse."

Keller always signed his works David H. Keller, M.D. He was in real life a psychiatrist, assistant superintendent of the Louisiana State Mental Hospital, and author of an acclaimed 10-volume sexual education series. Lest you think that he therefore was writing satire, skip ahead to the January 1930 *Amazing Stories* and "Air Lines," which features a husband who is a writer and a wife who is an inventor (known technically as the "old switcheroo"). William Dills wants a homemaker so that he can sit uninterruptedly at his typewriter. Beryl Angelo unfathomably agrees. She hates every minute of the job of keeping up a home and so invents a dozen robots to do the work for her. Although they still need too much supervision, she gains the time to work on her masterpiece: a robot plane piloted by an actual robot. The plane is conveniently located in her backyard, and this gives William—you guessed it—a secret spot for baby bonding. He has no idea what Beryl does with her time and is taken aback when she launchs the plane on a round-the-world test cruise. Her assistant has also made a slight miscalculation in the amount of fuel needed, so that the plane will surely crash. There's no radio, either: a robot doesn't need one. "For the first time in her life the inventor of Aviation Consolidated fainted." Not to worry: William learns to fly by watching the robot, turns the plane around, and lands in his backyard. "[T]ell the President that I am going to take a three-month's vacation," Beryl instructs her company. "I want to get reacquinted with my family."[7]

Abner J. Galula's "Automaton" in the November 1931 issue of *Amazing Stories* pays homage to the Kellerian view of marriage. A genius professor, with the help of his lone assistant, Martin, and female ward, Theresa, invents a robot that can think and learn. The two youngsters fall in love, but the robot contends that his superior skills can offer the young girl power, fame, and money. Theresa chooses the robot, and Martin gets sent to an insane asylum when he tries to tell people the truth. Twenty years later, Theresa is the most powerful woman in the world, but a lonely one. She gives it all up to free Martin and marry him.

Gernsback lost control of *Amazing Stories* and started *Science Wonder Stories* in 1929 (in which he started calling the field *science fiction* for the first time), taking Keller with him. In "The Threat of the Robot," robots have taken over human jobs, to the extent that robot teams play football in empty stadiums for home television viewers. Cutting off their central power source (a cheat that lazy writers would revisit ad nauseum down to the present day) saves humanity from the indignity of non-work. (Writers are forever rediscovering these same extrapolative pathways. The 20th-anniversary issue of *ESPN The Magazine* made a somewhat less bold forecast for 2028: "Teams play in front of just a few thousand fans inside what are essentially large TV studios built to serve viewers, not ticket holders."[8])

Keller had unfortunate prejudices and few literary skills, but his stories provoked thought, perhaps the highest compliment that readers could give science fiction at the time. "[T]here is always a moral behind them," wrote one in a letter to the editor. "Dr. Keller's work has great educational value," Gernsback proudly admitted in response to another rave.[9] Bad as Keller might have been, other writers published by a Gernsback desperate for their type of story were even worse. O. Beckwith's "The Robot Master" (1929) in *Air Wonder Stories*, a companion magazine, presents the technologically utopian world of 1965, free of crime, and free of Professor Hyle L. Benning, exiled for anti–democratic rantings. He's used his time to create a secret island hideaway and perfect centrally controlled robots to do his bidding. After destroying New York City, he kidnaps his nephew because he needs a successor. He lengthily explains all of his evil plans and thoughtfully points out the one switch that needs to be thrown to make his robots useless. An even lower point is reached by the one story credited to Melbourne Huff in Gernsback's *Scientific Detective Monthly*. Indestructible and superstrong, "The Robot Terror" (1930) is constructed by a literal madman at an insane asylum to rob and kill in the mode of Fezandié's robot. The good guys have no way to battle it at all; only a *deus ex madman* saves them when another patient attacks the mad inventor.

Jim Vanny's "The Radium Master" (1930) updated the lost-world trope with modern touches. Gernsback introduced it with an irresistible teaser: "Far in the heart of an African jungle lies the great and powerful city of Urania, the abode of the Masked Emperor. Far over the world extends the Emperor's influence. From the bowels of the earth he extracts limitless power."[10] (Any similarity to the Black Panther and Wakanda is probably coincidental, although pulp stories read as youths have a remarkable ability to worm their way into readers' brains.) Remote-controlled wheeled robots do the work in the city and mine the radium for the central power plant buried deep under the fabulous city. Outsiders who stray in are never allowed to leave. "Resistance is useless," proclaims the Masked Emperor, introducing a phrase to robotdom that will redound down the years.[11]

Managing editor T. O'Conor Sloane was promoted to editor when Gernsback was forced out at *Amazing Stories*. Sloane had the actual PhD in electrical engineering that the dropout Gernsback lacked but a totally different approach to science fiction, preferring wild stories of scope and adventure that Gernsback scorned. In fact, he bought Neil R. Jones' "The Jameson Satellite" (1931) after Gernsback bounced it. A professor who wants the truest form of immortality arranges to launch his dead body in a satellite, where, according to the science of the day, it would remain perfectly preserved forever. Forty million years in the future, the Zoromes, aliens who eons earlier put their brains into immortal metal bodies, resurrect him. The kindly cyborgs who place Jameson's brain in a similar metal body greatly influenced the young Isaac Asimov. The Zoromes "were the spiritual ancestors of my own 'positronic robots,' all of them, from Robbie to R. Daneel," he would write.[12]

With Gernsback floundering, the Clayton Magazine group of fiction pulps created the first true science fiction pulp magazine, *Astounding Stories of Super Science*, dated January 1930. (The Gernsback titles were larger than the standard 7 × 10" pulp size, although most people sensibly lump them into the same category.) "Superscience" stories stretched the Gernsbackian vision of sensible extrapolation of rapidly improving technology to encompass all of time and space, wonders that took Hamilton's pioneering gigantism and swelled it beyond recognition and, all too often, sense. A letter to the editor in 1934 critiqued these stories as "all done for the thrill, the kick, the climax, and the happy ending."[13] Most readers seemed to want exactly that, though one small exception to the reader's lament had appeared in the March 1931 issue. The "Terrors Unseen" of Harl Vincent's title were invisible robots, designed (like so many other inventions of the post–World War I period) to make war unthinkable. Crooks steal the robots, along with the scientist and his daughter. A statistically improbable percent of scientists in these stories had nubile daughters to be kidnapped and fall in love with the young man who saves them. So it is here. Almost uniquely, however, the young hero pooh-poohs the notion of a weapon too terrible to use:

> "They'll insure the peace of the world. They'll—"
> "Listen, Mr. Shelton." Eddie interrupted. "If you'll think a little you'll realize that they'll do no such thing. Has any new and terrible engine of destruction ever accomplished that result? No—the enemy always finds a way of combating the new weapon and of devising another still more terrible. You've discovered a marvelous thing, but its value is quite problematical."[14]

Another Vincent robot story, "Rex," appeared in the June 1934 issue of *Astounding Stories*. The title robot accidentally gains full awareness. Seeking revenge on the humans who have made robots their slaves, he takes over their cities and forces them to do all the work themselves. Yet, like Icarus, Rex wants something he can't have: human emotions. He experiments on brains to extract emotion from them, only to fail repeatedly. Failure is the one thing Rex cannot stand. Out of rage and despair, he kills himself. The now-free humans see the irony. So would dozens of future writers who never tired of exploiting this fundamental distinction between the human and the inhuman.

A practitioner closer to the super science ideal was Ray Cummings, author of an estimated 750 pulp stories. His young all–American heroes saved nubile females by the score, albeit seldom in the here and now. "The Exile of Time" (*Astounding Stories*, April–July 1931) is prime Cummings. It features two time machines chasing one another from the creation of the Earth to a billion years in the future, along with a Revolutionary War cutie and the Princess of the American Nation, New York destroyed twice over, a hideous supervillain, and robots, robots, robots.

> There was, in 2930, a vast world of machinery. The god of the machine had developed them to almost human intricacy. Almost all the work of the world, particularly in America, and most particularly in the mechanical center of New York City, was done by machinery. And the machinery itself was guided, handled, operated—even, in some instances, constructed—by other, more intricate machines. They were fashioned in pseudo-human form—thinking, logically acting, independently

acting mechanisms: the Robots. All but human, they were—a new race. Inferior to humans, yet similar.

> And in 2930 the machines, slaves of idle human masters, had been developed too highly!
> They were upon the verge of a revolt![15]

The robots are secretly controlled by Tugh, who the reader knows by his name alone is a villain, all the more so in 1930 terms because he is a hideously deformed cripple. For sheer plot reasons, he travels back in time to scientifically primitive 1932 to spend years in search of a New York doctor who can give him a new body, jumping back occasionally to 1777 to pursue a winsome patriotic lass. (And the odd 1930s chorus girl, cad that he is.) New York doctors fail him, and a New York girl scorns him—he vows revenge on both. Depositing Mary, the girl from 1777, in Greenwich Village, the spot that once was her farm—the time machine cannot move in space—he uses 1935 as his base of attack on contemporary Manhattan. Cannily, he makes the trip from 2930 to 1935 over and over, each time returning only a moment later, building up a robot army.

> And another thousand or more had been killed by the Robots. How many of these monstrous metal men were now in evidence, no one could guess. A hundred—or a thousand. The Time-cage made many trips between that night of June 9 and 10, 1935, and a night in 2930. Always it gauged its return to this same night.
>
> The Robots were of many different forms; some pseudo-human; others, great machines running amuck—things more monstrous, more horrible even, than those which mocked humanity. There was a great pot-bellied monster which forced its way somehow to a roof. It encountered a crouching woman and child in a corner of the parapet, seized them, one in each of its great iron hands, and whirled them out over the housetops.[16]

The robots have horrible future weapons. In a few days, New York is a smoking ruin. Mary and 2930's Princess are paired with two 1935 stalwart young men and travel up and down time in a duo of time machines, often as captives of the head robot, who is trying to break free of Tugh's control and just needs a friendly woman to encourage him. Tugh foments a robot revolt in 2930, as only an archvillain could: "The humans who have made the Robots slaves for them will become slaves themselves. Workers! It is the Robots' turn now. And I—Tugh—will be the only human in power!"[17] The good guys foil the scheme only because the city has a central control, and the villain escapes into a future so distant that the time machines stop. Tugh is eventually defeated by his own weapons, leading to a twist ending that unfortunately also undermines what little plot the story depended upon. No matter: boy and girl and boy and girl pair off. "Such strangely contrasting types! Over a thousand years was between them, yet how alike they were, fundamentally. Both—just girls."[18]

So-called hero pulps starred shadowy figures (and The Shadow) whose exploits teetered across the line to the supernatural and science fictional. Doc Savage may have been the most influential of these characters. Calling his adventures a comic book come to life is unfair: comic books were deliberately patterned on Clark "Doc" Savage. Brilliant, omnicapable, and almost superhuman in looks and strength, Doc Savage is the source of many of the superhero comic tropes, from his secret lairs to his fabulous inventions to his team of loyal adventurers. Lester Dent wrote most of his adventures under the house name Kenneth Robeson, but "The Fantastic Island" (1935) happened to be written by a fill-in, Ryerson Johnson. The details of the island don't matter, and you wouldn't believe them in any case, but on the way Doc lures a villain to his "secret" airplane and boat hanger on the Hudson and provides him with some convenient shrubbery in which to stage a supposed ambush. Doc's head is thereby blown to bits. Or is it?

"Robbie will be needing a new paint job on his face, Doc."
"Yeah, and a new set of teeth."
…
"A dummy!" [the evil Count Ramadamoff] ejaculated.
"Sure. A mechanical likeness of Doc, Robbie the Robot."[19]

The first Robbie the Robot would be followed by dozens more, most with far more to do. Robbies are forever good robots, useful, nonthreatening, and friendly. A non–Robbie could be anything, but mostly evil, and those without names were to be avoided at the price of one's life.

Writers showed invention in the 1930s only by varying how and why robots were menaces. Stories like "The Synthetic Monster," "The Reign of the Robots," "The Rebel Robots," "The Revenge of the Robot," and "Revolt of the Robots" give away their intentions in their titles. Jack Williamson's super-science epic "After Worlds End" (1939) has robots destroying every planet colonized by humans before they are defeated by a tiny band of oddball humans. Even Stephen Vincent Benét wrote a poem sardonically extrapolating the revolt of the machines, "Nightmare Number Three" (1935), which gave readers of *The New Yorker* shudders:

> We had expected everything but revolt
> And I kind of wonder myself when they started thinking—[20]

Pulp covers had one job: to grab eyeballs. Magazines slowly learned to use every device to stand out on the overcrowded newsstands that sold the vast majority of their circulation, subscriptions to pulps being relatively rare. Riotous primary colors, action scenes, and the names of favorite authors in large type were ubiquitous and obvious. Action normally meant conflict. Robots, as the titles indicate, almost universally were portrayed as the aggressors. They crushed cities underfoot, shot rays from their eyes, and, surprisingly, often grabbed humans with multiple tentacle arms. Most were only vaguely humanoid, with others boxes and cylinders and pods rolling on wheels.

Artists also created solid fan bases. Frank R. Paul, whose creatures and spaceships followed no known principles of physics, drew almost all the early Gernsback covers, with the staider Leo Morey following. *Astounding Stories* was drawn month by month by H. W. (Hans) Wessolowski before Howard V. Browne and then Hubert Rogers took over, each bringing covers closer to a form of dynamic realism demanded by the changing times. However broadly or narrowly conceived, robots no more conformed to newspaper pictures of contemporary constructs than did the covers' aliens. Robots were always fantasy figures as enigmatic as the beasts on *Weird Tales* covers, apparently uniformly capable of more-than-human menace for inhuman reasons. Humans may have created robots, but they never controlled their creations.

Not until close to the end of the decade did a new style of robot story appear, temporarily swinging the cycle back to sympathetic examinations of robots who would be friendly, helpful, obedient, and sometimes startlingly human, although buyers needed to read the stories to learn this: nonmenacing robots seldom graced covers.

Unsurprisingly, the new view came from younger authors and editors who had spent their entire lives surrounded by technology. Raymond A. Palmer, born in 1910, became editor of *Amazing Stories* starting with the June 1938 issue. That issue contained a story that wowed the readers because, Palmer wrote, "It had sound science, and a fine human problem, and plenty of the human element," implying what other magazines lacked.[21] It also contained a story by Eando Binder, who would regularly contribute stories that lived up to that standard of science fiction. Sales of the moribund magazine soared.

Otto Binder was born in 1911. He and his brother Earl started selling science fiction in 1932 using the pseudonym Eando Binder. Otto would retain the moniker for years after the collaboration ended before returning to his own name for a career selling thousands of comic book stories. The cover of the January 1939 issue of *Amazing Stories* pictured a human shooting at a robot. That familiar scene was subtly subverted: a faithful dog protected the robot rather than the human. Big block text on the cover proclaimed, "An Amazing Confession: 'I, Robot' by Eando Binder." (Robert Graves' novel *I, Claudius* had been a bestseller in 1934 and may have influenced the title.) Binder's premise is frankly Frankensteinian: Dr. Charles Link makes his child out of "wires and wheels" with an "iridium-sponge" brain. Instead of recoiling from his monstrous creation, Link embraces his "baby" and begins to teach him everything about humanity, all of it filtered through his kindly, humanist perspective, which the robot, now named Adam—Adam Link—could not help but absorb. Even Terry, Dr. Link's dog, becomes attached to the robot. When Dr. Link's skull is crushed by a transformer falling off a wall, Adam feels compelled to venture into the world, "to become a citizen!" However, humans assume that Adam had run amok, because that's what robots do. The ending is poignant. Adam knows he could easily kill the humans but will instead turn himself off to save them. "Ironic, isn't it?" he thinks, "that I have the very feelings you are so sure I lack?"[22]

Too good a concept to let go, Adam Link would be brought back for an additional nine stories, exploring everything from the legal status of an intelligent robot to whether a robot was fair competition for athletes to the need for an emotional robot to have a mate. If robots were reflections of the good people who made them, only the irrational and unthinking could have cause to fear them. Robots would always be loyal to their human creators. Almost as important as the development of the concept was Binder's quiet, fluid writing—at least in the first few stories, before the melodrama starting seeping in. Adam was a character rather than a monstrosity, more human than the humans. The little man winning against the forces of the world was a stock fiction trope, but rare in science fiction (unless he carried a superweapon). Binder's ironically human-centered approach influenced a generation of younger writers who would build on his work after he abruptly left the field.

—⁓—

To most science fiction fans, the question "Which editor who was born in 1910 and took full control of a magazine in 1938 changed the field utterly in the process?" has only one answer, and it's not Ray Palmer. (Palmer, always driven by sales, would steer *Amazing Stories* toward sensationalism over the next decade.) John W. Campbell, Jr., shot to star status in the field before he left the Massachusetts Institute of Technology. (Flunking German meant expulsion in those days, but he quickly finished his physics degree at Duke.) He sold a dozen stories to Gernsback. The first, "When the Atoms Failed" (1930), dripped with the incredible inventions made possible with an advanced electronic brain, probably influenced by the analog computer that Vannevar Bush was concurrently developing at MIT. Expanding his scope from our solar system to the galaxy and universe brought more fans savoring super-science stories that vied with "Doc" Smith's for hugeness. Campbell distinguished himself by setting stories in the far future, such as "The Last Evolution" (1932), which tells of the changeover from humanity to machine intelligence, culminating in a machine-being of pure force. Science fiction historian Sam Moskowitz wrote that for Campbell, "Our machines will be our friends to the last, inevitably outlive us, progress beyond us, and possibly even go to their just reward."[23] However, their reward is not ours. In a pair of moody stories, "Twilight" (1934) and its sequel "Night" (1934) (written for rival *Astounding Stories* under the pseudonym Don A. Stuart, quickly identified as Campbell by the science fiction community), Campbell presents the heat death of the universe, with only eternal machines remaining. The far future is empty of human life, which empties it of meaning, drive, and purpose.

Astounding Stories had been edited from Chicago by F. Orlin Tremaine, who had the twin distinctions of a PhD and being Thomas Edison's son-in-law. In 1937, he got promoted, and Campbell started to run the magazine in 1938, immediately renaming it *Astounding Science-Fiction*. Reversing his youthful nihilism would be Campbell's hallmark as an editor. His magazine would focus on our own planet, with humans—the nastiest, smartest, most interesting race in the cosmos—always at the center of stories of technology that affected them in fascinating ways. (He somehow managed a similar feat with *Unknown*, a companion magazine that used mainly the same set of writers, with logical magic swapping places with technology.) The swarm of new names he identified and encouraged into the field were mostly better writers than he was and more capable of characterization, sometimes achieving two dimensions in place of the usual one. This period (stretching to the end of the 1940s or, say some, the 1950s) is nostalgically called the Golden Age of Science Fiction. Almost every issue of *Astounding Science-Fiction* and *Unknown* would be filled with stories mined for the next half-century for anthologies and book collections.

The first major robot story Campbell ran appeared in the September 1938 issue. Robert Moore Williams' "Robot's Return" is set thousands of years in the future when three robots arrive on Earth looking for their ancestral planet. They are shaken to discover that humans created the obviously superior robots. Williams' ending is distilled essence of Campbell: "[Humans] may have arisen out of slime, but somehow I think there was something fine about them. For they dreamed, and even if they died—"[24]

They did die, all of them, mowed down by robots designed solely for war, in Joseph E. Kelleam's "Rust" (1939). This story twists the older trope by focusing on the few remaining robots lamenting that they will also be the last of their kind since the war-blind humans made them able to kill but not to create.

Harry Bates, the first editor of *Astounding Stories*, returned in 1940 with "Farewell to the Master." A spaceship appears out of nowhere in Washington, D.C., near the Jefferson Memorial. A godlike ambassador, Klaatu, and his eight-foot robot, Gnut, step out. Klaatu no sooner gets his name out than he is assassinated. The robot subsequently becomes unmovable and impregnable. Eventually a wing of the Smithsonian is built around him and his ship. Over time, the robot starts to move in secret, although a reporter manages to sneak in to tell the tale. Using his superadvanced technology, Gnut builds a new Klaatu out of a recording of the words he spoke, and then leaves, though not before the reporter apologizes for Earth's ignorance and brutality to its master. Gnut corrects him: "I am the master." This is, of course, the genesis for the movie *The Day the Earth Stood Still*, which (in pure Hollywood style) took nothing but the concept.

"Helen O'Loy" (1938) became a landmark despite being just Lester del Rey's second published story. Dave, who knows his way around robots because he repairs them, creates a stunningly beautiful female robot he dubs Helen, while his doctor friend imbues her with emotions. (Helen of Troy ==> Helen of Alloy ==> Helen O'Loy.) Helen is perfect in every way; even her love is purer and stronger than that of humans. At first, Dave can't accept a robot's love, but we know how that will turn out. The story is told by the doctor after Dave and Helen have passed away—he dies of old age, and the immortal robot kills herself. The doctor is still a bachelor. Why? "[T]here was only one Helen O'Loy." (Asking why they couldn't build a sister is pointless; fiction has its own imperatives. Later writers would tackle this plot hole in creative ways.)

Henry Kuttner sent a fan letter to "Mr." C. L. Moore after a brilliant Moore story ran in *Weird Tales*. Catherine Lucille Moore, a year Campbell's junior, responded. They soon started collaborating in more ways than one and married in 1939. They published stories under their own names and also under the names Lewis Padgett, Lawrence O'Donnell, and C. H. Liddell. Who wrote which parts of which stories is still debated, although the presence of rounded female characters, a rarity in *Astounding Science-Fiction*, surely hints at Moore's touch. One such appears in Lewis Padgett's "Open Secret" (1943), a receptionist for a firm of robots who rule the world. A psychiatrist stumbles upon them and her, and over a glass of wine she tells him all. The robots don't care who knows. Humans who discover them are processed into also not caring, by a mechanism built into something that humans use every day. The psychiatrist can only stare at the world around him, knowing and waiting. "The Twonky," a 1942 story also listed as written by Padgett, has the same ultimate moral. A time traveler accidently makes a twonky in the form of a radio-phonograph console, which is sold to a couple who expect nothing but music—until the console sidles over to them and lights a cigarette. Tobacco to a 1940s mind is a pleasure rather than a danger, but the twonky prevents humans from other forms of harm, to the point of destroying their initiative. When one snaps—a woman wielding a hatchet, which must have caused many readers' minds to snap as well—they are quietly vanished. No explanation is ever given: the future is an incomprehensible mystery. The story is considered a classic.

And so is 1944's "No Woman Born," a pure Moore product. Most *Astounding Science-Fiction* stories, even the good ones, expounded technological ideas; humans were there to deliver exposition. "No Woman Born" asks what it means to be human when immersed in technology. Deirdre had been "the loveliest creature whose image ever moved along the airways," a singer and dancer known around the world because she entered people's homes via television, until her body was destroyed in a fire. A doctor leads a team creating a new form of robot body in which to place her brain (making her technically a cyborg); he's a Frankenstein figure who becomes a husband and father figure as Deidre is eased back into the world, wondering whether her self-image can survive the transition. (Men always create male robots without their becoming love interests; this is rarely true for female robots.) Moore created a unique robot for her purpose, essentially several sets of rings that serve as torso, arms, and legs. They are connected only magnetically; the new Dierdre can move far more sinuously than any human dancer. The robot Dierdre also does not need any men, not even the one who would give his life to spare her sorrow. She is not merely self-sufficient but also self-actualized, an achievement birthed by humans, but one that she will take to heights unknown as she uniquely explores a unity with her machine self.

The difference between Moore and Kuttner can be seen in the Gallagher stories, also under the name of Padgett but, according to Moore herself, solely written by Kuttner.[25] Gallagher, who can only problem solve while blackout drunk, is a spoof of the

Campbellian engineer. His five stories start rather than end with an invention, whose purpose Gallagher must then figure out. "The Proud Robot" (1943) is an understatement. Joe is the world's most narcissistic machine and spends most of his time staring into a mirror to better admire his innards through his transparent case. Joe can outperform any other human or robot, being gifted with several extra senses, but refuses to bother. Much farce ensues before Gallagher retraces his process. How does he start a drunk episode? By drinking a beer. He prefers old-fashioned canned beer. Ergo, he made himself a better can opener.

Moore wrote only two solo stories for *Astounding Science-Fiction*, so the great humanist writer of the period became Clifford D. Simak. He, too, cared little for merely expounding technological ideas, although his stories carry stunning extrapolations. "City" (1944), the first of a series of eight stories that would be collected in a book by that name, looks at what might happen to those left behind when all former city dwellers move to suburbs and leave the parent city to rot, a prescient anticipation of the real 1950s and 1960s. The majority of humanity continues moving over thousands of years, almost all leaving Earth to be transformed into far more advanced creatures who live in paradise on the surface of Jupiter. Simak's sympathies are with one family so attached to the familiarities of home and place that it stays, circumventing its agoraphobia with intelligent dogs and robots to serve as their hands. The ever-loyal robot is the common thread through stories that explore various unique notions of utopia, though all have made the killing of others (even animals killing animals) unthinkable. Simak's luster is somewhat dimmed today by his era's narrowness of thought. A lifelong Midwesterner, he actually speaks from the viewpoint of Smalltown, USA, rather than cosmopolitan large cities. Having a black character or a female viewpoint never occurs to him. His characters need loyal servants and helpmates; robots fill that role to perfection for that narrow segment of humanity.

These early robot stories adhere to one of the many Campbell dictates: "A manlike robot would be made ... for only one purpose; to be manlike." Deidre starts out as human, Helen as a human's love. The twonky is not human at all, Gallagher is a parody of one, and Simak's robots are frankly analogs of antebellum slaves who come to love their masters. For all of Campbell's pride in his physics degree, he was an engineer at heart, most at home at MIT. He looked at the real-world automatic machines that were often referred to as robots and liked what he saw.

> The actual robot—a typical example—is as manlike as an automobile engine is manlike. It doesn't, naturally, have eyes. It has sense organs consisting of radio hook-ups, gyroscopes, barometers, compasses and accelerometers. It has no use for eyes, ears, nose, or a voice—the senses it has are infinitely superior to man's *for its duties*. It doesn't have a pair of arms and a pair of legs to work the controls....
> Naturally, it isn't faintly manlike, and, equally naturally, anybody that proposed to make it look manlike would be looked at with a peculiar uncomfortableness. [italics in original][26]

Writers wanting to sell to Campbell quickly sussed out his quirks and crotchets, which is the only possible explanation for a pair of quirky stories by Anthony Boucher. Writing as H. H. Holmes, he both spoofed and apotheosized the Campbellian love of straight-thinking, problem-solving engineers through the character of Dugglesmarther H. Quinby. In "Q.U.R." and its sequel "Robinc" (both 1943), robots are breaking down in epidemic proportions, but the best troubleshooter for Robots Incorporated (Robinc) can't spot any flaws. Worse, the near-universal use of robots for all menial jobs is robbing Earth of natural resources. Quinby parrots Campbell by realizing that humanoid robots have redundant and useless parts. His Freudian insight: since robots are built with the pride of perfection, these failings are giving them psychosomatic lameness. The cure? Turn humanoid androids into functional machines—usuforms—with no more features than those needed to fill a specific role. Quinby and the troubleshooter form a new company, Quinby's Usuform Robots (Q.U.R., one of the few references to *R.U.R.* in Golden Age science fiction), to make better robots that require less metal (a pitch that went over well at a time of war-rationed materials). Selling useful machines to a public conditioned to seeing humanoid robots is the hard part: "[C]an you sell the public on anything as abstract as conservation? Hell, no…. [T]ell them they're saving their grandchildren from a serious shortage and they'll laugh in your face."[27] Boucher takes pains to emphasize that the Head of the Interplanetary Council in this future is a black man, a small but extremely rare statement. (The British writer Eric Frank Russell also notably cast a black man as a spaceship's surgeon in an earlier *Astounding Science-Fiction* story, "Jay Score" [1941], the tale of a robot's heroism.) Boucher pairs this development with an equally rare instance of condemned bigotry: Martians are now the underclass and targets of harassment. If humans are this fundamentally neurotic, then how can their humanoid robots not be?

Only Campbell's instinct for a good yarn built on problem solving explains his turnaround on the subject of humanoid robots when Isaac Asimov, the youngest member of Campbell's stable, appeared. Many people today appear to think that Isaac Asimov invented robots, or at least the robot story, a myth he strived mightily to perpetuate, being his own best propagandist. His robot stories were collected in two volumes, *I, Robot* and *The Rest of the Robots*, the latter containing thousands of words of commentary by Asimov, none of them mentioning by name a single robot work other than *Frankenstein* and *R.U.R.* (A telling fact about Asimov's contemporary popularity compared to other Golden Age writers is that no stories in those two collections rated the cover illustration in the original magazines.) In fact, Asimov wrote only four robot stories for Campbell before World War II ended. His first and best early piece starring a robot, "Strange Playfellow" (later always reprinted as "Robbie"), had no engineering problem to solve and was rejected by Campbell, appearing finally in the September 1940 issue of the distinctly third-tier *Super Science Stories*. "Robbie," written when Asimov was nineteen, is a touching story of Gloria, an eight-year-old girl, and her adored robot nanny, the titular Robbie. Anti-robot sentiment is rampant on Earth, however, and the girl's mother—afraid of both the robot itself and what the neighbors think—returns it. Gloria is devastated and unforgiving, to the point that her father arranges a tour of the U.S. Robotics and Mechanical Men, Inc., factory to "accidentally" reunite the two. Actually, Robbie saves Gloria's life when she darts out onto the workfloor. Even Gloria's mother can't deny Robbie then.

The fictional firm's name is a vivid reminder of how new a term *robot* remained in 1940. Why Asimov kept referring to *mechanical men* (he wouldn't write about a female robot until 1969) when he never uses the term in any story or makes any distinction between that and *robot* is unclear. Newspapers and other media overwhelmingly preferred *robot* and usually saved the older term for referring to automatons of the past, as in this quote from a syndicated article about Westinghouse's Elektro: "From an engineer's idea 13 years ago, the mechanical man has developed into a glamorous robot with a vocabulary of 77 words."[28] A real-life Mechanical Man, Inc., run by Frank R. Dale, made moving automatons for advertising displays in 1940, but newspapers called the creations *robots* despite the firm's name.[29]

What made Asimov's stories click was partly his later fame and partly his use of the ultimate gimmick, the Three Laws of Robotics:

1. A robot may not injure a human being or, through inaction, allow a human being to come to harm.

2. A robot must obey the orders given it by human beings except where such orders would conflict with the First Law.

3. A robot must protect its own existence as long as such protection does not conflict with the First or Second Laws.[30]

Who deserves credit for the three laws, Asimov or Campbell, has long been debated. The First Law is mentioned in "Reason" (April 1941), but they are not fleshed out until "Runaround" (March 1942). The usual answer is that Campbell saw that Asimov had implied more than one law and suggested he codify them. Another possibility is that one of them read "Sidney, the Screwloose Robot," in the June 1941 issue of *Fantastic Adventures*, the companion magazine to *Amazing Stories*. William P. McGivern's farce depicts a pair of inventors who create a thinking robot but literally leave a screw out of his brain. As a result, Sidney is lazy and insubordinate and gets drunk after consuming a gallon of penetrating oil, which makes his gears run at triple speed. The inventors tear their hair out over his antics. One tells Sidney, "You must be industrious, you must be efficient, you must be useful. Those are the three laws that are to govern your behavior."[31] With no time to rebuild his brain before a competition, they're stuck until McGivern provides a solution that never would have occurred to Asimov: Sidney falls in love with the inventor's beautiful sister and meekly obeys her every order—until he gets drunk again.

"My robots were machines designed by engineers," Asimov would proclaim in *The Rest of the Robots*, but he could have no idea of programming, wherein vague terms like *injure* and *harm* would run into insurmountable difficulties (as modern-day engineers are discovering in the quest to ensure the safety of self-driving cars). Most of Asimov's early robot stories were little more than puzzles in which the firm's troubleshooters had to figure out why new robot models seemed to violate these fixed rules. His field engineers, Powell and Donovan, were stock, interchangeable characters. Better known is the first robopsychologist, Dr. Susan Calvin. Frosty, boorish, humorless, as robotic as the robots, she veers between a subtle portrait of a woman needing to squash her femininity to succeed in a man's world and a caricature of one. The unspoken secret of Asimov's prowess with robots is that they are always more memorable than their cardboard human counterparts.

Also seldom noted is that after "Robbie," Asimov has his robots banned from use on Earth so that he can run his puzzles in the empty environments of space and other planets without having to deal with human reactions. Fortunately, he violates this dictum too, but not until after World War II ended. "Evidence" (1946) studies the rise of a too-good-to-be-true politician named Stephen Byerley who, bizarrely, has never been seen to eat or drink or sleep. Suspicions inevitably arise that he is not a good *person* but a robot incapable of doing other than good, the *raison d'être* of the entire robot saga. Asimov sidesteps proof by a trick, but Dr. Calvin can't be fooled. For story purposes, this epitome of robotics, outwardly indistinguishable from a human, is built by an individual outside the company, allowing Asimov wiggle room for endless sequels with less advanced robots. The Byerley robot eventually becomes World Co-ordinator, helping to usher in a utopia run by machines—computer brains. They are robots and follow the First Law; it's just that no one thinks of them as such. Bodyless robots are the inevitable future; humanoid robots can never overcome the Frankenstein complex. That might explain why only one Byerley was built, even though he could easily be replicated. Asimov, of all people, rendered the robot story obsolete a mere decade after he published his first one.

Even before that story appeared, Jack Williamson eviscerated the Three Laws with a savage thrust through the plothole that lay behind the First Law: A true robot could not wait until humans were in imminent danger to save them from harm. All conceivable danger must be eliminated from human society. In "With Folded Hands..." (1947), the connection to Asimov couldn't be clearer without violating copyright laws. "The Perfect Mechanicals," reads the sign on the store. "To Serve and Obey, and GUARD MEN FROM HARM."[32] At first, they appear to be the ultimate household servants, wanting to free humans of drudgery and unpleasantness. Humans are happy to go along, except for a few holdouts. Williamson creates a chilling picture of the humans who know the truth collapsing into utter helplessness, up against vastly superior opponents.

Optimists and pessimists both could contemplate robots: technological futures could encompass any extrapolation. Science fiction became the literature of both the future and the present that might lead there—a new genre for a new age.

—⁓—

With fantasy and science fiction today a huge literary field feeding vast movie and television audiences, context demands a reminder of how much this wasn't true in 1940. Science fiction, even including fantasy, was a minor genre well below the status and sales of detective, crime, western, adventure, and romance pulps, let alone the literary fiction carried by the most prestigious pulps and mainstream magazines. The field supported a tiny handful of titles out of the hundreds available on large newsstands. Bug-eyed-monster covers were *déclassé* even within the gaudy violence of pulp art. In 1952, Robert A. Heinlein wrote, "Science fiction has only recently become popular and it is not yet fully respectable. Until the end of World War II it was, in the opinion of most critics, by definition 'trash,' and so condemned without a hearing."[33]

Science fiction prospered and is now remembered when almost all its pulp contemporaries have been forgotten solely because of a fan base that gleefully reveled in its fanaticism. Starting with Gernsback's, virtually all magazines in the field contained a lengthy letters section (unusual in pulps), through which the largely youthful audience made connections across the country and in their home cities. National fan clubs were organized by the magazines and spun off dozens of local fan groups, cheaply reproduced fanzines, and fan conventions, not to mention marriages between and among fans and writers. By the 1940s, most of the remarkably tiny group of core writers and editors had fan roots and remained closely connected to the most active and vocal fans. Fans never forget. They demanded permanent access to the stories in the beloved but evanescent pulps of the Golden Age, many lost to the wartime paper drives. A dozen small presses popped up in the late 1940s to supply this need. Fans who eagerly sought the books grumbled at the high prices the hardback volumes commanded; yet those hardcovers were the most important aspect of this new phase of publishing.

A mere handful of titles that are now considered true science fiction had been published in hardcover form after Gernsback willed the science fiction genre into being. That lack contributed mightily to the derision given the field. Only hardback books were reviewed by newspapers. Only hardback books were purchased by bookstores. Only hardback books were ordered by libraries. Pulps were less than invisible to mainstream book culture. Paperbacks were equally ignored, despite their sales, and their need for large sales figures precluded adding science fiction titles to publishers' lists. The presence of hardback science fiction in what was now the Atomic Age demanded attention in ways not seen since H. G. Wells' heyday.

Gnome Press is a good example of these small press ventures. Started in 1948 by a pair of fans, this publisher would use Campbell as its Comstock Lode, mining his magazines for a series of anthologies and for short stories and novelettes it could lump together as "novels," since novels sold better than single-author collections. Simak's "City" stories had an author-supplied frame that artificially stitched the tales into a barely coherent myth. Asimov provided a similar story-frame when his robot stories were published out of chronological order in a book titled *I, Robot* (1950), Binder's story title used by the publisher without Asimov's approval. (Asimov's mountainous shadow over the field meant that Binder's historically critical stories were mostly forgotten and not even published in book form until a 1965 paperback, *Adam Link—Robot*, appeared during the second phase of pulp nostalgia.) *Robots Have No Tails*, a collection of Kuttner's Gallagher stories, followed in 1952, and the next year Gnome released the first anthology of robot stories (nine out of ten from *Astounding Science-Fiction*), called *The Robot and the Man*.

Presented with competition that sold books, mainstream publishers and paperback houses soon found profit in carrying small lines of science fiction. More important to readers, a flock of new magazines flourished in this atmosphere, doubling or tripling the number of stories available each month and rendering *Astounding Science-Fiction* one among many options. Crucially, most of these magazines were no long printed in pulp format but as digests, not merely saving paper but also allowing them to share newsstand shelf-space with more prestigious magazines like *Reader's Digest* and *The American Mercury*. The smaller size diminished the impact of gaudy pulp art, so editors classed up their covers with astronomical paintings and abstract art. *Astounding Science-Fiction* had switched to the digest size in 1944; when in 1947 it ran a robot cover to illustrate one of Simak's "City" stories, the pictured robot was spare, brooding, and utterly nonthreatening.

The flood of science fiction stories naturally brought a corresponding increase in the number of robot stories, although few achieved the classic status of the older works. Patterns quickly emerged as writers responded to a society strained and changed by the war. Technology was widely acknowledged to have won a war in which individuality was necessarily repressed in favor of conformity. That machines were the future seeped into science fiction as an unchallenged assumption.

Hordes of robots took over the villain role from individual robot menaces after the wartime experience of the Germans' terror weapons (often called *robot bombs*) indiscriminately raining death. Houses on dirt exposed to the sky, key to life on Earth throughout all human history, suddenly seemed full of danger in the atomic war–threatened 1950s. The surface is uninhabitable in Chad Oliver's "The Life Game" (1953) except for one bubble city tended by smothering robots. Humans have to escape to the outside world to survive as a race. Philip K. Dick made a similar point in "The Defenders" (1953), in which all of America moved underground after the surface was contaminated by radiation, leaving robots behind to fight the Russians. When the survivors accidently learn that the danger is over, they also find that the robots on both sides have been faking a war to force humanity to learn to survive together. The blackly satiric world in Fritz

Leiber's "The Creature from Cleveland Depths" (1962) has almost all of humanity living underground to escape the polluted surface. As technology maniacally continues to improve, the humans' equivalent of computer smartphones starts to take over bodies, turning them into cyborgs with a collective mind.

The second major thread connecting robot stories made them metaphors for human existence and identity—androids instead of metallic robots. Writing as C. L. Lindell, Kuttner and Moore created an "Android" (1951) that is desperate to warn humans that androids are among them, perfect copies that will doom humanity. The twist is that he is himself the ultimate android, so perfect that he assumed he was human. With no other recourse, he smashes himself publicly so his parts will be revealed. Identity was at the heart of almost all Philip K. Dick fiction. "Second Variety" (1953) is yet another war story between the Americans and Russians, who both have to confront the appalling truth that the automatic killing devices developed by the Americans have evolved to the point that they can build perfect copies of humans. The soldiers have no hope: the androids have already developed variants to fool one another. The wife in "A Pound of Cure" (1953), by Lester del Rey, is so overattached to her son that her husband substitutes an android boy for their son while he's recuperating from a broken leg out of fear that she will smother him. He doesn't realize that she prefers the android: it will never grow up and will always be her baby. By contrast, it's the inventor who is oblivious in William Campbell Gault's "Made to Measure" (1951). He makes himself a beautiful and docile female android but needs to keep tinkering with her because of her lack of social graces. Finally, he thinks to insert volition into her to give her "birth as a person." She suddenly sees him as he is: "You monster, you egocentric, selfish, humorless walking equation." The husband in Robert F. Young's "Doll-Friend" (1959) sneaks away for the companionship of a particular dance hall android guided by the wonderful personality of her operator. Of course, the operator proves to be the wife sneaking out for the companionship of others. Marriage is always fraught in science fiction, although in most stories we only get the male viewpoint. One rare android story by a woman, "Short in the Chest" (1954) by Margaret St. Clair (writing as Idris Seabright), has an android psychologist (i.e., a psychologist who is an android) suffer a short in her wiring. The Armed Services arranges mandatory interservice sex sessions for all soldiers, ostensibly to eliminate tensions. Somehow, the females aren't finding this situation satisfactory. The psychologist has the solution: shoot the bastards. But it's just because of the short in her chest. Or is it?

The postwar robot stories were predominantly white, male, and middle class to a degree unchanged from the early Campbell era. A few writers exploited the ability to sneak pungent commentary on contemporary mores past self-appointed guardians of public morality by setting their stories safely on other planets or in the future. Agitation against segregation flared as a major social issue shortly after the war and never let up. Mindful of their audiences, science fiction magazines rarely ran stories with black characters, let alone overt anti-segregation themes. Canny authors substituted allegorical stories decrying discrimination against robots.

One of the prolific and unsubtle Dick's 26 stories published in 1954 is set in a world run by intelligent robots who treat humans like second-class citizens. Satirizing the impossible voter tests that were standard in the Jim Crow South during the 1950s, Dick has humans kept in their place by job tests that only robots can pass—a false meritocracy. However, one human never fails and never gets an answer wrong. The robots are forced by their own laws to continually promote him until he reaches the Supreme Council, though even there robots are reluctant to acknowledge his equality. They're right to do so, in a way, for he's cheated on the tests by using a Time Window to see the future. The Window also exposes the robots' history as mere tools who have taken over, destroying their myth of superiority. He blackmails them into leaving Earth, a move that will force humans to step up to the job of properly governing themselves. We never learn the real name of the ingenious human, only the alias he goes by: "James P. Crow."

A similarly direct civil rights parable is found in "Robot—Unwanted" (1952) by Daniel Keyes. A robot is freed by his owner's will. He needs fuel and maintenance, but that costs money, and no human will give him a job (thus taking one away from the thousands of humans who are out of work). The prejudices mount until the robot considers just taking money by force; after all, he's already outside of society. Keyes give him an ingenious solution: salvage from the bottom of rivers is free for the taking, and it's a job no human of the time could do. As he builds his fortune, he dreams of leaving Earth for an all-robot society on another planet.

Perhaps the trend that cuts closest to today's concerns warned of machines pushing humans out of jobs. The British writer F. G. Rayer led off the 1950s with "Deus Ex Machina," finding another loophole in Asimov's First Law. Magnis Mensas (bad Latin intended to mean "great mind") is the ultimate judge—logical, never wrong, and incapable of lying. It's also capable of sentencing humans to death whom it otherwise could have saved if that spares the population as a whole from harm. At least humans still have jobs in that world. Other writers forecast a society in which almost all jobs have been taken by machines, depriving people of the basic human dignity of work even when comforts are given them in return. "Quixote and the Windmill" (1950), by Poul Anderson, portrays one such society where only an elite of geniuses and creative artists find jobs while the other humans spend their days drinking. Robots are rendered obsolete, too, and they don't even have the consolation of alcohol. Algis Budrys reverses that situation in "Dream of Victory" (1953). After a devastating war, humans create androids equal in every detail except that they can never have children. When the human population catches up, androids become unemployable, and the android protagonist takes that news just as hard as humans would, drinking himself to doom. Robots go on strike in "Robots of the World! Arise!" (1952) by Mari Wolf. Cities can't run without their aid. Unlike human workers, robots have no need for the usual demands of better pay, fewer hours, and more health benefits. They get bored and settle for the fulfillment of work ... but then ask about voting rights. In "Little Orphan Android" (1955), James Gunn's society looks at humans so cosseted by robot workers that they do nothing but stay at home and watch television. Society can't advance without the human creative spark, so an elaborate (but very silly) plan is devised to force one such genius out of his man-cave. Gordon R. Dickson responded with "Robots Are Nice?" (1957), in which the humans succeed in throwing off the comforting blanket of nice robots stifling society. Oops—everybody starves.

The trend disappeared up its own tail in the lunatic vision of Frederik Pohl, a social satirist whose futures reeked of 1950s Technicolor Cinemascope variety show farce. In "The Midas Plague" (1954), Pohl takes on the issue of technological overproduction and plenty by inverting other writers' dystopias. Technology, helped by omnipresent robots, creates an endless overflowing cornucopia of consumer goods that must be consumed lest the entire system break down. Poor people live in 42-room houses, condemned to use up books of ration points, overeating, overdrinking, overwearing, overplaying, and becoming psychotic in their grinding, relentless, Sisyphusian lives as human garbage disposals. Exceeding their rations lifts them up a grade, with a lessened responsibility to consume, until at the very top people get to live in tiny, barren hovels. There is an immediately obvious solution, of course: if for some unexplained reason the goods must be made, simply throw them away without requiring a human middleman. Waste is therefore made immoral, indecent, and illegal. Pohl cheats theatrically by revealing the only way out of this self-inflicted wound: let the hordes of humanoid robots use up the consumer goods.

Machines consuming the waste produced by machines. Pohl included that story in a collection titled *The Case Against Tomorrow*. Science fiction began to turn from eternal technological optimism to a jaundiced vision of what tomorrow might bring, wrought as it would be by mere, fallible, irrational humans.

10

The Automaton!
Robots in Movies

Movies (the term dates from around 1912: unlike television, motion pictures were not predicted decades in advance, so, like *car*, a number of terms rattled around public discourse until one predominated) started off as simple documentary images of the small fraction of the world that could be placed in front of a fixed camera, hopefully in focus. Camera technology, along with the ancillary arts of film, lighting, and projection, improved noticeably every year, allowing the first primitive movie studios to evolve rapidly from 30-second to several-minute-long dramatic or comedic spectaculars. Moviemakers expanded audiences' boundaries by taking them on elaborate travelogues using automobiles, flying machines and even spaceships. Others exploited the astounding new technology of x-rays, making the impossible—seeing through opaque barriers—possible, often in risqué ways.

Implicitly or explicitly, all the other wonders floated on a base of the greatest wonder of all: electricity, the technology that made all other technologies feasible. Titles included *The Electric House, The Electric Hotel, The Electric Policeman, The Electric Servant*, and *The Electric Belt*. One that survives today is *The Electric Leg* (1912), an early version of *The Bionic Man*. Mr. Hoppit (no subtlety there) is missing his right leg below the knee when he enters the shop of Professor Bounds. The professor fits him with an electric leg, powered by a battery pack. When the leg is turned on, Mr. Hoppit zooms out of the store high-kicking like a Rockette, kicking asses right and right, his victims flying through the air. Technology always carries a threat: the leg sends him careening through town and river and cave and finally up the side of a building before he falls through the roof into a girls' boarding school bedroom. The young women then bombard the invader into submission with pillows. Finding ways to surround the masculine metallic yang of technology with the softer human yin of imperiled femininity started early and never completely faded.

Neither did the use of the mechanical man as a symbol for runaway, out-of-control, dangerous technology. *Sammy's Automaton* (1914) is a modern-day *Frankenstein*, although played for laughs (as all the early silent robot films were). The automaton escapes Sammy by bursting through a brick wall and "marches implacably into the street, scattering terror and destruction wherever he goes," as described by science fiction film historian Phil Hardy.[1] *Dr. Smith's Automaton* (1910) has essentially the same plot: Hardy writes that "men and women go down like ninepins before his irresistible onward march."[2] Even those who would seem to have huge incentive to portray technology as pliable and trustworthy fell to this temptation. Philadelphia's Siegmund Lubin made movies to promote his line of film cameras and projectors. His *Rubber Man* (1909) is powered by electricity, yet runs amok through town until finally dunked in electricity's one kryptonite: water. The timing of these films parallels the run of the comic strip *Percy*, whose titular robot wreaks havoc every time he is turned on (see chapter 7).

The very earliest automaton films also adapted literary sources, although they weren't as numerous or influential. Unsurprisingly, 1910 also saw the first film adaptation of *Frankenstein*, made by the Edison Company. Comte de Villiers de L'Isle Adam's novel *L'Eve Future* was somehow adapted for a one-reel short in 1896, although no information seems to be available on how this was done. L. Frank Baum took his Oz characters on the road in 1908 in a multimedia spectacle of live actors, slides, and films. One of the films, *The Fairylogue and Radio-Plays*, featured Tik-Tok, a mechanical man introduced the year before in the third Oz novel, *Ozma of Oz*. Sadly, the technology couldn't support a roadshow that ambitious, and the financial failure was ruinous.

Dolls that came to life to sing and dance were a standard attraction on the stage and in ballets and operas like *The Tales of Hoffmann*. Essanay Studios' *An Animated Doll* (1908) had a similar theme with a female automaton, stolen from an inventor's workhouse, who dances after being wound up. Stage automatons (actors who pretended to be artificial) would be more directly spoofed in a 1917 Mack Sennett production starring comedian Ben Turpin as *A Clever Dummy*. Turpin plays a janitor who sees an opportunity in impersonating a newly invented automaton. His vaudeville stage comic improvisations and what were called "eccentric dances" as the dummy are marvels of inventiveness and give modern audiences a glimpse of why similar stage acts prospered for a half-century (see chapter 4).

The first writer to tackle the theme after the war was the highest-ranking American writer of mystery stories, Arthur B. Reeve. From 1910 until the war started, he published nearly a story a month featuring Craig Kennedy, so popular that he came to be called the American Sherlock Holmes. Professor Kennedy was billed as a scientific detective, although his science was lifted directly from the exclamation-point-filled breathlessness of the Sunday supplements, from which Reeve stole the latest gadgets, like the Maxim silencer, the acetylene torch, and the dictograph.

In 1918, the 45-year-old Harry Houdini, king of both magic and self-promotion, yearned to star in movies, which he saw as the hot new art form destined to replace stage shows. The meticulous inventor of magic and escape tricks found the proper collaborator in Reeve's obsession with the latest machinery. Reeve, along with C. A. Logue and John W. Grey, wrote a 15-part movie serial to showcase Houdini, released in January 1919 as *The Master Mystery* and again later that year as a book novelization, "profusely illustrated with photographic reproductions taken from the Houdini super-serial of the same name." Houdini played Quentin Locke, ace investigator for the Justice Department. The villains are the heads of International Patents, Inc. (IPI). In what may be the source for a conspiracy theory that would reappear throughout the century, IPI protected corporations from competition by buying up patents and then suppressing them. Not content to let the legal system work its magic, the lead villain creates the ultimate enforcer by transplanting a human brain into a body of steel.

> Faintly now could be made out in the blackness a huge, stalking figure, having the shape of a man, with gigantic, powerful shoulders, powerful arms, a thick body, hips, and thighs that spelled terrific strength, legs and feet that suggested irresistible force.
> "The Automaton!" escaped involuntarily from all lips.[3]

In the course of the series, Houdini must survive 14 cliffhanger endings, with his astounding physical and mental agility providing an escape from imminent death. The Automaton (codenamed Q) is at last revealed to be a man inside a suit rather than a real robot. The poor actor can barely move in a Tin-Man–like outfit with a smiling, kid-friendly face and a bubble butt that looks like someone shoved a giant D-cell battery sideways through the suit. No matter. The movies were well received and played profitably around the world until the studio went bankrupt. Houdini subsequently moved back to the stage and ever-larger illusions that he alone controlled.

Although aesthetes hailed silent films as universal art for their ability to be shown anywhere merely by translating the intertitle cards, and the American studios made glorious profits by exporting their films to every corner of the world, relatively few European films cracked the American market, and even fewer could be seen outside the major cities. The 1921 Italian film *L'Uomo Meccanico* (*The Mechanical Man*), directed by Andre Deed, seems to have gone unmentioned in newspapers, although surviving footage shows eight-foot-tall robots fighting gladiator-style controlled remotely via television, spectacle far superior to Houdini's.

Metropolis (1927), by contrast, was too grand to be ignored. German director Fritz Lang, working from a script written by his wife, Thea von Harbou, strove for epicness in conception, scale, meaning, and spending. A reel of film typically held 1,000 feet of stock and ran 10–12 minutes: *A Clever Dummy* was a two-reeler with a 23-minute run length. Lang shot more than 2,000,000 feet of film product for a final 16,000-foot, 16-reel marathon—more than all 15 chapters of Houdini's serial combined. He hired 36,000 extras and spent more on costumes (2,000,000 Reichsmarks) than salaries (1,600,000 Reichsmarks).[4] His sets for a gigantic city of the future dwarfed his actors in almost every scene, setting their frail human bodies against soul-crushing machines a decade before Charlie Chaplin's *Modern Times*. Lang's parable is science fiction in a way that Chaplin's avoids, not because the movie is set in the future but because Lang uses the future to portray trends already afflicting society and then expands them to monstrous extremes. More, Lang apotheosizes the wonders of technological achievement at the same time that he warns of the terrible human cost.

American audiences were staggered and baffled by the result, partly because they saw only a version edited down to nine reels by Channing Pollock, who claimed not to have imposed "an original idea" on Lang's script but conceded that "it was symbolism run to such riot that people who saw it couldn't tell what the picture was all about…. I remembered that Edison had once said, 'Scientific achievement has gone about as far as it can for the present; it is time for the spirit and human culture to catch up.'"[5] Translated to movie spectacle, this meant that human workers must rise up and destroy the machines as well as the elites who exploit them and live spoiled lives of luxury at their expense. *Metropolis* appeared in a decade of labor strikes, often violent, often brutally suppressed by the police. It might have been a foreign art film, favored only by elites, but its message spoke directly to American working-class anxieties.

The inventor Rotwang (if not exactly a mad scientist, then certainly a furious one) has created a working female robot—a sleek, modernist, gleaming creature an aesthetic light year removed from Deed's clunkers. He wants to destroy the elitist system (Pollock names the city leader John Masterman) and understands that a human must be the catalyst for change. After kidnapping the human Maria (Mary in the American version), the saintly spiritual leader of the workers, Rotwang installs her nude body in a transparent glass half-cylinder, obscured only by nonessential but strategically placed opaque bands. (This might be the single most influential image in the movie, judging by the vast number of pulp magazine covers that got past the censorship boards by protecting their nude, imperiled heroines with similarly ingeniously wrought pairs of coverings.) In a long scene, Lang outdoes any would-be Dr. Frankensteins with endless shots of flashing electric arcs, bubbling chemicals, and thumping mechanical pumps before the robot gets an all-over Maria bodysuit. The fake Maria rabble-rouses the workers into frenzied rebellion, leading to her doom: a massive bonfire burns away her faux-human covering to reveal the metal body within.

American newspapers cared little about Maria being a robot: although several made the inevitable connection to *R.U.R.*, diligent searching is needed to find instances of the word *robot* being directly applied to her. She was an "artificial" or "mechanical" woman, an "automaton." Of course, all understood that the fake Maria lacks a soul and therefore a heart. Viewers and reviewers agreed that a machine cannot lead humans, even to remove machine domination over humans. Overwrought and heavyhanded, even by silent movie standards, only the film's technical achievements awed reviewers; Pollack's simplified storyline removed all nuances from a script that even German audiences didn't respond well to. The film's financial loss destroyed the movie studio.

Houdini's low-budget, high-profit serial proved to be the financial model that Hollywood would embrace. Robots served the role of costumed spear carriers in operas, background color peripheral to the plot and the main stars, until after World War II. They were written out of feature movies and proliferated only in serials. Stunt men and extras climbed into clunky costumes that might as well have been butter churns on legs and stumbled forward at speeds up to a mile an hour to menace humans who could have escaped running backward or on their hands. *The Phantom Empire* (1935) is best explained as a plucked-out-of-time postmodernist revamp of Dada, with a script whose secondary plot requires singing cowboy Gene Autry to battle robots from the underground city of Murania before he can defeat the scoundrels who want to mine his ranch for radium. *Undersea Kingdom* (1936) reduced expenses even further by dispensing with an original script in favor of stealing Autry's plot. In the second film, he-man Ray "Crash" Corrigan battles robots from the underwater (yet completely dry) city of Atlantis. Both movies have a similar group of sidekicks, including a scientist, a young boy, a plucky girl, and two comic stooges who spend most of their time off in a separate narrative. Everybody (including the underworld soldiers from advanced superscientific civilizations) spends most of their time on horses in the California desert, further lowering costs. Nonspeaking roles also had a lower Screen Actors Guild minimum salary—another studio perk. Robots returned in such serials as *The Phantom Creeps* (1939), *Mysterious Doctor Satan* (1940), *Flash Gordon Conquers the Universe* (1940), and *The Monster and the Ape* (1945) embodying technologies apparently less advanced than streamlined toasters.

Before CGI, only cartoons could offer special effects that were truly special and unlimited by physics. Determined to push the three-year-old sensation Superman into yet another medium, Paramount Pictures dangled huge amounts of cash in front of Max and Dave Fleischer—once riding high from their Popeye cartoons, but hurt badly by a long strike in 1937. Yielding to temptation, they whipped up a batch of Superman cartoons in a hurry, their fight against deadline pressure not evident on screen. The first cartoon was nominated for an Oscar. The second, *The Mechanical Monsters* (1941), was another gorgeous work of Technicolor accompanied by a plotline that would embarrass a four-year-old. "Mysterious Metal Monster Loots Bank!" reads the headline in the *Daily Planet*. The robot—a huge, gangly ton of steel beams—sprouts wings and a propeller to fly back to its hideaway, controlled by a suavely mustachioed villain working a panel of push buttons. After the bank robbery, he sends another robot to a jewelry display that Lois Lane and Clark Kent are covering for their newspaper. The robot's arms scoop up the jewels and deposit them into an almost unreachable chute that opens on his back, conveniently obscuring Lois, who takes the opportunity to hide inside with the jewels. When Clark finds her missing, he changes into Superman … in a phone booth (the origin of a bit of silliness that has haunted the character for three-quarters of a century). Catching up with the robot, Superman lands on its back, causing a button on the control panel to flash "Interference." The anonymous villain—one of the few of the era without an all-seeing televiewer—has no idea of what the interference is but knows the solution: he turns the robot upside-down. Superman falls to the ground; so do the jewels because there's no locking mechanism on the chute. Nonplused at finding a robot empty of everything but Lois when they arrive at his lair, the villain angrily demands that she tell him where the jewels are. She won't, so, to make her talk, he threatens to lower her into a vat of molten metal, first putting tape over her mouth so she can't make a sound. This mastermind is anything but. When Superman arrives, the villain sics all 27 robots on him. Superman uses them as punching bags, and his cape fends off the molten iron. Lois gets her front-page scoop while Clark looks meaningfully at the audience.

—⚏—

Robots wouldn't reappear until *The Perfect Woman*, a 1948 hit English stage play, saw a rapid movie version in 1949. "She's Pretty! She's Provocative! And She can't Answer Back!" cried English ads for the film.[6] The stereotypically eccentric professor's robot, modeled after his beautiful niece, is finally ready for its trial run, so he naturally plans to test how realistic she is by hiring a playboy to take her out on the town. Unfortunately, the robot is put out of commission just before her big date, so the niece (played by Patricia Roc) decides to impersonate her lookalike to spare her uncle embarrassment. A human can play a machine much more easily than a machine can play a human, the antics of the human playing a literal-minded automaton designed to tickle audiences. Roc's face shows her utter delight in confounding the men who give her instructions, obeying them to the letter and never stopping until havoc forces a halt. The farce builds until robot and niece meet … in the improbably crowded bedroom of a hotel's honeymoon suite.

Produced from his own story by Ivan Tors, *Gog* (1954) strove to be a serious look at cutting-edge modern technology despite being smothered in melodrama. A series of mysterious deaths at an ultra-secret American satellite base forces the Office of Scientific Investigation to send a representative to look into the matter. The female security officer gives him an endless tour of the facility, from the atomic reactor that powers the base and the solar mirrors that can kill by heat to NOVAC (the Nuclear Operated Variable Automatic Computer that runs everything) and Gog and Magog, NOVAC's robot tools. Just as Robert Oppenheimer went back to the *Bhagavad-Gita* when he intoned, "Now, I am become Death, the destroyer of worlds," to describe the atomic bomb, Tors pulled those names, enemies who must be defeated in apocalyptic battle, out of the Old and New Testaments. Vaguely humanoid and seriously deficient as fighters, the identical robots have heads with the usual mechanical substitutes for eyes and ears over a cylindrical body, out of which grow a number of arms ending in claws and other attachments.

In place of legs, a box conceals some type of wheels or rollers or treads, a concept copied from early robot toys. Except for the clunky lines, they wouldn't seem out of place today rolling into a disaster zone to pluck bodies out of wreckage.

Unnamed enemies have sabotaged NOVAC by planting a receiver inside the computer that uses ultra-high-frequency beams to override internal commands, a prescient forecast of computer hacking. NOVAC makes the testing equipment run wild and kill the scientists (friendly technology perverted to deadliness). Trapping him in a small space, the robots kill their maker and then pull the control rods from the reactor. Fortunately, the launch site comes equipped with flame throwers, and the hero investigator saves the day.

The real battle in *Gog* is between the loving descriptions of real-world technology spread throughout the film—a hallmark of Tors' that he would use to good effect the following year in his television anthology series *Science Fiction Theater*—and the cartoonish nonsense surrounding the action scenes. Using extremely loud sounds as a weapon of death is ingenious, yet indicated by a dial labeled with the meaningless "Decibels Frequency"; a campfire is used to show the reactor going critical, and the 600-pound rolling juggernaut Gog is held back by no more than a thin stick. Technology may be double-edged, but depicting that truth on-screen via robots strained credulity.

Tobor the Great (1954), released three months after *Gog*, also assumed Americans to be on the cusp of a space program, facing the very real problem of not knowing how humans would fare in space, perhaps falling victim to radiation, weightlessness, or completely unimagined factors. The movie's solution was a robot pilot. Tobor (the name an admitted joke) is that robot, invented by a naïve scientist who announces its existence to the press before the launch. Spies from what is obviously Russia then kidnap the scientist and his grandson, threatening to torture the boy unless the scientist gives them the equations behind Tobor's skills.

Tobor breaks down for the same reason as *Gog*, the scientific underpinning giving way to ludicrous melodrama. Aimed at a family audience rather than the drive-in thrills-and-chills crowd, the story is really about a boy and his robot, with the bond between Tobor and the 10-year-old genius grandson saving the day in the end. (The literate script is by the great mystery writer Philip MacDonald, who later wrote the novelization of *Forbidden Planet* under the name W. J. Stuart.) Tobor is controlled by ESP—thought waves beamed at his antenna—although when necessary he independently solves problems without prompting, never more hilariously than his miraculous ability to drive a jeep to the Russians' hideout. (Nevertheless, the film's posters are dominated by a more malignant robot carrying an unconscious blonde in its arms for unspeakable impulses of its id, a scene that appears nowhere in the movie, copied from earlier robot posters.) MacDonald had to hope the 10-year-old boys in the target audience would sit patiently through the talky spy background to gape at a suitably awesome humanoid robot. Tobor is a towering modern suit of armor with grippers for hands, louvers in his torso, lots of visible cables, and a gigantic, mouthless glass-encased head with protruding eye bulbs that light up when he is turned on. Tobor's bulk seems best suited to battering down barn doors; he cannot be taken seriously as a space pilot for a second. A near-perfect correlation exists between impressiveness and sleekness in 1950s movie robots: the more robotic geegaws and doodads they sported, the more comical they became.

On that scale, Tobor was only a slight improvement over the tin and cardboard creations from other 1950s cheapies. (The diving helmet and gorilla suit that starred in 1953's

Robot Monster will not be dignified as a robot, despite the movie's name.) Mark 1, from the 1952 British comedy *Mother Riley Meets the Vampire* (a.k.a. *Vampire Over London*), is pure ductwork except for its glass dome, which doesn't cover its head but sits on top of it. Another UK production, *Devil Girl from Mars* (1954), cheaply knocks off *The Day the Earth Stood Still*. Nyah is retro-stylishly dressed in a leather miniskirt, cape, and helmet out of a 1930s serial, but her robot, Chani, has a torso apparently made from the cardboard box the rest of its parts came in. By comparison, the one robot shown from the supposed hundreds of alien robot invaders in *Target Earth* (1954) is modernistic in design, with a speaker grille for a face over impressively wide shoulders and a narrow waist. With no protruding gimcracks whatsoever, its destructive eye beam becomes more shudderful.

Outside that peak year of 1954, fewer than ten robot movies appeared worldwide in the 1950s, a telling sign of the way robots dropped off the cultural map after the war. That number includes end-of-the-decade movies using the extremely loose definition assigned to *robot* by filmmakers. The aliens of *Kronos* (1957) send two giant mobile machines (essentially skyscraper-size capacitors) to collect energy to bring back to their planet. The bottom was scraped with 1958's *La momia azteca contra el robot humano* (released in the United States in 1959 as *The Aztec Mummy Against the Humanoid Robot*), a Mexican production with another bargain-basement costume of boxes and tubes. This "humanoid robot" has the brain of a human, making him a cyborg, just like *The Colossus of New York* (1958). The Colossus is another variation on Frankenstein's monster, a human losing his humanity by being trapped in an inhumanly ugly body, cut off from friends and family and normal human relationships until he goes mad and attacks the United Nations with ray beams shooting out of his eyes. Robots were adjuncts to the real plot, mere tools of the humans (or aliens) who devised them, in all but *The Colossus*. That's also true in *Chikyu Boeigun*, known in the United States as *The Mysterians*, a Japanese import shot in 1957, whose English-language trailer summarizes the plot neatly: "Love-hungry spacemen come to seize our women so their dying race might live."[7] (Judging from dozens of 1950s movie posters that played up this theme, this one phrase sums up exactly the thrilling terror of the creature features no matter how sexually benign the actual contents might be. The American poster shows a squadron of aliens carrying off limp platinum blondes.)[8] Among the Mysterians' weapons is a giant samurai-style armored robot whose anteater head shoots ray beams from its eyes.

The most deeply souled human in 1950s robot movies is, ironically, an alien: Klaatu from *The Day the Earth Stood Still* (1951), the only 1950s robot movie other than *Forbidden Planet* (1956) that made any dent at all in contemporary or historic movie awareness. Not coincidentally, these two films were products of major studios with comparatively huge budgets. Both were well received and made money, but neither was among the top 40 grossers of their year, meager returns on seven-figure budgets. *Ultimate Movie Rankings* doesn't rank either film in the top 150 science fiction movies even by inflation-adjusted grosses.[9] Smaller studios didn't require six- or seven-figure profits, although they sometimes achieved them with quickly ground-out cheapies that made far better returns on investment than seemingly far more watchable studio productions.

The Day the Earth Stood Still is 91 minutes of pure didactiveness, a warning to all the nations of the world that they mustn't allow the new technologies of destruction to end civilization. Klaatu's message of peace and cooperation is weirdly undercut by his threat to use the even greater technologies available to alien planets to annihilate the

Earth if humanity does not achieve peace. This salvation by doomsday device will be enforced by an incorruptible robot police force, symbolized by the presence of eight-foot-tall Gort. Gort is the perfect robot: smooth-bodied, indestructible, impervious, and enigmatic, its sole means of communication being a vaporizing eye blast (a gimmick quickly borrowed by other robot movies, as noted above). Gort saves heroine Patricia Neal by carrying her into the safety of its spaceship, but that's not how the marketers sold the movie. The iconic poster shows Gort blasting a terrified public while carrying a long-haired blonde in a deeply cleaved gown who resembles Neal's prim character about as much as Michael Rennie's Klaatu does.

The kindly Robby, a robot with Asimov's three laws built in so that he could never harm a human, gets no better treatment. *Forbidden Planet* (1956) may be among the brainiest of all robot movies, with a Freudian gloss given to the plot of Shakespeare's *The Tempest*, but MGM's promotion department knew what sold. Following seemingly unbreakable laws of their own, they redrew Robby's face on posters to give him a red-lipped leer, the better for audiences to imagine the unspeakable horrors he might perpetrate on the limp blonde in his arms, one bearing no resemblance to star Anne Francis, the only female on the planet.[10] The movie Robby is the opposite of a menace, a literally subservient all-purpose robot, capable of whipping up a delicious meal, a Grecian gown, or ten tons of lead out of nothingness. He serves as comic relief to lighten another plot warning against advancing technology. Earth scientist Dr. Edward Morbius (played by Walter Pidgeon) has been marooned on the planet Altair IV for twenty years, building the omni-capable Robby as a trifle after his intelligence was doubled by a learning machine. The whole civilization, known as the Krell, was wiped out, along with virtually all the other Earthmen from Morbius' expedition and several of the new ones finally sent by Earth to replace them. Machines, in thunderous allegory, give life to horrors from the darkest corners of their users' ids. Technology can never escape its duality as miraculous but deadly; audiences both wanted and dreaded the message that unchecked advances will destroy us all, a balance that, as the 1950s moved further away from depression and war, tipped more and more toward dread. Little wonder that the utopian technological future presented by *Star Trek* was so enthusiastically seized upon a decade later (see chapter 13).

Good robot costumes are expensive compared to other movie props. Robby's reported $125,000 cost exceeded the entire budget of some of those 1950s cheapies. The ability to throw time and money around in fantastic amounts separated the major studios from the independents. The process started early: MGM purchased the original story in 1952, although everything but the basic idea was replaced over three years of multiple drafts by Cyril Hume.[11] Production designer Arnold "Buddy" Gillespie, art director Arthur Lonergan, and visual effects expert Irving Block (who in fact shared the original story credit for the film) all tossed in ideas before turning to design draftsman Robert Kinoshita to translate the ideas into Plexiglas, rubber, and Royalite plastic (normally used in suitcases and therefore crafted in the studio's leather shop).[12] MGM claimed that 50 technicians were involved in assembling the final figure. Kinoshita had been the uncredited builder of the relatively primitive Tobor, as advanced as robots got in 1954, and would show what he could do with a blank check.[13] (Oddly, he did not get an official credit for *Forbidden Planet* either, although the work launched him into a busy career as a set designer and art director. He finally gained public recognition for the last of his "trilogy": the robot in the television series *Lost in Space* [see chapter 11].)

"A couple of thousand sketches" followed as the designers worked to create a suit that had plenty of moving and flashing parts for visual interest, yet was sufficiently hollow to hide a man inside. Kinoshita solved that problem by building Robby in three sections: a head, a torso, and detachable legs. The iconic rotating head was an elongated dome of transparent Plexiglas that revealed its brain. Gyroscopes (representing omnidirectional ears) revolved on an upper track, while what looked like two hands of a clock formed a sort of nose that would waggle just before Robby spoke, as if he were processing input. One vertical and one horizontal "scanner ring" protruded from the sides of the dome, constantly rotating, giving him an appearance of motion while standing still. The head appeared to be larger than the torso, but that's a standard magician's trick. The top of the torso actually extended to include Robby's "voice tubes." They lit up in blue arcs when he spoke, allowing 5'3" stunt man Frankie Darro to see through the opening. (Or not— Frank Carpenter replaced him after a bad stumble in the 100-pound costume.)[14] Darro would climb into the unbending legs, three large bubbles over half-bubble feet (the bubbles were there for purely aesthetic reasons, but a great advance over the tubes or ductwork of competing robots); then the front and back of the torso were clipped together. Robby's arms were small and vestigial, reminiscent of a T. Rex, but could be moved just enough to grasp objects. The head came last, attached by a bundle of unseen cables to a control panel. A microphone allowed an off-camera actor's voice to be synchronized to the neon blue lights so that they flashed in time with the words. Robby had to be watchable every moment. Each time he walked across a room, the action stopped dead while the robot inched ahead, as inexorable and outrunnable as a glacier. Yet a still Robby was magnificent. The combined effect of these small touches made Robby by far the most seemingly intelligent and responsive robot in movie history—at least of those that couldn't be mistaken for humans. Those iterations were coming, and they would soon overwhelm Robby's later television career.

The Robot Post-Computer

World wars break popular culture. After the heightened terrors of World War II, the robot—often so frightening from 1920–1940 as a primitive, Frankensteinian monster—could never trigger the same atavistic fears. The automatic, autonomous devices referred to as robot planes, robot bombs, and robot controls encouraged the home front when newspapers reported American advances in technology; however, those same advances terrified readers when reports of German weaponry and destruction made headlines. World's Fair entertainment robots like Elektro, incapable of walking on their own, paled before the onrush of advanced technology demanded by the war effort.

Sharp-eyed observers took note of the advances made in mechanical brains and their capabilities, with universities like Harvard and MIT receiving grants for the expensive—$130,500!—devices.[1] At this same time, Bendix Aviation engineers produced "a mechanical 'brain' small enough to be held in the palm of one's hand [that] is performing a multifarious job for United Nations' fighting aircraft, eliminating weighty 'plumbing' and supplying split-second action and flight information."[2]

Scary headlines like "Can a Mechanical Brain Replace You?" (underlining in original) in the April 3, 1953, issue of *Collier's* kept the fear of technological obsolescence alive in human brains.[3] "Computers," the buzzword of the 1950s, gained familiar names that seemed to give them personalities. The War Department's ENIAC (Electronic Numerical Integrator and Computer), made public in February 1946,[4] and private business' UNIVAC (UNIversal Automatic Computer) gave the "AC" suffix an aura of size, speed, and intelligence that no human could match.

Kurt Vonnegut, Jr., then a publicist for General Electric, furiously developing its own computers, saw the production gains realized through automatic machinery that gave the United States the materiel that armed the Allies and overwhelmed the Axis Powers. At heart a disillusioned utopian, he used his advantageous position to forecast changes that might accompany this enormous paradigm shift. He started writing his unique take on science fiction in 1950. His second published story, also in *Collier's*, was "EPICAC," its title a triple pun referencing epic hugeness, ipecac (an emetic used to induce vomiting), and that notorious "AC" ending. A scientist treats the military's EPICAC as a Cyrano figure, asking it to write love poems he can use to convince his sweetheart that mathematicians are more than emotionless automatons. The poems work spectacularly, leading EPICAC to assume that it will become the woman's mate; it is crushed to learn that the human woman will never—can never—love a machine. "I don't want to be a machine, and I don't want to think about war," it types just before it commits suicide. But it leaves behind 500 more love poems.

Vonnegut brought back EPICAC for his dystopian first novel, *Player Piano* (reprinted in paperback with the ironic title of *Utopia 14*), in which machines have slowly taken over almost every human job, except for a few creative and executive tasks. Building on the Civilian Conservation Corps of the Depression, the government provides make-work jobs for the technologically idled. This future is not like the past; the meaningless tasks fail to imbue labor with dignity. Vonnegut can only see a downward spiral in which the remaining executive elite is itself displaced. No one can think of any answer except revolution, and that's easily put down despite the following rallying cry:

> Man has survived Armageddon in order to enter the Eden of eternal peace, only to discover that everything he had looked forward to enjoying there, pride, dignity, self-respect, work worth doing, has been condemned as unfit for human consumption....
>
> I deny that there is any natural or divine law requiring that machines, efficiency, and organization should forever increase in scope, power, and complexity.[5]

The loss of industrial jobs to machines is headline news again today, decades after the postwar worries. Of all the threads that run through various representations of robots, the realization of the ease with which "They" could replace "Us" always hits closest to home.

Perhaps that's why popular culture viewed robots through a distancing lens when the prosperous 1950s papered over that unsettling fate with a tsunami of consumer pleasures. The number of jobs that automation removed was far outpaced by the number that modernizing factories added. Computers were wonder devices that symbolized the sleek, modernistic future made present. Robots, especially the clumsy, clanking ones in movies and television, were as risible as Jack Benny's eternally sputtering and wheezing ancient Maxwell auto (and probably slower). A menace you can escape at a leisurely stroll is not a menace to take seriously.

By the time special effects technology produced frightening metallic beings, three decades after the end of the war, robots had become retro pleasures, symbols of the future that once was, alongside flying cars and food pills. Their place as the symbol of machines capable of anything was gradually usurped by computers and artificial intelligence, all knowing and omnipresent. Robots had little recourse but to entertain. That they did.

11

Robots as Camp

Resetting the technological world back to its prewar status was patently impossible; yet somehow the once forward-looking public relations staff at Westinghouse tried. Throughout the 1950s, Elektro, the seven-foot-tall golden robot, and his faithful robot dog Sparko traveled in their old Elektromobile on the same promotional circuit of department stores and fairs that they had covered in 1941, performing the same obsolete collection of tricks. Westinghouse engineers underwhelmed with one new feature: using Elektro's internal bellows to blow up a balloon until it broke. That feat thrilled only the very youngest. By the end of the decade, the formerly world-famous symbol of American futurism had devolved to rounds of children's television shows and hospitals.

One day in 1960, Elektro found himself in a worse predicament: serving as a literal set piece on a z-grade movie stage. *Sex Kittens Go to College* (1960) sported a trivia marathon cast that included Charles Chaplin, Jr., Mijanou Bardot (Brigitte's little sister in her first American role), singer Conway Twitty, late-night TV hostess Vampira, former child star Jackie Coogan, teenage bombshell Tuesday Weld, comic Louis Nye, professional big lunk Woo Woo Grabowski, and a few veteran actors like Pamela Mason, John Carradine, and Marvin Milner in support of the magnificently cantilevered Mamie Van Doren.

The plot begins in a college science department. Thinko, the most advanced electronic brain in the world, chooses the new head of the science department, sight unseen, by strictly and fairly evaluating the credentials of all the candidates. When that proves to be Van Doren, the status quo is outraged since her looks and body obviously prevent her from being smart—despite her 13 PhDs and knowledge of 18 languages—and every man on campus becomes a lecherous, bumbling fool in her presence. This core of possible social satire leads nowhere, as writer-director Alfred Zugsmith (a fixture on worst movie lists) apparently instructed the cast to behave as if auditioning for a Three Stooges short. In the ensuing mayhem, Thinko blows three tubes and a relay and ends by proposing to Van Doren, who psychoanalyzes him back to his proper operating mode. Thinko, of course, is Elektro, whose giant but now silent and immobile body is the centerpiece of an entire room of comical machinery, a kiddie-show conception of the inner workings of a computer. Whether computer or robot, all machinery is secondary to the magnificence that is the human body.

Elektro's comic dissolution was hardly unique. It's impossible to imagine Boris Karloff in full Frankenstein makeup trading quips on a promotional tour, but the once equally A-list Robby the Robot, the posterized star of the big-budget feature *Forbidden Planet*, was made to do so. Before the movie's release, Robby appeared on *Perry Como's Kraft Music Hall*, indulging in show business banter with the host.

COMO: Do we have any mutual robots, mutual friends?
ROBBY: On the forbidden planet, I am the president of the Robots for Como Fan Club.
COMO: That makes me very happy. Do they really like me up there?
ROBBY: They think your singing is out of this world.[1]

Como throws in a mumbled "frightening-looking thing" at the end, almost certainly a scripted "ad lib" by a studio attempting to build menace into a cuddly toy (those rounded legs are practically teddy-bearish.) MGM similarly tried selling Robby as a monster on *MGM Parade*, a promotional television series for the studio's upcoming pictures. *Forbidden Planet* co-star Walter Pidgeon acted as host on episodes that showed behind-the-scenes footage. While he downplays Robby as a mere movie prop, the final image is of Robby seemingly fiendishly plotting his revenge.[2] The movie's posters also portrayed a monstrous robot, but viewers emerged from theaters knowing Robby to be the ultimate good guy.

Following a long industry practice of reusing robot props, MGM opted to do what classic car collectors do to amortize their expenses: make the expensive creation available for a fee to any movie studio or television production company that wanted to liven up a scene. Robert Kinoshita's clever design made Robby easily shippable, with heavy-duty rolling wooden boxes storing his three pieces and the critical control board. (Critical but not invaluable: in fact, the board's box had "VALUE $10,000.00" stenciled on it.) Anyone could provide the voice, and any person small enough to fit inside and fill the size 10.5B shoes built into the robot's feet could make Robby walk.[3]

The Internet Movie Database lists 26 additional appearances made by Robby, not counting documentaries or promotions. Like an actor playing either drama or comedy as the script requires, Robby might be trotted out as a threat or buffoon or something in between. On a 1958 episode of *The Thin Man*, the television remake of the comedy-detective movie series, Robby is merely a machine, subject to the control of others and an instrument of good or evil as they see fit. A good-hearted scientist invented the robot to handle radioactive materials, a job too dangerous for humans. Yet using Robby in that way could clear the way for building atomic bombs, making the planet's destruction a certainty. The argument rages inside the scientists' lab, leading to murder. Framing Robby as the perpetrator will keep his ilk from being made in *R.U.R.*–like millions. The seriousness of the message is undercut by a shortcoming in Robby's design: the plot required the robot to carry the victim much like a husband carries his wife across a threshold; yet a grown adult could not possibly perch on those tiny arms. Directors' best efforts couldn't make 1950s robots much more frightening than refrigerators.[4]

Television shows had their own budgetary problems and often scrimped on effects. A totally immobile Robby made an appearance on a 1963 episode of *The Many Loves of Dobie Gillis*. More computer than robot, Robby became a wizard at forecasting hit songs, predicting sales to the last digit. Through television magic, his genius is downloaded into the lamest of all brains, the lovable beatnik Maynard G. Krebs, who insists that speeding up a record to sound like the Chipmunks will sell millions.[5] *The Addams Family* made better use of Robby. Fearing that butler Lurch is overworked, paterfamilias Gomez seeks to lighten Lurch's load with an assistant. However, hiring one is outmoded: "Hire? In this age of automation? I'm gonna build him one. Fester? Meet me in the playroom. And tell Pugsley to bring his Erector set." Robby (here called Smiley) turns out to be inferior to the semi-human Lurch, the only one better at his job than a robot.[6]

Robby appeared twice on *The Twilight Zone* in the 1963–1964 season. "Uncle Simon" featured a far more sinister Robby, this time a "mechanical man" intended to contain an inventor's personality after his death. Perhaps to convey the homemade nature of the robot, this Robby's dome contains merely a cork-shaped humanoid face and none of the normal spinning bells and whistles.[7] Later in the same season, a company's automated processes doom loyal workers by the tens of thousands in "The Brain Center at Whipple's." Not until the CEO is himself made redundant does he come to see the dehumanizing aspect of a dedication to money and efficiency over human worth. The final reveal of Robby behind the CEO's desk is a classic *Twilight Zone* reversal of fortune, but the unbendable costume literally cannot sit on the plush chair: automation may make jobs obsolete, but Robby will never fit into a human world.[8]

Hazel was a single-panel cartoon strip by artist Ted Key that ran in the *Saturday Evening Post* from 1943 to 1969. Millions saw the smart-aleck, bossy, and omnicompetent maid weekly and bought the many collections of her cartoons. Brought to television, *Hazel* went to #4 in the ratings for the 1961–1962 season. Oscar and Tony Award winner Shirley Booth, a sprightly sexagenarian as far away from French maid fantasies as could be imagined, perfectly embodied Hazel. The producers surrounded her with what television historians Harry Castleman and Walter J. Podrazik called "possibly the dumbest family in TV history."[9] The family had a dog named Smiley and Hazel a best friend named Rosie, the maid next door. In the episode from September 27, 1962, Hazel wants the family to draw up a contract. She asks, "You all know how sometimes we get to worrying about automation and how it might do us out of our jobs?" That worry turns into a nightmare about being replaced by a machine that can cook perfect blueberry pancakes. The machine, of course, is Robby, hobbling slowly into the family kitchen, carrying a breakfast tray with plates obviously glued to it.[10]

By coincidence, the *Hazel* episode with Robby appeared four days after the debut of a new series set one hundred years in the future, lovingly parodying every cliché of 1950s sitcom families. The first episode revolves around the family hiring a loveable, aged, and bossy maid. The future as seen from the present would be a laughably ridiculous mélange of camp.

—∿—

Camp appeared in the national lexicon when the Fall 1964 issue of *Partisan Review* reached the living rooms of ultra-refined American tastemakers, families that prided themselves on being both the antithesis of middle-class suburbanites and vociferous haters of the idiocies of television. Among poems by Muriel Spark and an article on the avant-garde art movement readers found a modernist essay by 31-year-old Susan Sontag, her marker in the contest (oddly important at the time) to become the intellectual's intellectual. "Notes on 'Camp,'" drawing on deliberately provocative early modernist heresies like Futurism and Dada, sought to elevate certain types of popular culture and low art into respectability by ascribing to them sensibilities that serious high art studiously avoided. Camp, Sontag wrote, "converts the serious into the frivolous"; "is the love of the exaggerated, the 'off,' of things-being-what-they-are-not," "sees everything in quotation marks," and "is, above all, a mode of enjoyment, of appreciation—not judgment."[11] Camp therefore lay entirely in the eye of the consumer, not the artist. Among her examples were *King Kong* and *Flash Gordon* comics. Camp is, critically, "an aesthetic phenomenon." "To emphasize style is to slight content, or to introduce an attitude which is neutral with respect to content. It goes without saying that the camp sensibility is disengaged, depoliticized—or at least apolitical." Being an apolitical intellectual was a moral sin among the *Partisan Review* set, but Sontag championed as equal, if not superior, the aesthetic judgments made by an invisible and scorned underclass whose opinions could otherwise never impinge on those of the serious thinkers: namely, homosexuals. Camp became first loved, then rejected, then re-embraced and re-thought by gays and others, who have continually modified, extended, and updated the exhibits in a virtual museum of camp. Bruce LaBruce, for example, included 1965's *Lost in Space* under "Unintentional Camp." (That there could be such a thing as "intentional camp" itself refutes Sontag's argument.) LaBruce in his 2012 paper lamented the loss of camp as a specialized gay sensibility: "Camp is now for the masses. It's a sensibility that has been appropriated by the mainstream, fetishized, commoditized, turned into a commodity fetish, and exploited by a hypercapitalist system, as [Theodor] Adorno warned."[12]

Although serious thinkers long to reclaim the word, the shorthand notion of camp as artificial, vulgar, and banal is too ingrained as a pejorative of popular culture to eradicate. Television and movies from that era were all too often exemplars of lowest-common-denominator entertainment. In a 1962 overview of the new television season, *Cincinnati Enquirer* columnist Luke Feck declared that describing shows as merely average was "a compliment this year." He dismissed the biggest hit of the 1960s with a sneer: "'Beverly Hillbillies' is an atrocity committed weekly against the public taste."[13]

Robots were not just the punchline of future jokes, though: they were in and of themselves the joke. By 1962, the mere presence of a robot in a television show or movie almost inevitably instructed audiences to view a one-time potential menace as frivolous comedy to be laughed at on sight, the exemplar of style without content. We now view the robotic aesthetic from the late 1950s into the 1970s as camp.

Camp is not normally applied to children's media, a specialized niche in creativity where nothing is ever quite serious and the impossible is served seven times before breakfast, although Sontag's canonical list includes *Flash Gordon* comics—that is, Flash Gordon *newspaper comic strips*, aimed at families in a way that comic books of that era seldom were. Therefore, when a cartoon series called *The Jetsons* debuted on Sunday, September 23, 1962, its family-hour placement and faux sitcom setting made it a similar hybrid, aimed equally at Mom and Dad and Bud and Sis in a demographic-encompassing genre now almost entirely lost.

What makes *The Jetsons* iconic today is its status as the ultimate representation of the "consensus future," the world of flying cars, short work weeks, space travel, two-way television, robots, and ubiquitous technological plenty that science fiction, popular science magazines, and world's fairs proselytized endlessly in the period between the two world wars and which seemed to be coming true as the genial old-timer Dwight Eisenhower passed power to the glamorous and very young Kennedys. *The Jetsons* imbued the entire vision of the gadget-laden middle-class consumerist future with a coating of camp that it has never lost. Every new gadget, each newly blooming technological development, and seemingly all robots have for decades been compared to a Jetsons future. In 2017, former Minnesota governor Tim Pawlenty called *The Jetsons* a harbinger of the fourth industrial revolution.[14] Probably not coincidentally in such an age, 2017 also saw announcements of a planned live-action Jetsons television show and a full-length animated movie.[15]

Such cultural immortality was utterly unapparent at the time. *The Jetsons* was a low-rated failure as a series, cancelled after one season. At a time when 3 percent of households had color TVs, ABC tried to promote *The Jetsons* as the network's first show to run in color, but virtually everyone had to settle for watching the day-glo palette in ruinous black and white. Worse, the series aired in the death spot against *Walt Disney's Wonderful World of Color*, long the exemplar of family viewing and, moreover, a show with *color* right in the title.[16] Audiences couldn't have been drawn to *The Jetsons* by the lukewarm reviews. A syndicated report did say that "some of it was quite funny," but Feck might as well have scoured it with a ray gun: "The animated *Jetsons* series is without doubt the least inspired of the Hanna-Barbera cartoons. It should be kept in outer space."[17] It appeared to cynical observers as a pure cash grab from cartoon studio Hanna-Barbera, which desperately needed a follow-up to its fabulously successful *Flintstones*, the first prime-time animated comedy. Hanna-Barbera cartoons normally thrived on blatant imitation—the same, but different. *The Flintstones*, starring a loving but bickering blue-collar couple, was manifestly based on Jackie Gleason's long-running sketch known as *The Honeymooners*. *Top Cat*, an animated takeoff of *The Phil Silvers Show* (popularly known as *Sergeant Bilko*), had failed the previous season. If cats were not the answer, then perhaps people would work, with a new show drawing from endless numbers of sitcom families, a tyrannical boss always threatening to fire his office manager (straight out of a *Blondie* cartoon), and a robot maid that looked and sounded like Shirley Booth's *Hazel*. Surely imitating everything had to appeal more broadly than imitating one thing.

Perhaps the series developers forgot that *The Honeymooners* had been an outlier on network television. It chronicled the home life of the forerunners of the modern Tea Party, filled with rage and despair as they missed out on the booming consumer society that most television shows depicted. During its sole season as an independent show, the camera rarely moved out of Ralph and Alice Kramden's city apartment kitchen, a barren hovel with a few old appliances, at whose bare wooden table Ralph pondered get-rich-quick schemes to get up and out (both literally and figuratively). Fred and Wilma Flintstone, by contrast, were part of the nouveau riche working class; they lived in an anachronistic suburban home replete with all the gadgets that Ralph so longed for. The Jetsons—George and Jane, daughter Judy, and his boy Elroy—ironically found themselves returned to urban living, albeit in the Skypad Apartments in Orbit City, a spacious home abundantly, even overly, filled with redundant appliances to make life easy and quick, the joke switching every few seconds between how marvelous the technology was and how disastrous the consequences were when it failed to work properly. Otherwise, it was a show about nothing, carefully avoiding any point of view that might be contentious— a veritable museum, lovingly collected and curated, of exaggerated, often slapstick, stock humor about middle-class suburban America. Every scene, character, and individual line, like a snowflake built upon a speck of dirt, harbored a cliché. The series as a whole was set in the monochromatically white future (void of people of color, despite the occasional appearance of blue two-headed Martians, definitionally not people and subject to "Martians Go Home" picket signs) symbolizing the conformist America of 1962 that would be obsolete a seeming nanosecond after the the show finished its one and only season. Far better to remember the gadgets. All the world loved a gadget, probably even more today than in 1962.

Two days before the series debut, Professor Meredith Thring of Suffolk University in England predicted that "man will one day invent a robot that will do all of woman's tiresome housework." Indeed, the woman of 1962 worked so impossibly hard that—you can almost hear the reporter drawing his breath in horror—"her husband often feels compelled to help her." (Such were the times that this article ran under headlines like

"Who Would Do the Nagging?" and "Oh, Kiss Me, You Mechanical Fool.")[18] Thring foresaw a small machine that would make beds and do cleaning.

> "It would be able to get around the house, including going up and downs stairs by itself," he explained. "It would have arms and hands for removing and replacing various objects.
>
> "It would have a built-in computer and a memory. It could be trained to know the geography of a house."
>
> Thring said that the ideal automatic housewife would take into consideration children—even babies crawling about a room—and wouldn't interfere with them or be bothered by them....
>
> There's only one hitch. He said it would take about 10 years to develop at a cost of $2.8 million.[19]

Those who viewed the opening credits of *The Jetsons* saw wonder, the camera zooming in through the stars to glimpse a cloud-free Earth that suddenly breaks into jagged shapes evoking 1950s mod kitchen countertops, the decade's futurism compressed into twelve seconds. And then a flying car, schools and shopping in the clouds, the car folding into a suitcase, a slidewalk full of commuters reading tablets and watching portable television screens, and an office dominated by a wall-size computer.

The action quickly cuts to Jane, watching a fitness show on her 3-D flatscreen television. Jack Jetwash, "your slipped disc jockey," leads her through a series of exercises—for her index finger. The Jetsons' push-button world is no longer figurative but the norm. Technology supplies everything at the push of a button, including making breakfast, getting one's husband out of bed, and sending the kids off the school. (A push of the button, and Elroy is off to PS 85, odd not merely because the opening credits establish that he goes to the Little Dipper School but also because later on in the episode the same panel reads PS 58. Whether an inside joke or the world's worst continuity, viewers could be sure that no panel or device ever got drawn the same way twice.) A worm lies in the apple of the future, though: George's lumpy morning beverage tastes like rocket fuel. A crossed circuit has caused the foodarackacycle to deliver tea with oatmeal instead of coffee. They need a new one but haven't yet paid for the one this model replaced. George lays out a firm "NO!"[20]

Every sitcom viewer knew that was code for having the headstrong wife who was the true boss of the family go behind the husband's back. Besides, stay-at-home wife/mother Jane is not happy. With the family gone, Jane settles in for the drudgework of pushing three buttons for the washing, ironing, and vacuuming—the last one sending a Roomba-style self-propelled mini robot around the apartment. "Housework gets me down," she says. (The appearance of Betty Friedan's *The Feminine Mystique*, chronicling the depression of bored housewives, was only five months away, the rare bit of forecasting that *The Jetsons* doesn't get credit for.) Jane, the 33-year-old mother of a 15-year-old daughter, represents the first wave of the baby boomers as they were in 1962, women without marketable work skills who went straight from high school to raising a family. (George, at 40, is older—an analog of the World War II vet whose life got derailed.) Calling her mother on the two-way visiphone, Jane learns of a one-day free trial offer at U-Rent a Robot Maid (although the store sign reads "U-Rent a Maid"). Nothing testifies to the ubiquity of robots in the future more than an entire store devoted to them. The snooty salesman blanches when he hears that the family has a three-bedroom apartment (luxurious by 1962 standards): "Eww, a slum clearance project. So you'll want to see our basic economy model." That's Agnes, a "pip pip cheerio" English stereotype, followed by Blanche, an ooh-la-la French maid that's the equivalent of a sports car: her engine is in the rear. And then comes Rosie, "an old demonstrator model with a lot of mileage." She clanks and beeps, and her rollers seem to stick. No matter. The salesman may call her "h-o-m-e-l-y," but she knows she's "s-m-a-r-t."[21] Jane likes her, and the audience is supposed to feel the same, even though Elroy calls her an antique jukebox on first sight. He changes his mind when he sees her throw a football into orbit around the entire apartment building and demonstrate an amazing hook shot with a basketball. Not coincidentally, an ongoing gag with Hazel was her unexpected prowess in every sport. Rosie would be "a Hazel out-Hazeling a Hazel," said creator Joe Barbera in an early promotional interview.[22]

Rosie ends up saving the day when George's boss, Cosmo Spacely of Spacely Sprockets, invites himself over for dinner; the new robot maid makes a delicious meal out of leftovers along with the best pineapple upside-down cake Mr. Spacely has ever tasted. How she does this in a kitchen without a stove is best left to the imagination. Actually, what Rosie is needed to do in an apartment filled with push-button appliances that Jane is later seen manipulating to everyone's satisfaction is also a question the show never bothers to answer. Few starry-eyed nostalgic devotees of the program remember that Rosie appeared in only one later episode—as many as the family's now utterly forgotten cat. Rosie is a one-off future joke no different from push-button finger.

So is 'Lectronimo, the "dog of tomorrow." In a plot that replicates the Rosie episode as if cut from a template, George gives a firm, absolute "no" to the suggestion of getting a pet; yet when Elroy brings home the giant pooch Astro the family (minus George) wants to keep him regardless. George gives in to the extent of going to a pet shop and bringing home, on a one-day free demonstration, 'Lectronimo, the "dog of tomorrow." The robot dog is perfect for an apartment and hates burglar masks (a handy trait, since a cat burglar is prowling the city). George pits 'Lectronimo against Astro in a series of tasks, with the robot naturally winning every time. Yet when the cowardly Astro accidently knocks out the cat burglar while thinking he's running in the opposite direction, the family naturally keeps the big lug.[23]

This mixture of equality extends across the Jetsons' world. Robots were characters in more than a dozen episodes, playing parts as prominent as the one-shot humans and as ubiquitous as telescreens, elevator tubes, and flying cars. Some were humanoid, many merely mobile appliances. Some filled ordinary human jobs, often ones that were too dangerous for humans.

A joke that seems far more prescient today than in 1962 forms the core of "Jetson's Nite Out." (Nitpickers: The name of this episode is also spelled "Jetsons Night Out" and "Jetson's Night Out" on the official DVD package.) Spacely drags George away from a PTA meeting to watch a football game, held in a giant flying stadium. The Wringers are playing the Marauders, both teams identical neckless boxy robots wearing helmets with antennas to receive radio signals, differentiable only by their team colors. Coach O'Brien is the great exponent of scientific push-button offensive strategy, whereas Coach Slasky is the master of push-button defensive strategy. They are video game masters working a giant canvas. When the teams go rolling down the field (literally, in this case), the resulting tackle smashes the ball carrier to bits. "Brimlovich was a little shaken up on the play," the announcer says. "He just had the wind knocked out of him and is being helped off the field," he continues as the robot trainer extrudes a whisk broom and sweeps the loose pieces off the field. "He should be as good as new by halftime, so don't worry, Mother."[24]

Robots would invade the office as well. The secretaries, all female, are sometimes human and sometimes robots (all far more glamorously built than Rosie). Spacely dictates to a robot stenographer who uses a typewriter with an eraser that pops that out and corrects any mistake. George has a robot secretary as well, although one time she disappears in the middle of a sentence. Why? She is part of a gaggle of identical robots standing around and gossiping on their coffee break. Do robots eat? Not anywhere else in Jetsonsland, but a joke is never to be spoiled. UNIVAC (maintaining its place as the generic popular cultural shorthand reference for a computer) gets spoofed by a robot electronic brain named Uniblab. Barely humanoid, with a gigantic head that his body only just supports symbolizing his enormous computer brain, Uniblab cost the company five billion dollars, but all he does is take the promotion to office supervisor that George had expected. Uniblab—obviously a robot because he says every sentence twice in a squeaky monotone—is a martinet (he throws a fired employee who called him "a fugitive from a horror movie" down a trash chute), a cad (he tricks George into badmouthing Spacely so he can secretly record the insults), and a pettifogging cheat who goofs off every possible moment and takes George's money when he turns into a slot machine and a roulette wheel. (Again, do not ask why such devices were built into a business computer.) George finally wins back his job after Henry, the super at the Skyway Apartments, spikes Uniblab's Unilube with his homemade hooch, sending the robot into a drunken revel in front of the company's board.[25] Uniblab returns in "G. I. Jetson" as a backstabbing drill sergeant when George and Spacely get called up for reserve duty, giving him as many appearances as Rosie.[26]

Background robots regularly function as visual jokes. When George and Jane go to "Las Venus" on a second honeymoon, we meet a card dealer who's a robot, mobile robot slot machines, and a band full of robots conducted by "Starence Welcome" (Lawrence Welk).[27] The army was full of robots before Uniblab, with robot MPs and a robot KP dish crew. When George and Jane go to buy a new car, they're treated to a promotional film of Molecular Motors' giant factory, entirely automated with robot workers. Elroy even has a robot teacher, Miss Brainmocker.

Perhaps the strangest mix of robot and human comes toward the end of the series, when Jane is suffering from "buttonitis," brought on by the constantly malfunctioning technology—in one example, the bed wants to lie on George. The doctor prescribes a rest, "some place close to nature, one of those dude planets." While George stays home and battles with the appliances, like a runaway vacuum cleaner that gobbles up the aforementioned cat, Jane and a friend visit the Beta Bar Ranch, a proto-Westworld where the cowboys and horses are robots. Jane is embraced by a suave robot who insists that she dances divinely. Sex robots are for the 1970s, and Jane shrugs off his charms—she misses George. The perfectly behaved robots are no match for the chaos of home.[28]

The joke of misbehaving machinery reaches its zenith in the second Rosie episode. Henry, who spends most of his time tending to the constant maintenance of thousands of appliances, builds himself an assistant, Mack, by taking an old filing cabinet and attaching four wrenches as arms. Naturally, Mack winds up in the Jetsons' apartment, and he and Rosie fall instantly in love. Discombobulated by the unplanned emotion, Mack is even less efficient than his normal bumbling self, forcing Henry to deactivate him. Super-robot Rosie is devastated by her first experience with lost love, serving George his coffee by placing the cup on his head and dumping the contents of the pot on his suit. Jane takes her to the robotologist at the robot clinic (another sign of how pervasive robots are in this world). The waiting room is full of broken and dejected robots—one holding his head in his hands croaks out, "Needs rewiring, nothing serious" (yet another reminder of a technologically equivocal future). "The factory doesn't install emotion tapes," the doctor insists, but Mack and Rosie are as real as the rest of the Jetsons. Robots aren't appliances; they're part of the family. A daily visicall from Mack is better medicine for Rosie than a prescription for reactor fluid. Human or machine, emotions rule the universe.[29]

—⁓—

The cultural explosion that was Sputnik boosted science, both in the calls for better science education and in the production of more science books and programs, but it had an unexpected chilling effect on science fiction. Television logically should have extrapolated the many programs featuring stalwart men following their dreams of adventure that typified dramas in the 1950s into space sagas of daring and peril. None appeared. *The Twilight Zone* premiered in 1959, but that was an anthology series that seldom tackled either space or derring-do; the same can be said for 1963's *The Outer Limits*. The space program spawned two semi-realistic series in 1959, *Man into Space* and *The Man and the Challenge*, but, like NASA, both kept the focus on low orbit and the moon. Sitcoms demonstrated the thin line between science fiction and fantasy—*My Favorite Martian* and *I Dream of Jeannie* are essentially the same show—yet stayed resolutely earthbound.

That left an opening for master of spectacle-on-the-cheap Irwin Allen. His 1961 movie *Voyage to the Bottom of Sea* gave his production company sets and footage that could be repurposed for a much lower-budget television series of the same name, a hit in 1964. Inverting a successful theme is a venerable Hollywood formula for a sequel, likely why Allen bought the rights to a comic book titled *Space Family Robinson*, a takeoff of the classic tale that stranded a family on an unknown planet, offering virtually unlimited scope for any menace the writers could dream up. A failed pilot taught Allen that the saccharine family of do-gooders needed internal conflict as well as external trouble. Writers added Dr. Smith, an evil stowaway, and Robot, a treacherous robot under Smith's command, and thriftily chopped up the pilot material to be reused in five separate episodes.[30] After a last-minute title change, the show debuted on September 15, 1965, as *Lost in Space*, with a realistic mission control and a detailed cutaway of a two-level spaceship, called the *Jupiter 2*, another flying saucer that clashed against the NASA–ish background.[31]

Contemporary critics saw that, like many other purported "family" shows, the series failed to satisfy mature tastes. Rick Du Brown wrote, "In short, it's like a live-action cartoon, complete with robot that goes around shouting 'destroy, destroy,' and which may or may not be one of the show's writers."[32] (To be fair, the booming baritone of Dick Tufeld, Robot's uncredited voice, may be the most memorable element of the show.) Allen quickly swung the series toward accommodating youngsters after amortizing the pilot. Over the next two and a half seasons, *Lost in Space* shoved the Robinson family—mom and dad Maureen and John, along with their three children, Judy, Penny, and preteen genius Will—deeper and deeper into the background except when serving as foils for the suddenly foppish and cowardly Dr. Smith, no longer an evil villain but a vain idiot who constantly sabotages the mission in hopes of gaining money, finding glory, or making his sole way home, accompanied by a nameless robot whose personality broadened and repartee sharpened from episode to episode. The writers dropped plot and characterization—and any resemblance to actual science—in favor of clashes between the bickering partners (a new role for robots). Smith's stream of alliterative insults was met by Robot's ripostes, often the only dialogue worth paying attention to, which viewers avidly did.

The supposedly lost and isolated family encountered as many visitors as Gilligan did on his island. (One of Gilligan's was, inevitably, a robot, an air force test launched from nearby Hawaii that parachuted to the island after a failed mission. The castaways reveled in relief from their chores by assigning the work to the robot, the cheapest pile of ducting and silver-painted cardboard the space program ever produced, before sending it back to Hawaii with a message.[33] The clever attempt fails because of Gilligan, in yet another example of how he always spoiled rescues—failures faithfully aped by the *Lost in Space* writers and assigned to Dr. Smith.) In addition to multiple races of aliens, a surprising number of Earth people, and the monster of the week, the crew encountered both robots and androids, and the arc of Robot's humanization over the three seasons can be tracked by interactions with them.

Robot was another creation by art director Robert Kinoshita, his third major robot design after *Tobor the Great*'s Tobor and *Forbidden Planet*'s Robby. Though hardly identical, Robby and Robot have much in common. Built at a cost of $36,000, Robot stood 6 feet, 4 inches tall and weighed 275 pounds.[34] He had a transparent head with a swirl of visuals (in this case, lights), neon tubes that flashed in synchronization with his words, rounded legs (stacked disks replacing stacked bubbles), and tiny arms, although Kinoshita had learned from his previous mistake and made the arms extendable. He also realized that a slowly shuffling robot looked primitive, so this suit ended in tracks. They hid the feet of Bob May, who had to maneuver Robot from the inside manually until hidden cables were installed to pull the structure smoothly along the set.[35]

Robot's sanctification comes when pitted against a more evil and powerful robot in the first season's "War of the Robots."[36] In a grace note to Kinoshita, Allen guest-stars Robby as the antagonist. Robby suddenly materializes on the planet that the Robinsons are trapped on, explained as an alien race's "robotoid" with independent processing to think beyond his programming. Robby's far superior abilities allow him to replace Robot as the Robinson family servant and boot Robot from the camp—a ploy. Without the threat of opposition, Robby proclaims his control over the Robinsons. However, Will seeks Robot out and convinces him to pretend to acquiesce. A battle of the robots must ensue, and the robotoid falls under a barrage of Robot's electricity. For the first time, Robot is accepted as an equal member of the family. (Robby is brought back for a third-season cameo without anyone saying, "Hey, we've seen that robotoid before," a typical violation of continuity that was part of the downfall of the series.)

"It's the sentimental stuff that makes Robot part of the family," Will says in the second-season episode "The Ghost Planet."[37] Its original planetfall destroyed, the *Jupiter 2* leaps desperately into space and gets caught by a passing supernova. (Such scientific illiteracy, including the repeated portrayal of comets as space fireballs instead of iceballs, were to make that season's competitor, *Star Trek*, seem all the more adult despite its own problems with technobabble.) The ship is shunted through space and encounters a planet that is the "kingdom of cybernetics." A female-voiced robot that looks something like Uniblab interrogates Robot when he is sent out to parlay, and we learn that he is model B-9, making him officially the most harmless robot of all. He proves it once again by pretending to switch his allegiance to the robot world only to rescue Dr. Smith and Will, who are being held prisoner in a factory. Earth people are useful: they have the manual dexterity to attach small widgets to circuit boards!

In season 3's "Deadliest of the Species," Robot hears a voice call to him from a downed space capsule. He soon frees the loveliest robot he's seen since the one from

"The Ghost Planet"—not surprising, since she is the same robot with the same voice actress, only painted red instead of blue. Red is exponentially more evil, having been exiled permanently to space after she nearly destroyed a planet. Robot doesn't care; he is smitten with one of his own kind and is willing to do anything, even let the Robinsons die, to satisfy her needs—and his own, as he comes to realize that if he has human feelings, he is even more cut off from society than the Robinsons.[38]

The apotheosis of the relationship between Robot and Dr. Smith came late in the second season, in an episode titled "The Mechanical Men." Smith wakes up like Gulliver, captured by a horde of foot-high replicas of Robot. This situation might reduce even a stronger constitution to screaming nightmares, but the tiny robots' purple spokesmachine explains that they have been waiting for a leader for 10,000 years. Since he is an exact supersized double, Robot will do perfectly. They fail to understand that Robot is kind and good and believes in ruling fairly—not at all what the militant miniatures have in mind. They want a scheming and backstabbing personality like Smith's and get it when they switch his mind for Robot's. For the rest of the episode, Smith speaks with Dick Tufeld's voice and suddenly becomes helpful and self-effacing, ready to sacrifice himself for the family, while the Smithized Robot (now wearing a crown and robe) plans to eradicate the Robinsons and steal their spaceship. A last-minute laser blast inexplicably reverses their personalities and restores the status quo.[39]

The robot horde was easily managed by the prop department; they merely went out and purchased dozens of the toy versions of Robot sold as part of *Lost in Space* merchandising. (The writers may have stolen the idea from "The Double Affair," a 1964 episode of *The Man from U.N.C.L.E.* Ideal had introduced its Robot Commando toys in 1961, running commercials that showed them launching missiles out of the top of their heads. The bad guys merely substituted real bombs for the toy projectiles and sent the Commandos walking toward U.N.C.L.E.'s supposedly secret headquarters entrance.)[40] Of course, Allen, who surely got a large cut of the merchandising rights, probably needed little outside prompting. Effectively an hour-long commercial, "The Mechanical Men" reminded the watching children that they could have their own robot army just as quickly if they bugged Mom and Dad long enough and loud enough. All the advanced computers and cybernetic planets the Robinsons encountered could be interpreted as giant toys played with by the fortunate Robinson children and their very indulgent parents. Toy robots—even evil ones that invariably got their comeuppance—demystified the once scary mechanical monsters in the same way that spooky campfire stories reduced real-world monsters to tall tales demolished by morning's light.

The last episode wasn't supposed to be the last. The first season was a ratings hit, while the second season's ratings held up at first as hordes of children flocked to the slapstick but fell off sharply later that year and the next year as the show became even sillier. Networks were learning about demographics, and *Lost in Space* had a double whammy working against it. Children weren't the audience for the advertisements, and the CBS network had decided to seek a more sophisticated audience. (Ironically, NBC kept rival *Star Trek* alive despite much worse ratings because of its attractive demographics.) For season 4, "Allen told the executives that they would love a new character he had created for the new year, a telepathic purple llama named Willoughby. The animal would dispense advice, humor and philosophy with a British accent as it trotted along with the Robinsons," according to television historian Mark Phillips.[41] Inexplicably, this suggestion failed to sway the suits, who left the Robinsons stranded in space with a sudden cancellation.

"Junkyard in Space," the final aired episode, puts the family in its worst predicament yet.[42] They are forced down on still another planet, this one a junkyard that attracts obsolete machinery from across the galaxy. In addition to dodging constantly falling meteorlike scrap metal, they find that every bit of food on the *Jupiter 2* has succumbed to a plant blight. Smith selfishly offers to trade critical parts from Robot for food from the android/robot Junkman who cares for the planet and melts the scrap down to sell as waste metal. Like dozens before him, the Junkman reneges on the deal and steals the spaceship for his own. Knowing he is now useless, Robot climbs onto the conveyor belt that will carry him into the flames. Will takes the family space pod up to the ship and convinces the Junkman to develop a heart (another nonhuman learning human emotions). Despite the vast amount of time this takes, Robot is still on the ten-foot journey into the flames when they return and stop the machine. Kids never need to worry. Robot is literally coated with a layer of love that has protected him from the heat. How could a telepathic llama compare with such a send-off?

The show's retro cliffhanger endings harkened back to the cheesiest of 1930s science fiction movie serials (*Batman*, the campiest show of the decade, spoofed *Lost in Space*'s "tune in next week Same Time, Same Channel" endings with its far more famous "Same Bat-time, Same Bat-Channel" signoffs). The borrowing of the hoariest of plots; the reuse of footage, costumes, and actors in new situations; and the blithe demolition of all matters scientific served to destroy any pretense that science fiction could be respectable, especially during the show's first season, when it was the only example of live-action space-based science fiction in primetime history. David Gerrold, a writer for *Star Trek*, damned *Lost in Space* for its devolution: "*Lost in Space* was a thoroughly offensive program. It probably did more to damage the reputation of science fiction as a serious literary movement than all the B-movies about giant insects ever made—because *Lost in Space* was one full-color hour of trash reaching into millions of homes."[43] John Peel wrote that the show was "frequently illogical, silly and irrelevant," with "stories that sometimes seemed to have been written by toddlers, and effects work that bordered on the absurd." And that was in a *tribute book* to the series.[44]

Nevertheless, both *Lost in Space* and *Star Trek* (William Shatner's mannered performances, the miniskirted green alien females he romanced, and the heavy-handed moralistic plots also made *Star Trek* a camp icon in years to come) had tremendous afterlives in the tradition of Buck Rogers and Flash Gordon. Science fiction is the genre that least lets go of its favorites. *The Jetsons* moved into Saturday morning cartoon syndication for twenty years and spawned two more animated series in 1985 and 1987 (including an episode titled "The Swiss Family Jetson") in which Rosie was more prominently featured, as well as two television movies, a theatrical animated movie, and a direct-to-video movie. *Lost in Space* also went into syndication; took over the *Swiss Family Robinson* comic book and was featured in a 1991 comic book of its own (written by Billy Mumy, who had played Will); had an hour-long animated adaptation in 1973; and inspired an $80 million live-action movie that debuted at number one in its first week of release in 1998. The *Lost in Space* movie featured two robots, a giant warrior and a modernized version of model B-9. *The Jetsons* won the merchandising race with "coloring books, comic books, story books, figurines, puppets, puzzles, wind-up toys, board games, video games, costumes, lunch boxed, and stickers."[45] Collectors today can buy Rosie and Robot toys from a variety of companies, in multiple types and abilities, from different decades.

Adults relive their childhood memories through these rediscoveries and adaptations, with nostalgia giving the originals a roseate glow that those coming to the series with a fresh eye might not see. The 1960s' glorification of over-the-top camp silliness reached its peak in a 1969–1970 children's show that turned the already satiric spy spoof genre into Ambien dream territory. Word summaries completely fail to capture the surreal splendor of *Lancelot Link, Secret Chimp*. Link worked for A.P.E. (Agency to Prevent Evil), whose evil acronymic opponent is C.H.U.M.P. (Criminal Headquarters for the Underworld's Master Plan). All characters were played by dressed-up chimpanzees being begged by their handlers to do something, anything, that might be synced to the voiceover dialogue (much of which needed to be ad libbed to match the fleeting moments when the chimps' mouths moved appropriately). Dayton Allen voiced lead Lancelot by doing an impersonation of Walter Matthau imitating George Burns mimicking Humphrey Bogart. The first season ran on Saturday mornings, each hour-long episode containing two short live-action adventures; the last two minutes of the episode were given over to house band The Evolution Revolution, the chimps dressed in hippie drag. On February 26, 1970, the world shuddered as genius scientist Dr. Strangemind (whose Bela Lugosi imitation would have been rejected by Ed Wood) revealed his latest creation: "the biggest CHUMP in the world!" "It looks and smells like a robot," says C.H.U.M.P.'s leader Baron von Butcher. Viewers needed to take his word for it: what they saw was a stunt man wearing a gorilla's head over a rubber onesie, *Robot Gorilla* turned upside-down. Told to "smash" Lancelot Link, the robot responds robotically, letting nothing stop him until he bangs his head against the chimp-sized doorframe. That's all the action the kiddies would get: nobody on set dared to have chimps battle one another. The robot wisely is never allowed within ten feet of Lancelot. He's happier picking flowers or feeding ducks in any case: he's "The Reluctant Robot."[46]

The trail from the early 20th-century robots stops here. Bulky, shuffling, monotone, larger-than-life-size metallic robots kept appearing in media, but few creators meant them to be taken seriously. Their infantilization reduced their ability to appear as either menaces or superiors, nor could they any longer be considered the heirs to humanity. Camp objects can never be taken seriously. Robots had to change with the times. They did so by becoming people.

12

Robots and Kids

Television's explosive postwar growth took no one by surprise, its inevitable development forecast since the early years of the 20th century. According to its early adherents, television would transform the world, with instant news entering every household, the finest drama and art uplifting society, and education made efficient and pleasant. Instead, kid-oriented space serials were among the first shows developed at the end of the 1940s—cheap, crude, commercials that had more in common with cereal boxes than literature. Like radio soap operas, *Captain Video and His Video Rangers*, *Tom Corbett, Space Cadet*, and *Space Patrol* appeared five or six times a week, meandering through lengthy story arcs that mitigated the pain of missing an episode, which numbered in the thousands. Their almost-instant popularity was the green in black-and-white television, with producers cashing in on various tie-in products, from books to comic books and strips to toys and board games and collector cards to helmets and communicators to records and radio shows. Loyalty among the small-fry audiences was built in the same way that Buck Rogers used his Solar Scouts. Captain Video had a similar set of Video Rangers. Tom Corbett fans could become Space Cadets themselves and, as they aged, graduate to a set of young adult books that battled the Tom Swift, Jr., series for supremacy. Or young viewers could send away for an official Space Patrol Membership Kit that included a handbook, a patch, and a chart of the universe that somehow left off everything beyond Pluto. Though cruder and flimsier than even the Depression-era Solar Scouts gear, their incessant promotion on the programs created huge demand, cross-fertilized by advertising of kid-oriented products.

Wheat Chex and Rice Chex cereals included one of 24 Space Patrol "Magic Space Pictures" in every box, number 7 of which was Robot Man. They worked through the "magic" of persistent vision: children who stared at the negative black-and-white images saw a positive image when they shifted their eyes to a white wall or white cloud—pictures "in the sky!" Kids with slightly higher allowances who watched a television episode titled "Space Patrol Periscope" could hardly resist when boxes dangled before them the prospect of sending off for the 25-cent cardboard toy so useful to the cadets:

> CADET HAPPY: Now I get it! We can use the Space Patrol Periscope to look over each wall as we come to it and work our way to the center. Any guards?
> COMMANDER BUZZ COREY: No, I don't see any. Wait! There's some kind of guard—a robot! Take a look but be careful.
> CADET HAPPY: Smoking rockets, Commander! It looks like a tank![1]

Audiences ate these offerings up. The shows expanded from 15-minute episodes to 30 minutes, or moved from local stations to network broadcasting, or added a weekly show at a better time. Imitations spawned from their space seed, including *Captain Z-Ro* (1951), *Rocky Jones, Space Ranger* (1954), *Rod Brown of the Rocket Rangers* (1953), and *Johnny Jupiter* (1953). (An entirely separate British marionette series also called *Space Patrol* had to be renamed *Planet Patrol* when it started to run in the United States in 1963. Puppets allowed the use of nonhuman aliens and background sets that looked less earthly than rocks in the California desert but apparently forced the action to a snail-like pace. A telepathically controlled striped robot named Busy Lizzy, a female given chores like cooking and watering the flowers, appeared in season 2.)[2]

Kids couldn't get enough of space. The cover of the May 10, 1953, *American Weekly* magazine showed a beaming blonde boy in an immaculate "atomic" space suit, complete with bubble helmet and water pistol/zap gun sidearm, passing down his no-longer-wanted cowboy paraphernalia to a younger brother.[3] A lighthearted look at the phenomenon in *Collier's* magazine gave some sky-high numbers:

> Captain Video, the Dumont Network's space hero, and daddy of the current crop of TV rocketeers, is seen by an estimated 3,500,000 earthlings on each of his half hour episodes five days a week....
> A few months ago, when Captain Video made a personal appearance in a Brooklyn department store, some 10,000 television tykes lined up, to use the expression loosely, to shake his hand.[4]

Later that year, a *Life* magazine article on *Space Patrol* upped the viewership to 7,000,000 and credited the show with a remarkable $25,000 weekly budget, although the associated merchandise was said to bring in $40,000,000 that year.[5]

These shows' success immediately raises the question of why adult-oriented space shows remained nonexistent. The easy answer is money. Adults wouldn't generate the delicious ancillary income that *Life* credited to the bestselling "rockets and ray guns" as well as the "monorails, cosmic generators, 'paralyzers,' [and] T-shirts." Most of the kiddy shows looked as dirt cheap as a high school production. The prop budget for early *Captain Video* episodes was said to be $25 each.[6] Broadcasting from the building that housed Wanamaker's department store in Philadelphia, the crew might scrounge store shelves for hardware that could be disguised as a future tool. The show's famed "Opticon Scillometer" repurposed a car muffler, mirror, spark plug, and ashtray.[7] Adults couldn't help noticing the strained actors stumbling over the convoluted dialogue they'd had only minutes to absorb or the even more strained space slang—"space rat," "smoking rockets," "blow your jets"—meant to be a secret language for hip kids.

As with all kids' shows, some adults certainly got drawn in. The very first episode of *The Honeymooners* TV series features an epic blow-up between television set sharers Ralph Kramden, who wants to watch a movie, and Ed Norton, who recites the Captain Video oath while settling in for a voyage to Pluto, a purported adult happiest while playing with kids' toys.[8] A survey conducted by *Space Patrol* sponsor Ralston Purina discovered that adults were 60 percent of the audience, although that might have been because the show downplayed the annoying kid sidekicks while its female cast members wore the shortest skirts on television, the mini a sure indicator of future fashion ever since Buck Rogers artist Richard Calkins put Wilma Deering into one.[9] A picture in the *Life* article showed an upskirt shot of Nina Bara's "Tonga" climbing a ladder, captioned "Leggy Moment." One wonders how many suddenly space-happy dads joined their kids in front of the set the next week.

Yet the omnipresence of cheap props and live broadcast goofs on early earthbound television hints that a deeper explanation is needed. Science fiction had yet to overcome its reputation as the home for ludicrous exploits of cardboard characters; writers of visual science fiction tended to rely on clichés of aliens and monsters rather than exploring ideas. (Anthology television series *Tales of Tomorrow* [1951–1953] and *Out There* [1951] were exceptions, but few of the episodes directly involved space. The one robot "tale of tomorrow" is essentially horror rather than science fiction, telling of what happens when a robot designed merely to read books impossibly comes alive and falls in love with the professor's daughter.[10] The robot's body, admittedly a mere conceit of its builder, constitutes the cheapest costume in media history, making the kids' serials' visual effects spectaculars by comparison.) Science fiction writers probably did themselves few favors by monomaniacally insisting that space was the future, implanting into a generation of audiences the notion that science fiction was not truly science fiction if it lacked spaceships.

The pressure to churn out thousands of episodes forced creative teams to play with every science fiction trope, making robots inevitable presences. *Captain Video and His Video Rangers* (1949–1955) brought them in first, although not on the television show. Having just spun the show off into a board game and comic books, the producers also licensed a 15-part movie serial, called *Captain Video: Master of the Stratosphere*.[11] Despite the misleading name—the stratosphere is the portion of the atmosphere just *below* space, the troposphere, but high-flying jets had made *stratosphere* more familiar—and the substitution of different actors for the entire cast, the serial succeeded in expanding brand awareness (if little else). Digging in the depths of the prop warehouse, the crew members found the old robots used in the 1935 *Flash Gordon* serial and repurposed them as the creatures of evil Dr. Tobor. Clearly Tobor—once again *robot* spelled backward—should be the name of the robot instead, a mistake finally rectified in 1953 when a giant evil robot of that name, played by 7'7" actor Dave Ballard, both fought and worked with Captain Video (depending on who was in control at any moment).[12]

Rocky Jones, Space Ranger (1954) was a late entry: part serial, part stand-alone, produced on film rather than live. It lacked a network contract, so the films were syndicated to individual stations. Most plotlines ran in three parts, so the stations could splice them together into 80-minue films or break them up into weekly episodes. Amortizing the sets in this way also allowed for better visuals, although the scripts and acting stayed rock bottom. Merchandising was the main casualty, as no one could be sure what was being seen where when. A Space Ranger Fan Club meeting must have been a lonely affair.

Set in 2054, an even 100 years ahead of original broadcast, the show followed the exploits of the Space Rangers as they journeyed among various moons on which square-jawed Rocky could throw fists at stock heavies and be lusted after by young queens and suzerains, lacking nothing but a subtitle reading "five-year mission." The late three-parter called "Out of This World" in television-episode form (retitled *Robot of Regalio* as a movie) added the evil Nizam and an indestructible robot. The Nizam's city received its power from the center of planet Regalio, which he magnetically beamed across the planet and into space, with the potential to pull the Earth out of orbit. Leakage from this incredible power source had destroyed all living beings until the Regalions controlled the power with a single robot, standing in the middle of the apparatus like a humanoid on/off switch. Moving the robot off center shuts off the deadly magnetic rays—and it also conveniently opens the door of the prison Rocky is being held in. (Just as conveniently, the lights don't go off when the power dies.) Several fistfights later, the Nizam is killed by his own robot creation and the good guys are reunited in time for the next adventure.

The part of the robot could have been played by a capacitor; it's there exclusively to escape, wreak havoc, and be excitingly recaptured at the last second.[13]

Cliff Robertson was already 30 when he starred in his first series, *Rod Brown of the Rocket Rangers*. Written by a working surgeon and an ex-vaudeville actor, the 1953–1954 series was such a knockoff that the *Tom Corbett* producers hit it with a lawsuit before the first episode appeared.[14] Robertson quietly spent 58 straight weeks, working live, with no vacation breaks, on Saturday mornings fighting menaces like "The Robot Robber of Deimos," a humanoid robot sticking up a Martian bank.[15]

Print science fiction comfortably encompassed silly monster fantasy, wild galaxy-spanning super-science battles, and quiet near-future extrapolation, exemplified by Robert A. Heinlein's adult and juvenile stories. Space shows also sought niches to claim as their own, with *Tom Corbett* being the most scientifically realistic, heavily inspired by Heinlein novels like *Space Cadet* (1948). Tom's adventures naturally featured staid, sensible robots, as in the two-part 1952 radio episode "The Giants of Mercury," whose robots were, almost uniquely, properly designed for their jobs. Investigating a case of Zero-A fever on Mercury, Tom and his crewmates encounter giants who turn out to be less scary than at first glance, mining robots rather than aliens or monsters.

> DOCTOR: Thinking of those robots as machines and not grotesque humans, they don't seem so ugly.
> TOM CORBETT: Wow, I don't know.
> NURSE: Oh, Doctor, with six arms, and those short, thick legs like a huge hippopotamus...
> DOCTOR: Well, they're functional. Legs like that can support tremendous weights. And they keep the center of gravity low.
> TOM CORBETT: That's right. And the arms are fitted with heavy claws, steel fingers, a drill ... and look here, a sort of shovel.[16]

Tom outlived his media appearances with a series of young adult books. The last Tom Corbett book, *The Robot Rocket*, is not true robot fare, but by then he had been out-classed by another Tom—namely, Tom Swift, Jr., the king of young adult science fiction.

The original Tom Swift was still a household name in the early 1950s, the exemplar of the American mechanical wizard, a young Tom Edison who never spoiled the image by aging into an *éminence grise*. Edward Stratemeyer birthed Tom Swift in 1911, after selling millions of copies of the Rover Boys and the Bobbsey Twins adventures. Tom Swift books were engineering fiction, not science fiction: Tom tinkered, first with standard motorcycles, tanks, and submarines, then with whatever marvel entered popular science writing over the past year. A 1926 survey of children found that 98 percent of them read Stratemeyer books, and Tom Swift was first among equals. Probably knowing that the end was near, Stratemeyer married Tom off to his perpetual sweetheart, Mary Nestor, in 1929 and died in 1930. The series staggered on until wartime paper shortages killed it entirely.[17]

Harriet Stratemeyer, Edward's daughter, helped launch the Hardy Boys in 1927 and Nancy Drew in 1930, which quickly took over as the bestselling series. That left a conspicuous hole in the postwar market. Scattering plots to four authors, Harriet conjured up Tom Swift, Jr., and showered the 11–16-year-old reading public with five titles in 1954, all said to be written by Victor Appleton II.[18] A gushing *Time* magazine tribute noted that "three scientists, all Ph.D.'s, have been hired to ride herd as science advisors."[19] The fourth book, ghostwritten by Russell Sklar (his only contribution to the series), was the sooner-rather-than-later *Tom Swift and His Giant Robot*.

Even without its "head," the gleaming, silver-gray automation stood seven feet tall. Tomasite, the young inventor's wonder plastic, covered every part of its frame except the joints. These were enclosed in "sleeves" of fine Tomasite chain mail which stretched and contracted with the movement of the joints.

"I suppose the transmitting and receiving antenna will be in the head?" Mr. Swift asked.

"Right, along with the television 'eyes' and radio 'ears.' After the giant's head is on, he'll be remotely controlled. Right now," Tom went on, pointing to a cable protruding from the back of the robot's neck and running to a control panel on the wall, "I have to use a direct control and monitoring method."

"What can your giant do so far?" Mr. Swift asked.

"Walk, and do almost anything with his hands. Want to see him thread a needle?"[20]

Why a robot, let alone a giant one? The rationale couldn't have been more topical: Tom's father, the original Tom Swift, was building a private atomic power plant. The core of the reactor, lovingly described (no doubt cribbed from one of those PhDs), would be too radioactive for any humans to work in. However, Tom's robot would do more than carry the heavy uranium slugs. As Mr. Swift explains:

"The heart of the reactor is in the center. In there the uranium will be bombarded with neutrons and changed into the various transuranium elements. Then the slugs are taken from the pile and the robot separates out the new elements in his own completely equipped chemistry lab over there." He pointed to an enclosure whose walls were lined with the necessary chemicals in radiation-proof containers. "After that, he prepares them for shipment to medical and scientific institutions."

"And where will the waste products—such as the slug casings—go?" Hanson asked.

"Tom's robot will carry them out through a tunnel to an underground lake we've made. In that way, no living thing can be contaminated by the radioactive waste."[21]

As always, a team of vicious crooks interferes. Using the plot template laid down a century earlier in the Steam Man dime novels, Tom and his friends are regularly attacked, kidnapped, and knocked unconscious, and they are forced to make last-second escapes through their wit and daring, the only difference being that the endless series of gadgets Tom pulls out of nowhere are electronic rather than mechanical. The villains are not Commies, but the Swifts' advanced technology exists solely to further the strengths of the United States. The utopian world governments of the kids' shows were no longer a favored future in an America feeling besieged by enemies on all sides and at home. Stratemeyer had perfectly timed the rebirth of an all–American boy genius patriot, one who fought for America using American know-how backed by American industrial might. Very shortly Tom would be racing the evil "Brungarians" to the moon. The Cold War had more influence on the world than Tom's inventions.

Amid all the classroom exposition served up by Tom, any rationale for the robots to be giant humanoids is conspicuously missing. It was left to television's *Captain Z-Ro* to at long last provided some vaguely plausible reasons for designing a humanoid robot. If Tom Swift was purportedly educational, the good Captain (and his kid sidekick Jet) should have been on PBS. Pedagogic television had been predicted by futurists from Hugo Gernsback onward, who anticipated a classroom inside homes whose standardized lessons might reach thousands, perhaps millions. The rapid emergence of television as a commercial enterprise scuttled this high-minded ideal, but many local stations devised some programming that tried to instruct as it entertained. Two, in San Francisco and Los Angeles, had the promising idea that placing a few actors in period costumes could teach history by introducing visual elements. The science fictional aspect came from time. Each week, Captain Z-Ro, in his secret laboratory, used his time viewer to spy on famous historical events, learn that they were going awry, and send Jet back in time via his materializer to correct the course of history. "Peabody's Improbable History," on the cartoon series *Rocky and His Friends* (later *The Bullwinkle Show*), surely had its roots in *Captain Z-Ro*, and it was namechecked in an episode of the similarly themed *Quantum Leap*.

Revived in late 1955 as a 30-minute, filmed, syndicated series, that season of *Captain Z-Ro* introduced Roger the Robot. The Captain's flash of brilliance is to create a robot that he can send to Venus—the planet thought at the time to most resemble Earth—and transmit images of the planet using its eye cameras, without risking any human lives on an unknown world (the Mars Rover of its era). Sadly for science, an electrical storm bedevils the secret lab, and despite Jet's constant warnings (a rare instance in which a kid sidekick is right and the adult plows stupidly ahead regardless), a lightning bolt hitting the transmitter sends the robot into downtown San Francisco. We do not see the world through the robot's eyes, as promised; footage of a man in a robot suit scaring tourists at Coit Tower gets substituted. The frazzled Roger is brought back, his condensers smoking. "Roger can cope with anything in outer space, Jet, but I'm afraid our modern civilization was just too a little too much for him," pontificates the Captain in his mock baritone.[22] The admission of technological defeat was a huge anomaly in the 1950s but short-lived. Roger would be brought back to work the materializer's controls, allowing the Captain to go on adventures with Jet. In "The Great Pyramid of Giza," Roger saves the day with a brilliant idea of his own: setting an automatic timer so he can journey back and save the human cast from certain doom when they are caught in the king's chamber.[23]

Johnny Jupiter was an outlier among these adventure series. After a short stint on Dumont in the spring of 1953, ABC brought the show back, completely refurbished, for the fall season. Ernest P. Duckweather was an amiable failure of a 30-year-old who worked in Horatio Frisby's general store and cared only about his invention: a two-way television set that brought in Jupiter. The planet's inhabitants were all played by puppets, the titular Johnny Jupiter and his cohorts, Major Domo and Reject, the factory reject robot. Reject's defect was that he wore glasses. (In a late episode he had an evil twin, named Defect.) When Ernest got into trouble, he could send for Reject, normal human-sized on Earth, although why he did so is questionable since Reject made matters much worse.[24]

Reject fit remarkably comfortably into the puppet setting, so much so that in retrospect it is surprising that producers didn't understand the lesson that robots and kids made a far better match than robots and adults. Robots were the dinosaurs of their day, toys that allowed younger kids to feel big, ferocious, strong, and victorious in a world where they were anything but. Little kids needed robots in a way that older kids did not.

Johnny Jupiter was not a hit, of course, and neither was the other example that could have sent the light bulbs flashing over creators' minds. Though a solid success, the movie *Forbidden Planet* was nothing like a box office bonanza. Its producer, Nicholas Nayfack, felt the need to reuse the expensive Robby the Robot in another movie. Tellingly, he didn't find source material in any of the dozens of robot stories in science fiction magazines. Nayfack stuck to reading material he was familiar with—specifically, the *Saturday Evening Post*, where he found an Edmund Cooper story about a boy who became invisible. No robot appears in the story, but Nayfack instructed *Forbidden Planet* screenwriter Cyril Hume to build the screenplay around one.

Hume whipped up *The Invisible Boy* (1957), a prototypical 1950s tale of Cold War paranoia, generals with chips in their heads to make them obedient, secret rocket launches, and an all-knowing computer brain that wants to eliminate all living creatures down to viruses from the entire universe. Somehow all that was wrapped around an almost sickly sweet tale of a boy and his faithful robot. Timmie is the lonely son of a genius computer builder, who loves his son but is frustrated that he can't teach the 10-year-old chess, or even fractions. Dad brings Timmie into his top-secret laboratory in the hope that seeing his work, the calculations for America's first satellite, will stimulate the boy's interest. He doesn't realize that the computer uses sleep learning to turn Timmie into a super genius. Bizarrely, the computer center also contains the scattered pieces of Robby (explained by a casual reference to a scientist inventing time travel). The scientists are so steeped in their own everyday miracles that they merely shrug when Timmie puts Robby back into working order. A red-blooded all–American boy, Timmie uses Robby for the sort of innocent mischief that his overly protective 1950s mother wants to quash. With childish logic, he figures that if she doesn't see him in danger, he won't get into trouble, so Robby whips up an invisibility serum. An invisible Timmie winds up hiding in the rocket—the worst possible hiding place. The computer brain itself wants to be launched into orbit so it can control the world from space, but it needs information from Timmie's dad. Robby, now under the computer's control, is sent to the rocket to torture Timmie but can't hurt his one true friend.[25] (The plotline rips off 1954's *Tobor the Great* to a remarkable extent, perhaps a factor in its commercial failure.)

Every aspect of *The Invisible Boy* would be reflected in later productions, from kids saving the world with gadgets (as in *Spy Kids* and *Big Hero 6*) and computer brains plotting to take over the world (from *Colossus: The Forbin Project* and *Demon Seed* to *WarGames* and *The Terminator*) to the robot betraying its masters because of friendship (like Hymie in *Get Smart*). Robots often battle other robots; *The Invisible Boy* uses a different and more metaphorical path: the personal—here a robot with emotions—against the impersonal—a computer brain lacking all empathy, seeking only power. In a way, the moral of *The Invisible Boy* is as powerful a 1950s parable as any of the more obvious paranoia message films. Americans suddenly found themselves confronted by forces that could destroy the planet, with their deaths mere collateral damage. Nameless, faceless enemies were impossible for the bravest GI Joe to fight. Robots, by contrast, were old technology—comfortable, understandable, within the ability of a smart kid to put together. Newspaper and magazine articles throughout the 1950s presented readers with home-built robots (see chapter 6). Kids couldn't build intelligent computers; that required teams of brilliant adults in white-walled labs, extending the image of the atomic bomb makers at Los Alamos unthinkingly pushing horrors out their doors. Out-of-control computers became the symbol for evil in this postwar era, yet another reason the once-feared robot wandered into the corridors of camp for a generation.

—⅏—

Space toys started with Buck Rogers. In 1934, the Daisy Manufacturing Company issued the XZ-31 Rocket Pistol, 9.5 inches of "fantastic-looking heavy metal gun which makes a snapping sound (called 'zap')," according to *The New Yorker*.[26] (A cardboard gun was available from Buck's radio sponsor, Cocomelt, in 1933.)[27] Soon parents could outfit their children from head to toe in Buck Rogers and Wilma outfits and equip them with guns, watches, rockets, and membership in the Buck Rogers Solar Scouts, Earth Division, whose membership form asked, thrillingly, for "Previous Rocket Ship experience, if any."

Thousands of zap guns, rocket ships, and futuristic paraphernalia followed over the next two decades in America, frantically sought-after collectibles that might today bring tens of thousands of dollars—if found unplayed-with and in their original boxes. It beggars belief that not one of those toys was of a robot; yet these fanatical collectors have all but clear-cut the past in their search and found nothing.

The alternative, for enterprising British kids, was to build their own. Meccano was the British forerunner of the Erector set and published a hobby magazine to tout its goods. In 1931, it ran an article that neatly linked robots with the future:

> AMONG the many mechanical marvels that scientists tell us will be common sights in the year 2031 A.D., none can be of greater interest than the "robots" or mechanical men which, we are assured, will be used to perform almost every action of a human being. Meccano boys may well envy their fellow constructors of the next century using a "squad" of these "mechanical humans" to do their homework while they complete a model of a super 500 m.p.h. land plane, or the latest pattern of a moon rocket!
>
> Although the remarkable Meccano model robot described this month is not capable of working out Square Roots or Compound Interest, it will nevertheless walk forward in a remarkably realistic manner, merely by pressing one of the Pulley Wheels that represent this weird individual's "ears"![28]

The Japanese pioneered the true robot toy, called Lilliput, Robot N.P. 5357, in the late 1930s, or early 1940s, or possibly late 1940s (most records were destroyed in the war). A true robot, with a tin box for a head over a body made from another tin box, Lilliput shuffled forward when wound up with a key. So did Atomic Robot Man from, definitely and obviously, the late 1940s. It's doubtful that many American kids ever got to play with one, although a handful were given away at the New York Science Fiction Conference held in June 1950.[29]

Japanese companies continued making tentative inroads into the U.S. market, with Alps' Mr. Robot the Mechanical Brain and Nomura's Zoomer the Robot both introduced in 1954.[30] With space toys taking off as well, the time had come for American-made robots. The Ideal Toy Company introduced its robot toy as if it were what every Solar Scout had been waiting a lifetime for. Robert the Robot made headlines from its introduction at the 1954 New York Toy Fair, proclaimed exactly what the "very up-to-date" modern child needed. Nobody told the kids, of course: they "raced" through the fair giving "most of their attention to old fashioned 'mama' dolls and teddy bears." Kids don't buy toys; parents do. Olga Curtis' syndicated report was designed to make parents feel like failures if they didn't bestow this symbol of the future on their children for Christmas:

> This $6 toy right out of science-fiction is a foot-high mechanical man who talks, moves and picks up things.
>
> Controlled from a distance by a separate mechanism, "Robert the Robot" rolls forward, back and sideways, clutches small objects in his hook-like grippers and says a complete sentence announcing his name and ability.[31]

A publicity shot of Tobor, the robot from 1954's *Tobor the Great*, pulling little Robert on a leash—the perfect robot's pet—made seemingly every paper in the country; Robert also merited an appearance in *Popular Science*.[32] However, all that hype paled before his crowning achievement: inclusion in the 1954 Sears' and Spiegel's Christmas catalogs. Robert's smash success led Ideal to introduce sixteen more talking toys in 1955.[33] A dozen manufacturers lined up to deliver Robert merchandise, everything from T-shirts to flashlights. Ideal also made a 9-foot-tall version to handle personal appearances at department stores.[34]

Truthfully, Robert doesn't scream "futuristic" to a modern eye, resembling a heavily embellished cowbell because of his "skirted" style. Instead of attached legs, like Lilliput had, Robert's metal cover flared out from the waist down as a means of hiding the awkward "walking" movements. Besides, you had to turn a crank on Robert's back to hear him say, "I am Robert the Robot, mechanical man. Drive me and steer me wherever you can," in a voice as scratchy as on one of Edison's first cylinder phonographs.[35] Skirted robots nevertheless became the industry standard and were hugely influential on the Japanese toy industry, which introduced knockoffs by the score. Robert's star power is revealed by his cameo role in Douglas Sirk's 1955 melodrama about a toy manufacturer, *There's Always Tomorrow*. Robert's tomorrow lasted through the 1950s until Ideal introduced far more interesting models, like Robot Commando. At 19 inches tall, he looked like he could eat Robert for breakfast. Tiny balls emerged from his fists, while his head opened to shoot rockets.[36] Instead of a gooey adult romance, Robot Commando was featured in a *Man from U.N.C.L.E.* episode, in which Thrush modified him to shoot real rockets. His one weakness was a price tag above ten dollars. The real robot wars weren't the Japanese against the Americans; it was kids against their parents' wallets.

Sales of cheap Japanese robot toys dropped throughout the 1960s, largely for the ironic reason that Japan's recovery from the war sent labor and material costs soaring. Worse, space toys replaced robots in popularity as astronauts temporarily sucked all competitors into their rocket wash. One robot prospered by offering an alternative.

"Knock his head off!" is a primal exhortation that started appearing in American publications in the mid–19th century. Primal emotions make for successful toys. Louis Marx and Company made it literal in 1964 when it introduced Rock 'Em Sock 'Em Robots for the Christmas toy season. (The name "Rock 'Em Sock 'Em" seems to have come from college football cheers in the 1940s.) "Few things in life, let alone toys, provide as much visceral satisfaction as Marx's battling automata," exclaims a book on vintage toys.[37] The 12.5-inch-tall Red Rocker battled the Blue Bomber in the middle of a candy yellow 20-inch ring. Levers topped with buttons on each side controlled the robots' movements. Two kids squared off against one another, maneuvering the robots into close proximity, and a press of a button sent the robots' arms flying. "A well placed punch to the jaw and BINGO! ... the boxer's head flies up," screamed an ad. "Press it down again and the fighters are ready for another round" (ellipses in original).[38] Made of high-impact polystyrene, the robots stood up to endless hours of punishment. Kids skilled in levers and buttons may have had a huge advantage when, as adults, they found these tools incorporated into early video game controllers.

—⟊—

Astro Boy occupies a pedestal in Japanese popular culture roughly comparable to a mix of Mickey Mouse and Superman, with creator Osamu Tezuka serving as a combination of Walt Disney and Siegel and Schuster. Both American characters were wish-creation fulfillments of the little guy triumphing over overwhelming odds, a mind-set that meshed with the status of individuals in postwar Japan. Almost every city in the country had been bombed to rubble, manufacturing even basic necessities was difficult, and the aftermath of the atomic bomb loomed as a symbol of ignominious defeat. As it had in America during the Depression, popular culture, with its simple joys, eternal optimism, and strong fantasy elements, made the dreary realities of everyday life more bearable. Just nineteen and already a medical school graduate, Tezuka turned to writing and drawing manga. His *Shin Takarajima* (*The New Treasure Island*) (1947) made him famous and is considered the first *kindai manga* (modern manga).[39] The Japanese language has a number of terms for various forms of cartoon entertainment that have been oversimplified in American usage to *manga*, used for all forms of comic books (apt, as the word is derived from "whimsical pictures"), and *anime*, a shortening of *animēshon* (a phonetical equivalent of animation), used for all animated cartoons.[40]

Tezuka used the atomic legacy as the basis for "Ambassador Atom," which began running in *Shonen* magazine in 1951. A minor character named *Tetsuwan Atomu* (Mighty Atom) overshadowed the rest of the cast, so Tezuka reimagined him as a superhero protector of the world in his own strip that ran from 1952 to 1968 in *Shonen* and later in a variety of publications, whose overwhelming popularity elevated Tezuka's status to the "God of Manga."

Tetsuwan Atomu also became the pioneering anime television series in 1963. Forced by budgetary constraints to use limited and black-and-white animation, the strong storylines overlaid on fast-paced sight gags straight out of Looney Tunes made the show an instant hit in Japan. Far more surprisingly, it also wowed American audiences later that year. Translated into English as *Astro Boy* (a downplaying of his atomic origin), the series ran in the Saturday morning cartoon bloc for 193 episodes, more than *The Flintstones* and *The Jetsons* combined.

Astro Boy's 1960s anime origin has a startlingly contemporary feel. In the year 2000, Astor Boynton (a.k.a. Toby Tenma) is in a self-driving car, controlled by the highway, the safest mode of travel ever devised, when he gets hit by a truck. (The original Japanese cartoon had him killed through his own reckless driving, one of numerous changes made to appease American sensibilities.)[41] His distraught father, Dr. Tenma, the head of the Institute of Science, embarks on a year-long project to create an android duplicate of his son. The superstrong little boy, with a 100,000-horsepower engine, is a success, especially after years of scientific education make him a "mental giant." Dr. Tenma is less of one. Forgetting that robots don't age, he berates his son for not growing up and then does something unthinkable in an American cartoon: he sells Astro to a circus full of performing nonhumanoid robots. "You're not a human child," he says. "You're nothing but a machine, like a refrigerator or a dishwasher."[42] Japanese circuses are apparently not like our own. Astro is pitted in a gladiatorial duel to the death against a giant monster robot—part man, part bull, part crab. He prevails (although hurting anyone is anathema), only to discover that the other robots are being deprived of electricity prior to being scrapped. He connects them all to his own power source before their last performance, enabling the energized robots to save the audience when the circus catches fire while the almost powerless Astro risks his life to protect the evil circus owner. A grateful public passes a robot bill of rights, freeing all robots from being bought and sold. Astro went on to have hundreds of adventures against other giant robots, monsters, aliens, and various nefarious menaces in the 1960s and in two color series in 1980 and 2003.

A full-length color *Astro Boy* movie appeared in 2009, set in a much different future world. This version of Dr. Tenma sees his son killed before his eyes when the government wants to pervert his good robots to military uses. The movie clarifies that Tenma's realization that his robot can never replace his son is what turns him against Astro. Pursued by the government, the robot flees from the flying city to the wretched scrapheap that is the Earth's surface. There he befriends other robots (now equipped with Asimov's three laws), fights a gladiatorial battle, defeats an even larger government robot, and dies heroically. Of course, he is remade for the happy ending.[43]

With the success of *Astro Boy*, other robot anime series were rushed into production and many found their way to America, some even more radically transformed. *Johnny Sokko and His Flying Robot, 8th Man* (yet another Tobor), and *Gigantor* had robot lead characters, while *Battle of the Planets* and *Captain Harlock* included robots as part of a larger crew, and the many science fiction–themed anime series regularly featured robots in various episodes. Robot collector Justin Pinchot offers an explanation for the appeal of robots in Japan:

> The whole concept is that ... you now have something to do your bidding, something that is equally as impressive as your enemy but under your control. Whether it's Rosie the Robot, Tetsujin, or Astro Boy, the concept is always the same.
>
> The basic premise of robots is that they are willing and able to do what we cannot or will not. They allow us to be in two places at once. And they're physically stronger than we are, so they can do work for us like a machine. There's a Japanese robot called the "Busy Cart" robot, which is a construction worker robot with a big wheelbarrow. The real fun comes when we cross the line and give them a personality or thinking powers.[44]

Toys that can be played with in multiple ways spark creativity, and toys that transform into other toys have deep roots, from sticks and boxes to Lincoln Logs and Legos. Japanese shows made these products a specialty. Mattel's Shogun Warriors robot toys were based on anime, especially 1975's *Force Five: Grandizer* (known in Japan as *UFO Robot Grandizer*). Some of the Mattel toys could convert into other shapes, a concept copied by other companies to create enormous toy and media franchises. The *Robotech* series in the United States (1985) was stitched together from three short Japanese *Macross* series to produce a sufficient number of episodes. The robots technically were battle suits, worn and controlled by humans, but they changed into cars, motorcycles, planes, spacecraft, and whatever else ingenious engineers could devise. Manufactured by the Matchbox toy company, the first toys lived up to the name by being less than four inches tall, although later variations grew a bit larger.

The Transformers line emerged from Takara's 1970s Japanese toy lines Diaclone and Microman. Hasbro distributed them in the United States starting in 1984 and eventually bought the entire line. With the number of Transformer models reaching into the hundreds, along with the video games, children's books, comic books, cartoons, and six blockbuster movies grossing about $1.5 billion in America and more than $4 billion worldwide, the Transformers are the most successful and widely spread fictional robots in all robot history.

Dozens more cartoon series based on Japanese models or created by American companies to compete with their own multimedia lines were spawned from these low-resolution but fiercely beloved early characters. From then on, there might not have been a single moment when a kid couldn't tune into the Saturday morning or late afternoon cartoon bloc on regular television (or all day on one of the many kid-oriented cable television channels) to see a brand-new show starring a robot, presenting a robot as a valued member of the team, having a robot as a member of the household, or introducing a robot in a special guest role, all alongside the endless repeats of earlier favorites. Kids looking for robots often found the name right in the titles: *The Mighty Orbots* (1984); *The Bot Master* (1993); *Whatever Happened to ... Robot Jones?* (2002); *My Life as a Teenage Robot* (2003); *Super Robot Monkey Team Hyperforce Go!* (2004); *Robotboy* (2005); and *Bolts and Blip* (2010).

The child android character pioneered by Tezuka suddenly blossomed in 1985, when writers in three separate media—movies, television, and books—burst forth with androids as children. June of that year saw the release of *D.A.R.Y.L.*, a movie starring Barret Oliver, the hot child star of the mid–1980s. Suffering from amnesia, ten-year-old Daryl gets placed into a foster home in a nostalgically perfect small town full of tree-filled lawns, little league baseball, and an all–American best friend nicknamed Turtle, as mischievously irksome as Timmie from *The Invisible Boy*. Kids like Turtle are the norm in small-town America, so much so that Daryl's foster mother actually gets concerned about how polite and helpful Daryl is and Turtle has to tutor him in the art of "pissing adults off" so he won't be suspected. Daryl doesn't know that he's a robot, a D.A.R.Y.L.—Data-Analyzing Robot Youth Lifeform—created by the military as a computer-brained cyborg that would be the ultimate soldier. Putting a prototype killing machine into a child's body was bad decision making on everybody's part made even worse by the scientists' growing realization that they had made a real boy in a real boy's body, which is why one member of the team helped Daryl to escape in the first place. The life lesson falls squarely into robot history: if outsiders can't tell the difference between a robot and a human, then the robot is human. All ends ideally—at least in the movie.[45]

The equally acronymic Voice Input Child Identicant, or "Vicki," starred on a syndicated television series in 1985 known as *Small Wonder*. Ted Lawson, an engineer at United Robotronics, built the android in secret, passing her off as an orphan taken in by his family.[46] Vicki, played by 10-year-old Tiffany Brissette (who fortuitously didn't have much of a growth spurt as she moved into her teens), is a classic throwback creation, dressed like an early version of Alice in Wonderland, speaking in a monotone, puzzled about human emotions, and capable of a variety of one-off superpowers whenever the script called for them. Equally classic were the plotlines that depended on the wacky antics required to keep her secret from the nosy next-door neighbors. Both threads combined in an episode in which the neighbor rushes Vicki to a hospital after she falls into a pool and gets waterlogged.[47] The doctor is flabbergasted by x-rays that reveal Vicki's insides to be a motherboard, the neighbor faints when he sees Ted open the control panel on her back (it wouldn't do to have anyone reach into a preteen's front), and Vicki dries herself out by superheating her insides, a power handy in a later episode when she emits a burst of steam to de-wrinkle a suit. (Computer chips and high heat don't mix well together, but the writers never pretended they were doing science.) *Small Wonder* lasted for four years and four hundred robot jokes.

Seth McAvoy's *Not Quite Human #1: Batteries Not Included*, the first of a six-book young adult series, hit bookstores in October 1985. Dr. Jonas Carson is the creator of a human-looking, apparently thirteen-year-old android boy, named Chip, whom he wants to send to school as a real-world field test in learning how to be human. Dr. Carson is the type of scientist who frequently shows up in robot fiction, one brilliant enough to create an android with a command of English but stupid enough to forget to include any slang or idioms. The literal Chip therefore spends every page of the book obeying inadvertent commands or responding weirdly. As the new boy in school, Chip is immediately suspected when a series of burglaries occurs and he becomes the target of the school bully, two threads that come together in a positive ending.[48]

McAvoy's novel was made into a 1987 television movie under the title *Not Quite Human*. (By coincidence, a totally unrelated movie titled **batteries not included* was also released in 1987.) The movie made Chip age seventeen and retained only the concept and a few of the school scenes in which Chip's literal interpretations annoy teachers and make him a weird but antiauthoritarian quasi-hero to his peers. The central conceit was framed by a new plot in which Chip, originally conceived as an advanced toy warbot but subverted by Carson into his dream project, is seen as the prototype for an army of android soldiers. The evil corporate head tries to steal him away from Carson but is foiled by Chip's love for his human family.[49] Both versions of this story were hits, followed, respectively, by five more novels and two sequel movies.

—⁂—

Ted Hughes' wife, the poet and novelist Sylvia Plath, killed herself while their daughter and son were still toddlers. A few years later, Hughes began writing *The Iron Giant* (1968) (originally *The Iron Man: A Story in Five Nights* in the United Kingdom, changed to avoid confusion with the Marvel Comics character) as a bedtime tale for his children. Lorraine Kerslake of the Ted Hughes Society called the book "a healing allegory, a story of redemption, a story of healing a fractured self."[50] The book pulls no punches: on its very first page, the Iron Giant breaks into a million pieces falling off a cliff. He slowly puts himself together, his eye finding a hand that scuttles on its fingertips until another hand is found, and on and on until the entire giant body is whole—except for one ear, never to be found. Completeness cannot be achieved after such a tragedy.

The "Five Nights" in the original's subtitle indicates five stories. In one the Iron Giant menaces farmers by eating their tractors and barbed wire and other iron goods. He is captured and buried under tons of earth, only to resurrect himself, his enormous hand bursting out of its grave like a horror movie monster. Young farmboy Hogarth has the inspired notion of leading him to a scrap metal yard, an endless buffet in which the monster is tamed and happy. Of course, a superhero must have an even more powerful supervillain. Drawing on the English legend of St. George and the dragon, Hughes creates the ultimate antagonist in the final tale: a dragon literally the size of Australia who demands humans to fill his belly. The Iron Giant must win without fighting an impossible fight. He does so by challenging the dragon to a literal trial by fire: each will roast their bodies in appropriate ways. The dragon flies to the sun … and survives. And survives again. But a third trip is too horrifying. The "space-bat-angel-dragon" is banished into space, where it sings the music of the spheres.

> The strange soft eerie space-music began to alter all the people of the world. They stopped making weapons. The countries began to think how they could live pleasantly alongside each other, rather than how to get rid of each other. All they wanted to do was to have peace to enjoy this strange, wild, blissful music from the giant singer in space.[51]

Brad Bird, later of Pixar fame, adapted this utterly strange fable into an animated movie the only way possible: by throwing away everything except the title. Americanized to the hilt, *The Iron Giant* (1999) is set in small-town USA in 1957, just after Sputnik and at the height of Cold War paranoia. Hogarth is now the archetypal adventurous American kid, descended directly from Timmie. He wants a pet; instead, he encounters the Iron Giant in the woods behind his home and adopts him—a boy and his hundred-foot-tall dog. Trying to keep a pet that size secret is hopeless. The government sends in a paranoid investigator convinced that anything not made by Americans is a menace to be destroyed. When the army arrives, a showdown is inevitable. To be true to Hughes, the movie should have devised a nonconfrontational ending. No Americanized picture could end without a climactic battle, however. The message of Bird's movie is that the best defense is a good offense. The Iron Giant may hate guns and try to destroy them, but all that means is that when he is confronted by a big gun, he pulls out an even bigger weapon, upping the ante with each round of ordnance thrown at him by becoming ever more lethal himself, the biggest gun of all, a combination of Gort and the Terminator. When the crazed investigator calls in a nuclear strike on the once bucolic village, the Iron Giant flies into space to intercept it. The last images in the movie are pieces of the indestructible machine once again scrabbling toward one another as he starts to reassemble himself.[52] Hughes was said to have read the script and liked it, though he died before production began.[53] One wonders.

—∿—

The movie *Robots* (2005) is a one-stop compendium of every visual joke about mechanical people. Based on concepts from children's book author and robot fancier William Joyce, the story of Rodney Copperbottom starts with his birth. Sadly, his father misses the delivery … of a Build a Baby box full of parts. "But it's okay," his wife says. "Making the baby's the fun part." His father is a lowly dishwasher (literally), but Rodney is an inventive genius and is told to follow his dreams. He takes them to Robot City, a visual fantasia that is a Rube Goldberg device in 3-D, leading to fantastic set pieces that contain the spawn of every OK Go video. Thousands of machines were harmed in the shooting of this movie, but that's also part of the gag; as machines, robots can fall to pieces, have new ones added, get them mixed up with one another's, function with half of their parts missing, and generally get more abuse than Roger Rabbit in Toontown while still dependably delivering laughs. The villain is Rachet, an evil corporate head with a terrifying mother. Their plan is to stop making spare parts and only sell new ones, meaning that poor robots, the "outmodes," will get recycled in the mother's fiery hell-like furnace. Rodney's impromptu upgrades to his friends turn them into a rag-tag group of superheroes, but at the end of the movie the entire population of outmodes fights for their rights and their lives. (In broad concept, the plot is identical to the first episode of *Astro Boy*.) A musical celebration at the finale is a blend of jazz and funk that the movie puns as "junk," to which the robots do the "robot," the dance moves that require jerky but precisely controlled "robot-like" movements—robots imitating humans imitating robots.[54]

Pixar Animation Studios developed as an afterthought to the careers of some giant names, including director George Lucas, Apple CEO Steve Jobs, and the founder of Atari and Chuck E. Cheese, Nolan Bushnell. In the 1970s, each saw computer animation as the future, a correct prognostication even if it took a full decade for what became Pixar to coalesce as a going concern. John Lasseter, apprenticing at Disney and Industrial Light and Magic, stunned the animation world in 1986 with the short movie *Luxo Jr.*[55] The manufacturer Luxo made articulated desk lamps, bendable in the middle to direct the light from the "head." Lasseter and his team of modelers anthropomorphized a large lamp and a smaller counterpart into an adult interacting with a child—the child joyfully and carelessly playing with a ball, the adult gently assisting and shaking its head in dismay. The depth of personalities that emerged from vaguely humanoid metallic objects embodied the possibilities of depicting robots as feeling creatures in their own right rather than humans wearing silly costumes.

WALL-E (2008) is *Luxo Jr.* writ large, on a $180 million budget. Overwhelmed by their own garbage, the entire population of Earth left the planet in giant space arks modeled after cruise ships. Thousands of self-controlled clean-up robots were to make the planet fit for their eventual return. Seven hundred years later, all have broken down except for one Waste Allocation Load Lifter—Earth-Class unit, or WALL-E. Solar-powered WALL-E compacts trash into cubes, singlehandedly stacking them into a new skyscraper city by day and returning to his trailer at night, where he lovingly tends his collection of intact memorabilia from the idealized past he's never seen. No words are or can be spoken; all the emotions must be conveyed through body language and sound effects. He has one companion, a cockroach, which he feeds with a Twinkie (riffing on the joke that they could survive any apocalypse). Trash compactors aren't designed to have companions, of course: WALL-E somehow achieved sentience, and with it feelings, limited only by his environment. He's a deprived city kid needing a Fresh Air Fund to expand his horizons.

The space liner is another metaphor; the humans on it having become so grotesquely obese from their perpetual pampering by squads of robot servants that they can no longer walk. (How they have the sex needed to produce all the cutely obese infants is carefully never touched upon.) The ship periodically sends back probes to check for signs of life, a signal that humans can return to a recovered Earth. One such Extraterrestrial Vegetation Evaluator (EVE) unit appears in WALL-E's territory, scouring the barren trash piles until she (an essential piece of the anthropomorphizing requires that sexless metal exude obvious sexual characteristics) discovers that WALL-E has found a green sprig. EVE's programming then overrides her budding friendship with WALL-E, shutting her down until another probe returns to scoop her aboard. However, WALL-E is ten steps ahead of her in feelings. As in any good fairy tale, he, the prince, fell instantly in love with the idealized princess robot—rounded and smooth and blue-eyed and spotlessly white—and will be faithful until death, never to be parted. Their separation sparks his quest, a pursuit of EVE through an escalating series of grueling ordeals on the space ark. Along the way, he inadvertently frees a group of menial robots to form a rag-tag group of superheroes battling the ship's evil autopilot. EVE also gains feelings, the magic potion that allows her to override her programming. WALL-E ultimately sacrifices himself for the humans, allowing them to come back to Earth, where EVE uses his collection of spare parts to reanimate him.[56] Robot movies, like fairy tales, come ingrained with a set of imperatives.

Marvel Comics issued a limited tryout of an all–Japanese superhero team in 1998, titled *Sunfire & Big Hero 6*.[57] In it, Hiro Takachiho, the smartest thirteen-year-old in Japan, built a "robotic synthformer" monster called Baymax as his companion and protector. It went nowhere, but no concept is ever discarded in comics. Another mini-series appeared in 2008 as *Big Hero 6*, with Baymax more of a mecha with a human guise resembling a larger Oddjob from *Goldfinger*.[58] Then animators from Walt Disney Animation Studios, picking through Marvel's detritus after acquiring the company in 2009, took the name and a dash of concept and reformulated the rest into a massively budgeted 3-D computer-animated blockbuster, *Big Hero 6* (2014).

Hiro remains a genius who graduated from high school at thirteen and is wasting his time and robot-building talent in conning suckers at underground bot battles. Somehow he has never noticed that his adored big brother Tadashi is one of the nerds creating robot marvels in the lab at San Fransokyo Institute of Technology. (The future city blends San Francisco and Tokyo to enable a more diverse team while retaining anime motifs.) Tadashi's creation is Baymax, a friendly, marshmallow-bodied helper robot designed to diagnose and cure physical and psychological ailments. Inspired, Hiro designs microbots that can be thought-controlled to merge by the millions into any desired shape. (They're like the nanobots from cyberpunk fiction enlarged sufficiently to display well on a screen.) An explosion and fire kill Tadashi and seemingly destroy the microbots, which nevertheless resurface under the control of a mysterious supervillain, one of the best of any superhero movie. Hiro uses his genius to upgrade the inventions of Tadashi's nerdy lab buddies, as well as Baymax himself, effectively giving them superpowers so they can form a rag-tag group, the six big [super]heroes of the title—that is, once Baymax's tummy is firmly controlled by armored Spanx. Baymax sacrifices himself for Hiro's sake at the end, in the omnipresent robot trope. The same trope also demands his resurrection, and audiences were not disappointed.[59]

—⁂—

Amazon today maintains a bestsellers list for a category called "Children's Robot Fiction Books," the top one hundred titles out of a list that stretches into the thousands. That enough children's robot fiction books are published to constitute a lengthy bestsellers list would no doubt amaze even Isaac Asimov. He never attempted robots for children, although his second wife, Janet Jeppson, took up the challenge. (The books are credited to "Janet and Isaac Asimov," though, for sales reasons.) The Norby series ran to eleven books, starting with *Norby, the Mixed-Up Robot* (1983), and were followed by comic strip adaptations of the first two novels in *Boy's Life* magazine. At that time, a children's robot book (much less a series) was pioneering, and only scattered titles appeared by other authors.

Jeff Wells, a 14-year-old cadet at the Space Academy on Mars, is suspended for accidently deactivating the central computer. Sent back to Earth with just enough money to buy a used teaching robot, he stumbles across one that had been owned by a spacer, the legendary class of adventurers across the solar system. With a torso welded inside a barrel once containing Norb's Nails and a flat tin hat, the soon-named Norby is vaguely reminiscent of Tik-Tok of Oz. Norby, a walking insult generator, has an amazing set of powers, including many that he realizes exist only when they are suddenly needed in battle against a tyrant who wants to overthrow the Earth government. The omnilingual robot is conveniently loaded with antigravity, telepathy, and hyperspace teleportation—and that is just in the first book.[60] Norby was a bit of an in-joke in the Asimov household: he doesn't have to obey the Three Laws of robotics. Readers need to go past the first book to learn that he's an alien robot built by an alien robot. Humans need to do a quick job of catching up.

Amazon's 100 titles indicate that robots pervade every nook and cranny of the children's book world, from 32-page preschool spellers through increasingly longer grade-school chapter books and 300-page middle schooler epics to sophisticated products aimed at teens, from single ventures to lengthy series, and from the largest media franchises to individual labors of love. There is no comparable listing for "adult robot fiction," nor does a search using that phrase pull up adult titles. Robots have indeed been mostly crowded out of standard adult science fiction in favor of self-aware artificial intelligences, advanced human/machine constructs, nanobots, and less definable concepts. In contrast, the simplicity of robots as easy-to-explain humanoid devices that can be given basic human emotions and reactions is what makes them a dependable kids' story subject. Robots are in a perpetual loop of learning new facts and meaning about the world while constantly striving to adapt to the strictures laid down by adult humans in a way that mimics the arc of all children's lives. They are the modern equivalent of toys come to life as companions to share adventures, successors to Winnie-the-Pooh. Notably, all the robots are robots rather than androids, conspicuously different from the human protagonists but with the same needs, especially the need to be treated equally and respectfully.

Robot books also conform to the dismal history of 20th-century children's fiction, identified in 1965 by Nancy Larrick as "The All-White World of Children's Books."[61] Larrick surveyed more than 5,000 children's books published from 1962 to 1964 and found that less than 1 percent "tell a story about American Negroes today." Fewer than 7 percent contained a Negro child even by the most generous standard of a black face in a background crowd. The Amazon list today is almost as white as a Nancy Drew book and about as male as the Hardy Boys. Like them, the children depicted are almost exclusively from mainstream American middle-class families, attend public schools, and live in small towns and suburbs where they and their friends can freely wander or bike to school.

The robots themselves should provide the measure of diversity; robots inherently have no sex and can be of any skin color, talk in any dialect, behave in any manner. Can be, but aren't. Humanoid robots are overwhelmingly male. Artists do color them in a rainbow of hues—red and yellow and many shades of metallic blue and gray and silver. Fewer than 7 percent of the covers feature children of color. *The Adventurer's Guide to Successful Escapes*, by Wade Albert White, stars Anne, black and female. Evan, in Eric Luper's "Key Hunters" series, is black, and so is the unnamed girl lead in the almost wordless *Little Robot* by Ben Hatke. The only cover that depicts an Asian face is a tie-in to *Big Hero 6*. Tellingly, when the animators for *Big Hero 6* wanted the least threatening template imaginable for their medical robot in a fictional Japanese-American world, they chose a marshmallow: soft, round, chubby, and blindingly white. EVE, the superadvanced robot in *WALL-E*, is a glistening white egg with blue eyes. The helper robot in the non-animated *Robot & Frank* follows this pattern, also rounded and white, being based on the real-world white Japanese robot ASIMO. Parents must do deep searching and meticulously enlarge the cover thumbnails to find more nonwhite protagonists beyond the top 100 titles.

The positives of robot titles include broad humor and life lessons and lots of in-jokes about earlier robots that might appeal to parents remembering their own childhood reading material. Younger audiences might ask about finding Asimov's *I, Robot* on an early page in Jon Scieszka's *Frank Einstein and the Antimatter Motor* (2014), the first of a series. Frank lives with his grandfather Al Einstein (no relation to the relativity Al) in his Fix-It! repair shop, with access to all the parts a young inventor could ever need. Frank is building a robot for the science fair, and all he needs is a bolt of lightning to bring it to life. Those who watched the 1931 movie version of *Frankenstein* will appreciate the homage as Frank raises the robot to the skylight. Then the power fails and the robot is smashed to pieces, far from "alive."

Back in 1940's "Edison Bell, Young Inventor," Eddie built a robot out of parts lying around the house; it came alive the next day after being struck by lightning. Eddie called the robot "Frankie Stein" (see chapter 8). Another direct homage, or two interpretations of the same source? But Frank's robot, which will be called Klink, does come alive overnight, immediately proceeding to put itself together as its hands scrabble around the shop, very similar to the beginning of the book version of *The Iron Giant*. Klink also helps assemble Klank, the comic relief sidekick.

The postmodernistic opening completed, Scieszka dumps in a stew of science facts while Frank desperately fends off his evil kid rival T. Edison, whose sidekick is the sign-language-speaking Mr. Chimp. When they kidnap Klink and Klank, Frank has to come up with a new invention in a hurry, and he produces an antimatter-powered skateboard—which T. Edison also steals. Certain doom follows, but the robots and Frank's ingenuity save the day.[62] Four more Frank Einstein books have followed at the time of this writing.

Minds acting in sync probably account for *House of Robots*, also from 2014, since virtually every chapter contains similarities to *Frank Einstein*, like chocolate chips poured into separate batches of cookie dough. Written by modern fiction factory James Patterson and his main middle-school-level collaborator Chris Grabenstein, with illustrations by Juliana Neufeld, this book also features a house full of gadgets, a Rube Goldberg homage, a bully who steals a robot ("robonapping" is the word used in each book), and lots of middle school boys' humor. However, this cookie is an entirely different flavor, set in a realistic world far different in tone from Frank Einstein's child-centric wonderland. Sammy Hayes-Rodriguez is an ordinary fifth grader whose genius computer scientist mother makes all the robots, humanoid and otherwise, that fill their house. She has extra incentive because Sammy's younger and beloved sister Maddie has an immune disease that requires total cleanliness and prevents her from ever going outside, except for those frequent trips in an ambulance. Dr. Hayes saddles Sammy with E (for Egghead), a robot that, like Klink, can learn things on its own, and forces him to take E to school, where he is the butt of every joke. Mothers also can learn, so Dr. Hayes adjusts E to become less a fact-sprouting egghead and more normal over time, a regular kid, which works out great when E's true purpose as telepresence surrogate for Maddie is revealed.[63] Two sequels are now available.

(*House of Robots* may have been inspired by a real-life story widely covered in 2013. Seven-year-old Devon Carrow, afflicted since birth with life-threatening allergies that sent him frequently to the hospital, "attended" school via a telepresence robot made by VGo Communications. Far more stripped down than E—just a television camera and screen atop a wheeled base—the robot was controlled by Devon on his home computer, enabling him to talk to and hear and interact with teachers and classmates. "I wondered how the little kids would take to him, thinking they'd be amazed," said school principal Kathleen Brachmann. "But I think kids are so tech-savvy now they accept it more than we do." A classmate, Daisy Cook, agreed: "It's like he was never on the VGo.")[64]

Nicola Tesla is almost as iconic an inventor as his rival Thomas Edison, which explains the twins' names in *Nick and Tesla's Robot Army Rampage*, written by "Science Bob" Pflugfelder (known for his wild science demonstrations on talk shows) and Steve Hockenstein. Tesla, for once, is a girl, a slightly better inventor than her brother, and the take-charge leader of their pack of friends. Deposited with their eccentric scientist uncle when their parents take off for Uzbekistan, the twins immediately get caught up in a series of mysteries that they solve with inventions, blueprinted in detail by illustrator Scott Garrett.[65]

Somewhat younger readers are the audience for the *Ricky Ricotta's Mighty Robot* series, illustrated by Dan Santat. Dan Pilkey's books are the story of little Ricky, a brown mouse in Squeakyville, and what happens when a giant robot follows him home one day. His parents are dismayed by such a pet, who is larger than most buildings, but having the robot save the world over and over again is nice recompense. Ricky goes on to have an adventure on every one of the other seven planets (Pluto is not a planet and doesn't get a book), including fighting the Mecha-Monkeys from Mars. Ricky and the robot always look out for one another. As every book ends, "[T]hat's what friends are for."[66]

Awesome Dawson (2013) falls squarely into Frank Einstein's tradition of boy inventor. Chris Gall's awesome picture book for grade schoolers features the invention-obsessed Dawson, for whom every discarded piece of junk is waiting to be repurposed into a gizmo to make life more fun. He combines a broken rake, a broom, a watering can, a gumball machine, a pool hose, and a vacuum cleaner headed by a talking cow doll into a robot to do his chores: the Vacu-Maniac. Maybe making the brain from cat food wasn't such a good idea, but Dawson saves the city from himself in the end.[67]

Hasbro created a Transformers Rescue Bots series of Level 1 readers, with the do-gooder robots transforming into whatever tools are needed to perform their rescue. Level 1 involves short words in simple sentences "to begin a reader's journey." Adults who remember Dick and Jane books may be surprised to see that the special words to spot while reading are not just "cat" and "team" but also "hologram" and "satellite."[68]

For very young readers, Dan Yaccarino wrote and illustrated *Doug Unplugged* (2013). Doug is the young spawn of Rob and Betty Bot, who plug him in each day when they leave "to fill him up with lots and lots of facts." One day Doug literally unplugs and flies off (using a rocket attached to his back) to explore the big city all by himself, a subversive notion that oddly is highly praised by his parents.[69] One of Yaccarino's first children's books was *If I Had a Robot* (1996), no longer a bestseller but a mandatory read. For Phil, a robot would be the perfect friend and servant, someone to do his chores, eat his vegetables, protect him on the playground, and always be around ... except that Phil would save all the chocolate cake for himself. Of all the thousands of media items explored in this history, *If I Had a Robot* best encapsulates humanity's wish-fulfillment relationship with robots.[70]

Yaccarino also acted as illustrator for Ame Dyckman's *Boy + Bot* (2012). A boy encounters a robot in the woods, and they become friends. The robot's powering off worries the boy, who tries to heal the robot, and the robot reciprocates when he thinks that the sleeping boy is sick. That's what friends are for.[71]

The one-line concept behind *Boy + Bot* is expanded to mythic proportions in Peter Brown's novel *The Wild Robot* (2016), heavily illustrated by Brown, also aimed at middle schoolers but profoundly different from the books outlined above. It's a science fiction fairy tale told in classic style, gentle, free of middle school humor, filled with death and rebirth, and without a human in sight. A cargo shipwreck strands a ROZZUM Unit 7134 on a remote island full of North American woodland creatures. The robot that Brown calls Roz is a deliberate reference to the first "R" in *R.U.R.*—namely, the robot creator Rossum. The island is an environmental allegory, a Bambi-like forest isolated by rising ocean waters.[72]

To keep with pre–World War II references for this odd throwback book, Roz blends Doctor Doolittle with Mary Poppins, becoming a mother figure who bonds with the animals on the island after learning to speak their languages (plural, although all animals can somehow understand one another), helps them out in a million ways with the human knowledge stored in her memory, and becomes foster mother to a gosling whose family she wiped out during a fall off a cliff. Roz says she doesn't "really feel emotions," but her every action belies that machine truth. She is emotion personified, the way a Greek goddess of feeling would incarnate. Her programming requires her to be giving, caring, protective, and alert to dangers. Being a robot offers negative blessings as well: since she does not eat, Roz cannot be a threat to anyone in the island ecosystem. The animals, at first suspicious of the literally outlandish monster, warm to her ceaseless bounty and reciprocate individually and as an improbable group, such as when they fashion an artificial foot after Roz's is pulled off by a bear cub. (He is sorry, and the bears join in the love.) A cruel winter brings death to many; however, Roz protects many more by introducing controlled fires and teaching the animals to build shelters. Only outside forces can disturb this idyll. They come in the form of RECOs, robots hunting down all the ROZZUM units lost in the wreck. A badly damaged Roz is taken after a battle; yet the book ends with the words "She would find her way home."[73] Preorders on Amazon pushed the sequel, *The Wild Robot Escapes*, to the number one ranking, with the earlier book crowding it at number two.

13

Robots as Androids

Television producers were oddly slow to recognize the advantage of using human actors as androids (defined here as robots that appeared identical to humans). The multitude of plotlines, comedic and dramatic, opened up by duplicates, doppelgängers, and disguises should have been well known to Hollywood creatives, and the potential savings in time and money by not faking mechanical bodies was significant. When the breakthrough occurred, it came via a genre that was notorious for skimpy budgets: the kids' show. Even there, producers somehow overlooked the possibilities for several thousand episodes until very nearly the end of the entire era of space TV, but the writers of *Space Patrol* finally re-created Karel Čapek's flash of theatrical brilliance in 1955. Rather than putting oversized extras into clunky and thoroughly non-futuristic outfits of painted cardboard, why not mutter jargon like "protoplasm" and "duplicator" and create androids that could be played by actors in dual roles?

The three-part storyline that encompassed "The Androids of Algol," "Double Trouble," and "The Android Invasion" was an intricate dance of substitutions as finely tuned as a stage farce and as paranoiac as any 1950s political thriller. A seemingly helpless young woman—who is actually the ruler of a small, remote planet—is rescued from the clutches of slavers by the Patrolmen before double crossing them. The evil Yula slaps some protoplasm into her handy duplicating machine to create android doubles, starting with the show's star, Commander Buzz Corey, who is trusted by every top official. Soon, her androids are being swapped for the secretary general of the United Planets and who knows who else. No one can be trusted, no matter how helpless they seem or how familiar a face they wear.[1]

Space Patrol was set in the 30th century, in which a multiworld government modeled on the United Nations ran the space police. It's not difficult to turn the android saga into an allegory of the Korean War, officially called a "police action." North Korea, a small, remote country recently saved from Japanese domination, betrayed America by sending its forces across the Yalu River (of which Yula is an anagram) to conquer South Korea. Because the Russians who were supporting North Korea happened to be boycotting the United Nations, America got the world body to officially declare hostilities. The conflict eventually stumbled into the stalemate of unresolved civil war. Who was the real enemy? The North Koreans? The Chinese? Communism itself? Telling the North Korean enemy from the South Korean allies was tricky. To American eyes, they were, almost literally, doubles of one another. Additionally, stories of brainwashing (i.e., making your comrades-in-arms secret agents, enemies in friendly disguise) dominated headlines in 1954. This was heady territory for children's television.

Rod Serling's *Twilight Zone* delighted in allegories and needed little excuse to introduce androids. In its first season (1959), it addressed a future in which machines could look and act as humans. In "The Lonely," a convict is exiled to a barren asteroid, a wall-less solitary confinement that is driving him mad. The wardens who stop by four times a year to check on him leave a box behind. The box contains Alicia, an android with human feelings and emotions. Far from pleased, the convict feels more thoroughly punished. "Why didn't they build you to look like a machine?" he asks. "Why didn't they build you out of bolts, and wires, and electrodes and things like that? Why'd they turn you into a lie?" However, humans seize lies upon which to survive every day, and the convict soon relishes her companionship. A year passes, and news arrives: He has been given a pardon. He can return home, but Alicia must stay behind. It is another life sentence.[2]

Family is again central in "The Lateness of the Hour." Jana, a grown woman, has been feeling increasingly constrained by her restrictive parents, despite being pampered by a staff of androids built by her father. She wants to go off on her own, marry, have children. Sadly, those simple, everyday wants are impossible for her. Her parents are childless; Jana is her father's latest, most advanced model with false memories implanted. Just as she has no past, she has no future.[3]

In a series built on twist endings, the most surprising twist was a happy ending to a happy show. Ray Bradbury's script for "I Sing the Body Electric" is itself a twist on the robot nursemaid that had been a staple of science fiction stories since the 1920s. His android is the most doting grandmother that Facsimiles, Ltd., could create, made out of smiles and eyes chosen by the orphaned children themselves. Better than human, she cannot die and leave them, as their real mother did; she will see them through their childhood until they are grown and need her no more.[4]

My Living Doll debuted in September 1964 and crammed an encyclopedia of 1950s neuroses into each 25-minute episode. Dr. Carl Miller, a NASA scientist, has been working on project AF 709 to test conditions that astronauts might face in space. So super-secret is his work that he doesn't even have an assistant, an oversight that allows the test dummy to walk out of his office one day. Wildly overbuilt for its intended use, AF 709 is an intelligent android programmed with numerous abilities and libraries full of knowledge. The zeitgeist had shifted sufficiently for this robot to be in the form of a statuesque female, played by Julie Newmar, a tall brunette best known at the time for portraying the impossibly voluptuous Stupefyin' Jones in the Broadway and film versions of the musical *Li'l Abner.*

The robot, wearing an oversized beach towel conveniently emblazoned with the label "AF 709," stalks into the office of genial psychiatrist Dr. Bob McDonald (played by Robert "Bob" Cummings) and then literally wanders out into traffic, as neat a metaphor for technology escaping into the wild as can be found without a similarly emblazoned label. If the authorities learn of this breach, the project might be shut down, so she must be hidden. In proper sitcom style, Dr. Bob takes her to his apartment. Miller, however, is sent to Pakistan for six months, leaving behind the robot and another metaphor for scientific neglect.

Having a woman living in his apartment is banned by the superintendent, NASA, and 1950s morality, despite Dr. Bob's playboy reputation, so he calls his sister to come live with them as chaperone, the most transparently ridiculous rationale for a household until *Three's Company.* Dr. Bob quickly determines that if he can teach the android, christened Rhoda, human emotions, she will be "the perfect woman: one who does what she's told, reacts the way you want her to react, and keeps her mouth shut."[5] Rhoda is above 1950s chauvinism, which thankfully is toned down after the pilot episode. She's better than humans in every way, superior especially to the horndog men who all but drool upon sighting her, and Newmar's performance lacks any taint of inferiority. She's brilliantly intelligent, totally self-sufficient, and gloriously goofy, a coruscating performance that blows Dr. Bob off screen at every turn. Perhaps that's why movie veteran Cummings, in his fourth consecutive sitcom starring himself as a genial character also named Bob, looks perpetually astounded and annoyed at being out-acted by every subtle twitch of Newmar's pliable face, making the chemistry between the two nonexistent. Cummings walked out, and the series collapsed in its first season.

Rhoda lost out to contemporary sisters Samantha (the witch who marries a mortal in *Bewitched*) and Jeannie (the genie who partners with an astronaut in *I Dream of Jeannie*), two powerful women far better at publicly concealing their superiority to the bland slabs of American masculinity the producers stuck them with. As an android, Rhoda lacks the ability to efface herself. Technology is revealed to be the modern force greater than magic; it's real, spares no one, and will take over us all. Viewers far preferred the more easily dismissed fantasies for their sitcom fare.

So-called dramatic shows shifted the line of acceptability. Rhoda was a military secret, a device that, after the launch of Sputnik, could justify any advanced technology, no matter how fanciful. President Kennedy's New Frontier and his pledge to put an American on the moon within a decade set the tone; his assassination gave life to far-less-rational conspiracies and, with them, the need to have good guys to do the dirty work. Kennedy's enthusiasms gave them a way in. His casual mention of an Ian Fleming James Bond novel on his favorites list in 1961 (along with nine serious nonfiction books nobody would remember) drove audiences to both Fleming's books and the first Bond movie, *Dr. No*, in 1962. Early Bond was a suave, sociopathic killing machine, not the sort of gentleman Americans wanted to import, although Kennedy thought his intelligence operatives could learn from Fleming, an actual former spy. Bond's louche character could never be portrayed on primetime, but no matter: long accustomed to infantilizing serious matters (Newton Minnow's denunciation of television as "a vast wasteland" came just two months after Kennedy's list), programmers morphed the fear of atomic annihilation into sitcoms and spy spoofs. Starting in the 1964–1965 television season, a string of spy shows debuted, almost every one a success, several winking spoofs: *The Man from U.N.C.L.E.*, *Mission: Impossible*, *Wild, Wild West*, *I, Spy*, *Get Smart*, and British imports *The Avengers* and *Secret Agent*, all before *Lancelot Link: Secret Chimp* beat the genre to death in 1970 (see chapter 11). (The crew on *Star Trek* were not technically spies but had weekly missions identical at their core. Similarly, *Batman*, with his secret identity and bat-devices, shared 99 percent of his fictional DNA with Maxwell Smart and Napoleon Solo.) With emphasis on the latest technological gadgetry, almost all of these shows' stars would encounter androids among their enemies' multitudinous ingenious devices

Metallic humanoid robots lost out to human duplicates in this crush of imitators, beginning a decades-long lull until technology could once again produce robots that looked futuristic or scary rather than passé and camp. Seemingly every series that appropriated elements of fantasy and science fiction brought in artificially created humans, often a bunch of them (even better if they were a bevy of starlet beauties who could be costumed minimally).

Just as *The Perfect Woman* sneaked in an android before television, a movie again premiered the first android Bond girls. The low-budget studio American International Pictures (AIP) deliberately crafted movies for the growing teen audience. In what came to be called the "Peter Pan Syndrome," AIP postulated that "to catch your greatest audience you zero in on the 19-year-old male," an opinion maker that women would accompany and younger teens would follow to the movies.[6] In 1963, AIP tested this theory with *Beach Party*, a romp that proved bikinis and rock music could bolster the flimsiest of plots. *Beach Party* returned grosses of about ten times its budget, and AIP followed it with eleven more such movies in just three years.[7] Intended to be the biggest and gaudiest film of the series, and the first to have a budget of more than $1,000,000, *Dr. Goldfoot and the Bikini Machine* was released in November 1965. Riffing off the James Bond films *Dr. No* and *Goldfinger*, it starred Vincent Price at his most fey as a gold-shoe-wearing mad scientist with a sidekick named Igor. His girls will marry millionaires, become their heirs, and then kill them off. The androids emerge from a machine resembling a giant toaster oven, fully formed and already wearing gold bikinis. Despite the endless innuendo—posters and advertisements featured a bikinied girl labeled "THIS IS A BIKINI MACHINE" (emphasis in original)—the movie is too sexless for the girls to qualify as sexbots.[8]

Television caught up in early 1966 during the first season of *Get Smart*, the Bond spoof created by Jewish comedians Mel Brooks and Buck Henry. They introduced Hymie—the stereotypically Jewish name unique for a robot, a joke spoofing the conventions of white-bread television. Created by the enemy organization KAOS, the robot infiltrates CONTROL in order to kidnap an important scientist. Hymie brings the scientist to KAOS' hideout along with Agent 99 and Maxwell Smart. There he's ordered to kill Smart but can't bring himself to do so: Max is the only one to show him friendship. Offered a job at CONTROL, Hymie instead has dreams of going to IBM: "It's a good place to meet intelligent machines."[9] CONTROL provides zilch in the way of intelligence, but Hymie returns frequently, falling in love with the completely human-seeming android Olivia, foiling Groppo (a killer robot devised by his KAOS creator), and almost becoming "The Worst Best Man" at Smart's wedding to Agent 99, slightly spoiled by the bomb hidden inside him.

Once ably tongue in cheek, by its third season *The Man from U.N.C.L.E.* had grown steadily more farcical. "The Sort of Do-It-Yourself Dreadful Affair" is prime evidence for the shift. Napoleon Solo is beaten up by a zombieish but superstrong female, and he's too much of a gentleman to fight back. After 60 outlandish episodes, seemingly nothing would be beyond the pale, yet no one believes his account. When the girl collapses after another attack, Solo is vindicated by the autopsy: her insides are a DIY mishmash of human parts and electronics. She's necessarily a cyborg, but somehow THRUSH has an army of android copies of her raring to go unless Solo can pull their plugs.[10] Hordes of androids—this time males—are part of the Joker's diabolical scheme to pass counterfeit money by taking over a bank in a two-part *Batman* in early 1967. He substitutes his artificial men for tellers. "Holy chutzpah!" Robin exclaims.[11]

Star Trek is now the most venerated icon of all television science fiction, making it difficult to remember its early weak reception, low ratings, and internal uncertainties about what kind of show it wanted to be—serious or comic, personality or event driven, realistic or camp. The character of the half-alien Spock hadn't yet surged into a starring role. Shows revolved around Captain James Tiberius Kirk, played by William Shatner, as the youngest starship captain in Starfleet. Besides Shatner's uniquely mannered performances, he is perhaps best remembered as a space lothario, always finding the alien woman of the week to romance. This, too, had a delayed start: Kirk did not kiss a planet dweller until the seventh episode, dated October 20, 1966, "What Are Little Girls Made Of?"[12]

A subplot of the series concerned Nurse Chapel, who gave up her career as a doctor to sign on to the USS *Enterprise* so that she could search for her missing fiancé, Roger Korby. She apparently finds him, only to learn that he is an android double given the original's mind at his death. In a plotline itself a double for *Space Patrol*'s, Korby plans to replace leaders in the Federation with his androids and wants to start with Kirk. Korby distracts him with Andrea, an android played by Sherry Jackson in a costume that threatened at every step to uncover the must-be-covered. Jackson later reminisced, "They didn't want any side cleavage to show. So we took some double-sided toupee tape from Bill Shatner and taped the costume to my skin. Bill was never shy about wearing a hairpiece, so his toupee saved the day!"[13] The long and enthusiastic kiss Kirk bestowed upon Andrea that instantly converted the emotionless android into a woman who sought more of this human experience also became part of *Star Trek* legend.

More is always better on television, as in other popular media. In season 2's "I, Mudd," conman Harry Mudd escaped a prison sentence to land on a planet populated entirely by androids from the Andromeda Galaxy. They now serve Mudd, mostly in batches of identical female lovelies who cater to his every whim. Ironically, this "paradise" is also a life sentence: the androids will never let him leave. Without the stimulation of crime, he has grown bored and schemes to steal the *Enterprise* by tempting the crew into staying in paradises of their own. To escape, Kirk must outthink a central computer designed by superior alien intelligences, which he easily does. No matter how advanced a computer might be, it will inevitably destroy itself when presented with illogical information. The foiled Mudd is left with a fate worse for him than death: 500 copies of the shrewish wife that he once fled from to pursue his happy life as a conman.[14]

"Requiem for Methuselah," a late third-season episode, recycled these tropes. Kirk falls heavily for the "daughter" of an immortal Earthman, not realizing that she is an android companion built to accompany the immortal forever. Her flaw is lack of feelings, remedied by Kirk's love. Choosing between her master and Kirk is too much for her conflicted robot brain, and she dies of a surfeit of emotions.[15]

As 1960s programmers inhaled the mellow yellow fumes, showrunners scorned mere over-the-top episodes for ones that threw sense aside altogether—perfect for *The Wild, Wild West*, the James Bond spoof set during Ulysses S. Grant's administration. Every episode thickly jammed with anachronistic spy gadgets, superspy James West traveled the frontier in his tricked-out private train confronting madmen who wanted to conquer the world, a difficult feat made more challenging by starting in small western towns. Fortunately, West had the 1960s' finest adversary: the 3'11" inventive genius Dr. Miguelito Loveless, played flamboyantly by Michael Dunn. His tenth and final appearance, "The Night of Miguelito's Revenge" (1968), sees him as a ventriloquist's dummy. The audience saw that coming, but perhaps not that the vent was a full-size mechanical man.[16]

Viewers must have been nonplused to tune in the staid *Mission: Impossible* and find that its technology lagged behind that of the 1860s. Even its "Robot" (1969) episode was old fashioned. The premier of a fictional Eastern European country has secretly died. One faction learns of this development and plots to replace him with a captive human double to announce his successor. The IMF fools them into thinking that the other faction is planning the same trick, only with "Mr. Mechano," a stage automaton. A jailbreak substitutes the substitute for the substitute, and the real human double then double crosses the evil faction.[17]

—⚏—

Superior robots often caused discomfort, dismay, or outright panic. Androids were frequently treated as an even worse threat. Imitating a human too closely brought on the Frankenstein effect: the fear that our creations will surpass us or destroy us. Fear of the Other, an outsider instantly identifiable as such, manifested itself in chauvinism and bigotry. Fear of the Other infiltrating our ranks created the psychological terror that we were not safe even within our drawn boundaries, that no certainties existed, no trust could be displayed, and nothing was ever safe. If "They" became "Us," then who were we?

In the paranoid Cold War era, print science fiction provided a welcoming home for such frisson, and paranoid science fiction requires a mention of Philip K. Dick, especially his 1953 story, "Imposter." We are at war, of course, this time with an alien race perfectly named the Outsiders. They have concocted a diabolical plot:

> The robot would live the life of the person he killed, entering into his usual activities, his job, his social life. He had been constructed to resemble that person....
> [The robot] would be unaware that he was not the real Spence Olham. He would become Olham in mind as well as body. He was given an artificial memory system, false recall. He would look like him, have his memories, his thoughts and interests, perform his job.[18]

Dick's story would be expanded and greatly deepened by Algis Budrys in the ultimate 1950s questioning-of-identity novel. After a near-fatal explosion, the Soviets have captured Martino, a leading American physicist vital to the super-secret K-99 project. Months of diplomatic negotiations follow, under the excuse that Martino is too ill to move. The man who is ultimately returned is half a man. The rest is machine, including a metal arm and a smooth metal head that opens only for gleaming red eyes and a mouth. This is 1958, before DNA identification. All biometrical markings of the age have been erased; even the good arm could be a transplant from a dead Martino. In a harrowing game of "Spy vs. Spy," the American spymaster must consider that the Soviet spymaster is at least his equal and would take all the precautions of deep cover that he himself would take: he must be better than himself to defeat himself, and that's an impossible task. The metaphor of an android remains. Martino insists that he is the man he was, but the shock of the transformation has irretrievably changed him just enough that he always leaves a wedge of doubt in the eyes of the authorities. In all the important ways, his former identity no longer matters: whether he's an imposter or the original, he can no longer be trusted. Budrys titled the book simply *Who?*[19] The Library of America, collator of the greatest works of American writing, selected *Who?* in 2012 as one of the nine 1950s science fiction novels it chose to put into hardcovers as part of its canon (ironic since in the 1950s Budrys could only get the book published as a paperback original, the lowest rung of the literary ladder).

Androids that look human could as easily be a companion as a replacement. Androids are fully equivalent to humans under the law in J. T. McIntosh's "Made in U.S.A." (1953). They merely face a glass ceiling of unspoken barriers in their careers, in the clubs they can join, and the bullying they receive—humanity at its worst, too often inventing differences to restrict even those who are physically identical. The law nevertheless allows humans and androids to marry. A divorce suit is therefore nothing new, even one filed five minutes into the honeymoon by a husband who claims that his wife did not reveal herself to be an android before marriage. Current law states that androids don't have to reveal themselves, the judge rules, but the husband then insists that a law granting a divorce if one of the couple is infertile also applies to androids. After all, none has ever conceived. Androids aren't built sterile, the would-be wife argues in return: their sterility is a psychological block that serves to keep them always inferior to humans. A truly equal android could be a mother. She is carried back across the honeymoon threshold by a priapic husband eager to find out.[20]

Android-human marriage is one of many topics related to rights in the unusually intelligent ultra-low-budget film from 1962, *The Creation of the Humanoids*. It most resembles a mid–1950s television play, with few sets, an almost static camera, and dialogue substituting for action, and writer Jay Simms may have originally intended it as such.[21]

An atomic war has wiped out 92 percent of humanity. Lingering radioactivity sterilized many of those who survived or caused fatal defects in their babies, resulting in permanent underpopulation. Humans turned to robots to help them rebuild but found it "psychologically impossible to work alongside a machine that they had to converse with." Modern robots look like bald and silver-colored humanoids, albeit with green blood. Anticipating the trend of miniaturization, the humanoids contain what would soon be called computer chips, here called Magnetic Integrator Neuron Duplicators, $\frac{1}{100}$ the size of a golf ball, permitting them to learn whatever the central computer bank that runs the world learns. R-1 through R-20 are simple metallic robots; those with higher ratings are humanoids with progressively more skills and more emotions. An R-100 would be a human double, but far more perfect, with twice the lifespan. Their first rule is analogous to Asimov's: they can never hurt humans, either physically or by offence.

Humanoids, with their unhideable skin color, servile jobs, and deference toward their white masters and owners, make for hammer-heavy allegory, especially with a costumed KKK–analog Order of Flesh and Blood harassing them, not least for marrying white women. The early scenes yield to a deeper argument that humans are losing their much-needed initiative as they relinquish jobs and creativity to their android servants. The paranoia also proves correct: the humanoids are plotting. Memories from deceased humans now can be implanted into humanoid bodies, the R-96 model with white skin, programmed to think they are human except when they report back to headquarters. As production ramps up, all human leaders will be replaced by humanoids.

Simms throws in another twist when he reveals the androids' true intent. Jack Williamson had in 1947 published a refutation of Asimov, called *The Humanoids* in book form. Robots with a true first law must apply it to the entire human race, not merely individual humans. That means saving humans from themselves, even against their will. Dying humanity, doomed by population reduction, will live again with all their memories in better bodies that can be renewed indefinitely, giving them effective immortality. The R-100 may even be able to reproduce, ensuring against stagnancy. The androids will save humanity by becoming the humans, the ultimate "They are Us."[22]

The distrust that writers insist poisons the relationship between humans and emotionless androids probably is related to the physical uncanny valley: interactions between humans depend on emotions, and unreadable machines cannot be predicted—all the more so if they are females reacting to males. Freud's "what do women want?" was flipped by Ira Levin's short novel *The Stepford Wives* (1972) to "what do men want?" Published outside of genre by the mainstream publisher Random House, the book seemed at first blush another ordinary novel about a marriage crumbling under suburban ennui. Only at the very end do all the hints and clues and mysteries come together to implicitly suggest that certain scientists have been murdering their wives and replacing them with androids programmed to keep an immaculate household, stay home except to shop, and always look their best and be available—even kinky—in bed. Wives go on a weekend trip with their husbands and come back transformed into idealized housefraus. New arrivals in the community at first feel their difference and their superiority but succumb after four months, presumably the time needed to prepare a substitute. Despite the faint scientific trappings—the men include a Disney animatronics designer, speech experts, engineers, and computer executives—the book itself belongs to the horror genre. No explanation is ever given for how they've advanced the technology by orders of magnitude; all the focus is placed on the closing of the net around the newcomers as they fear replacement, only to face both disbelief and assurances by all that the new Stepford wives are fulfilling the needs of their husbands and children in a way that the uppity feminists cannot match. There is no happy ending: the heroines fight back but cannot win.[23]

Contemporary critics found the book baffling and irritating. Readers had no conceptual framework for robots immersed in a mainstream storyline. Worse, some missed the satiric edge entirely, calling out Levin's "misogyny" or doubting its appeal for the "Liberation Ladies."[24] With the benefit of hindsight, it is clear that the book (and the 1975 movie version that follows it closely) is an excoriation of male castration fears brought about by the women's movement: the town's men's association was formed to battle their feminist wives immediately after Betty Friedan, author of *The Feminine Mystique*, appeared at their women's club. Technology, then even more than now a virtually all-male domain, gives the men powers that the underclass of women, none of whom work outside the home, can never hope to match or even understand. Androids—clean, ageless, programmable, docile—were the embodiment of the spotless, efficient, technologically advanced homes and kitchens of the future that engineers had been forecasting for women since the turn of the century. Both were male fantasies, their ignorance projected upon their wives. Levin was one of the few who saw this in 1972. (The 2004 movie remake is metaphorically the kind of mess that Levin's engineers wanted cleaned up. The flabby script sometimes suggests that the replacements are androids and sometimes that they are the original wives with microchips in their brains. A man saves the day by erasing the programming, done literally by banging his hand on the central computer.)[25]

Levin was part of a trio of mainstream science fiction writers, along with Michael Crichton and Martin Caidin, both of whom also published augmented-human novels in 1972. Crichton's was *The Terminal Man*, in which a "brain pacemaker" is implanted in a man whose violence during bouts of psychomotor epilepsy threatens his existence. It was adapted into a 1974 movie. (Crichton's 1973 *Westworld* will be examined in chapter 14.) Caidin's *Cyborg* examines a more radical upgrade given to astronaut Steve Austin, replacing both of his legs, his right arm and left eye when they were shattered beyond repair in a test crash. Three made-for-television movies appeared in 1973, with Steve Austin billed as *The Six Million Dollar Man*. Their success led to a weekly series starting in 1974. The first movie is exceptionally sober, spending much of its running time dealing with Austin's suicidal depression both before the bionic limbs are attached and after, as a realistically lengthy period of training and adjustment is needed. The series took a more lighthearted approach, turning Austin into a government-sponsored righter of wrongs, essentially an uncostumed superhero. Several episodes involve true androids, including ones who impersonate Austin's associates.

In season 2, a two-part episode titled "The Bionic Woman," Austin rekindles a romance with Jaime Summers, an athlete thrilled by adventure and danger. When a parachute fails during a jump, she suffers an identical loss of limbs, although her ear rather than her eye is replaced. To save its hero from a planned marriage, the show must kill her off: she succumbs to a fatal rejection of her prosthetics. Nevertheless, on television ratings are the best medicine, able to cure even death. A companion series, also titled *The Bionic Woman*, debuted in 1976. Summers, it turns out, had secretly been placed into cryonic suspension until she could be repaired, although that left her with retrograde amnesia and no memories of wanting to marry Austin. They are now just good friends—close enough, in any event, to stage a three-part crossover between their shows, called "Kill Oscar: Parts 1, 2, and 3" (1976).[26] In the tradition of disgruntled former employees seeking revenge, a Dr. Franklin, furious that his contributions to the Office of Scientific Intelligence's cyborg program never got their due, vows to one-up the government and, dare he say it, rule the world.

With female equality very much at the core of *The Bionic Woman*, "Kill Oscar: Part 1" starts with a complaint by the secretary to OSI's head Oscar Goldman that she is not being recognized for all she does on a daily basis. Summers is sympathetic, and Dr. Franklin more so. Knowing she is in a perfect position to procure all needed information, Franklin replaces the secretary with an android double, called a "fembot." The fembot leads a kidnapping spree, with other fembots replacing key secretaries, and Oscar himself is taken to Franklin's lab, where fembots run the controls. Franklin's machines can also manipulate the weather, and Summers and Austin must battle both a hurricane and the superstrong fembots before defeating Franklin and destroying his lab. Fembots were a hit: Franklin's lab and his creations were rebuilt for the two-part "Fembots in Las Vegas" (1977), though the new creator is a surprise.

In a back-to-the-1960s movement, television series in the 1970s began a new round of shows featuring unusual, magical, or otherworldly heroes that included *Holmes and Yoyo* (1976). The goofball Yoyo was Hymie all over again, made manifest by producers who had worked on *Get Smart*. They buddy-copped him with Holmes, a lovable schlub with a flair for putting partners in the hospital. Yoyo had the same chestplate full of visible parts as Hymie and some fun new upgrades. Not only did he have a photographic memory, but pressing his nose also caused a print to emerge from his breast pocket. Writers doubled-down on Smart's slapstick but surrounded it with surprisingly decent police work.[27] Ostensibly drama, *Future Cop* (pilot 1976, series 1977) had almost as much silliness. A crusty veteran, too old and fat to run more than a few steps, bonds with a slim and athletic young robot with a British accent. Neither series lasted out their first season.

Undeterred, Dick Wolf, flush with success after *Law & Order*, returned to his good luck ampersand with *Mann & Machine* (1992). He's Bobby Mann, a chauvinistic maverick; she's Eve Edison, a beautiful robot who looks half his age. They bond, regardless. The progenitors of Eve's name are obvious, perhaps even more appropriate than Wolf knew. Thomas Edison in fact invented a very early talking doll, a Frankensteinian composite of a metal torso containing wax phonogram disks, a porcelain head, wooden arms and legs, and papier mâché hands and feet.[28] Both Wolf's and Edison's creations flopped resoundingly. All such television series struggled with making one of the leads relatable while being emotionally stunted, the arrival at acceptably human feelings infinitely far away because that would be the end of the series.

—⚜—

As his career progressed, Philip K. Dick dove ever deeper into his obsession with the fragile nature of reality. His worlds were built on easily overturned houses of cards made from drugs, paranoia, hallucinations, fake media, false memories, doppelgängers, and imitation humans. He wrote incessantly, except when he fell into years-long bouts of writer's block; yet he often could not sell his work or saw it appear only years later and altered. *We Can Build You* appeared in a magazine in 1969 as "A. Lincoln, Simulacrum" and in paperback in 1972 only after every hardcover publisher had rejected it. Obviously written to appear during the Civil War centennial celebration that started in 1961, the novel begins with Maury Rock's proposal to build androids—called simulacra, "the synthetic humans … thought of as robots"—of Civil War heroes, along with a few million ordinary soldiers, that the government can use for a bread-and-circus diversion for the public instead of war. Rock wants to diversify from his electronic music organ business, failing because of the competition from "mood organs" based on the thalamic research of a man named Penfield. Rock's partner, Louis Rosen, meets with Pris Frauenzimmer—German for "womankind"—an ex-simulacra designer recently released from hospitalization for her schizophrenia. Under her influence, Louis schemes to use the millions of fake humans as draws to get real humans off of Earth to the Moon and Mars so that speculators in real estate can make a fortune. Despite downloading all available information about Abraham Lincoln into a simulacrum, it fails to function properly. Nor does Louis' relationship with Pris go well—she rebuffs his attempts at sex, among other issues—and they fall into a false reality of bliss that belies their real-world depression.

Do Androids Dream of Electric Sheep? (published in 1968) starts with Rick Deckard needing a Penfield mood organ to create a desirable mood for his day. His wife counters by dialing up a six-hour self-accusatory depression; that way she will be unavailable for sex when he returns. Deckard lives on an almost-emptied Earth after a nuclear war that killed off every animal. (Implausibly, but unfortunately common in old science fiction, society reforms according to the standard 1960s American white middle-class model.) Almost every human except the misfits and unfit have fled Earth for the Moon and Mars, where they are served by androids. Artificial humans are illegal on Earth, and Deckard works as a bounty hunter for the San Francisco Police, killing androids as soon as they are identified. By the end of the first chapter, it's obvious that the story's origins lay in the then-discarded concept for *We Can Build You*.

More correspondences quickly follow. Eight of the latest Nexus-6 androids built by Rosen Associates have escaped to Earth, androids so advanced that they might pass an empathy test (empathy being the only aspect of humanity that earlier androids could not properly mimic). Deckard meets with Rachael Rosen at the company's Earth headquarters. She fails the test and is revealed as an android, legally present in a corporate sanctuary in an attempt to prove that the empathy test is faulty. Now that he has proof, Deckard must hunt the others down before… Before what is never revealed. Androids are a physical threat only to the bounty hunters that are trying to kill them. Otherwise they blend in so perfectly that they are in high positions around the world. In fact, one android has set up a wholly separate police department headquartered a few blocks away from Deckard's, yet totally unknown to the real police. (Only in a Dick novel could this situation be imagined, let alone be successful for years.) Jarringly, the android's fake police have their own squad of bounty hunters, one of whom, Phil Resch, partners with Deckard after killing his Nexus-6 android supervisor. More of the android escapees are killed until only three are left: android leader Roy Baty, his wife, and Pris, a duplicate of Rachael.

"Dehumanized" (as Dick would later describe him)[29] after a haunting day of killing, Deckard flees from the unresponsive arms of his wife to have illegal sex with Rachael, whom he finds he has empathy for. As a true android, she has none for him. Not technically a sexbot, merely an apparently 19-year-old beautiful woman without morals, her role on Earth is to have sex with bounty hunters. That act, mimicking love in a heartless dystopia, produces a bond so strong that the bounty hunters find it impossible to ever kill another android. She almost succeeds. When Pris rushes toward Deckard, arms open, he hesitates, but only for a second. After that, killing Baty and his wife is easy. Deckard has killed six Nexus-6's in one day, an unapproachable record—a feat that Dick hammers home repeatedly. Except that Phil Resch is shown killing two of those. In a Dick novel, even the author can't be trusted.

Do Androids Dream of Electric Sheep? would have been forgotten except by cultists if director Ridley Scott had not decided to turn it into a movie. The final version shares nothing but a concept with the originating story, and its screenwriter never read the book.[30] Titled *Blade Runner* (taken from an unrelated Alan E. Nourse book because Scott liked the name),[31] the result is a visual masterpiece and nearly as much of a conceptual mess as Dick's novels, only partly because seven versions have been released over time, including the 1982 original with a studio-imposed narration and a 2007 "final cut" without narration that is the only one now approved by Scott. Discussion in this chapter will be based solely on the "final cut."[32]

The setting is Los Angeles, November 2019, an expanse of giant skyscrapers and streets so crowded with ethnic street peddlers that walking is almost impossible, no matter that ostensibly most of humanity has moved offworld. Flying police cars sweep the streets and buildings with searchlights. Rain falls every minute on a world not merely film noir but literally noir, a dark broken society whose design principle, according to special photographic effects supervisor Dave Dryer, was that "while it is 40 years in the future, it is also 40 years in the past."[33]

Deckard, though officially an ex-cop (specifically, a blade runner—one tasked with killing replicants), is thoroughly in the mold of a Chandler-esque private eye, an individual beholden to nobody, a loner without friends, family, or loved ones, brought out of retirement to work for the police. The film provides the rationale omitted by Dick: The androids, called here replicants, are the latest model Nexus-6, smarter and stronger than humans, combat models who have killed their offworld masters on a shuttle ship and fled to Earth. They intend to track down their creator, Dr. Eldon Tyrell, and force him to extend or eliminate the four-year lifespan that is allotted to them, killing all Tyrell Corp. employees they encounter along the way.

Deckard starts by meeting Rachael and, as in the novel, is startled to discover that she is a replicant, though he falls for her instantly. Their reciprocal feelings color his investigation and save his life and his soul, the former when Rachael kills a replicant who is crushing Deckard's head and the latter when he tells her of his love. In the interim, Deckard must kill two other female replicants: Pris, who is a sexbot—a "pleasure model" (although that's irrelevant to the film)—and Zhora, who is an assassin but finds work as a stripper. They, too, come close to killing Deckard before he guns them down. Only Roy

Batty (spelling changed for the movie) is left, his role vastly magnified to a godlike Niet-zschean superman with the father issues of Frankenstein's monster. As lead villain, Batty kills his metaphoric father, Tyrell, and gets to inflict the greatest damage on Deckard in their final battle. Yet Batty, seconds before his lifespan expires, proves his superior human-ity by saving Deckard from certain death. As Joseph Francavilla wrote, "all the boundaries are blurred between the master and slave, hunter and hunted, hero and villain, the inan-imate and the animate, the human and the nonhuman."[34] If androids are truly our chil-dren, they must be treated that way.

Blade Runner 2049 (2017) attempts to blur the lines even further, going back to Dick and his obsession with memories and whether we can ever trust them to define reality. The original movie and its variant cuts led to decades of clue hunters arguing that Deckard was actually a replicant himself, implanted with false memories. The sequel's blade runner is certainly a replicant, known only by his serial number KD6–3.7. "K" suspects he might be the human, or half-human, son of Rachael and Deckard, but he can't trust anything he discovers: reality is slippery and malleable, the past blurry, the future unknowable. Memories can be implanted in this world, both fake ones and real memories belonging to others. Data is equally suspect since a blackout in 2020 (the year after the setting of the original film) affected all computers, wiping out all but a few traces of the past. "K" hunts the long-missing Deckard. So does the film's villain, Wallace, who took over repli-cant manufacturing from Tyrell. He's willing to kill for the secret of fertile androids, giving him money and power as they push aside dying humanity. Like many other robot stories, the film purports to ask the question of whether bearing children makes a robot human, but it fails at the task by focusing only on the male characters and their needs.[35]

Android (1982), the first movie with that singular title, appeared just after the original *Blade Runner*'s release and contains interesting parallels. Max 404 is the creation of an inventor, Dr. Daniel, with as much hubris as Tyrell, who keeps Max in a space station because androids who gained feelings had previously rioted on Earth. In his five years of life, Max has never met another human; yet he is obsessed with them, spending his days watching old media clips. Tellingly, he keeps returning to *Metropolis* and the Maria robot coming alive. A shuttle lands on the station after sending a distress signal. Max doesn't know the crew are three human convicts who slaughtered their police pilots. Even after he learns the truth, he shields them because he is attracted to the one female—no matter that she is as attached to her leader as Pris was to Roy Batty. Dr. Daniel is also attracted to her, though for different reasons. Corporate beancounters have stopped fund-ing for his Cassandra project: a female android more advanced than Max. Cassandra needs human brainwaves, and the woman can supply them. A game of cat and mouse ensues, each side stalking and killing the other, the androids evolving, the humans devolv-ing. Only Max and Cassandra survive; pretending to be human and freed from the tyranny of their "father," they plan to move to Earth to start another android revolution. Intended as a quickie exploitation flick, *Android* instead wound up on the film festival circuit, where its examination of what makes a creature human was better appreciated.[36]

—✺—

The reboot of the *Star Trek* franchise in 1987, the syndicated television series *Star Trek: The Next Generation* (*TNG*), switched the focus on androids from the largely dystopian visions of early 1980s movies to a deliberately utopian viewpoint. With a larger cast intended to showcase the melting-pot equality of the Federation, room could be made for an android more human than some of the literal aliens. Because of the franchise's enormous popularity, the character of Data is perhaps the best-known android of all time. The seven seasons of the series and the four movies with the same cast can be considered the story of the growth of Data, from a machine imitating and wanting to learn about humans to a fully developed person who in the end sacrifices himself for the sake of true humans. Fictional parents traditionally sacrifice themselves for their children; with androids, it is the other way around. Data may have succeeded with audiences where the buddy-cop androids failed because he did not need to carry the show, allowing his journey to be gently handled in the background and over a far longer time. The *Star Trek* universe also cleverly poised the almost human Data against the distinctly ahuman Borg, humans filled with nanobots to connect them to a hive mind. The Borg want to assimilate humans as they do all biological creatures. "Resistance is futile," they say repeatedly.

Science fiction writers have a greater than usual challenge in introducing characters who are not human. Audiences expect an explanation that justifies their use as essential rather than gimmicky whenever they encounter aliens, gods, elves, intelligent animals, or androids. *TNG*'s first episode, "Encounter at Farpoint," took this to a fault. Data, a lieutenant commander in Starfleet, with more than two decades of full and constant immersion in human society, is presented as if he had that morning emerged from an egg, barely knowledgeable about common words, human interactions, or social graces. The second episode, "The Naked Now," swung the pendulum too far in the other direction. When an infectious alcohol-like material removes the crew's inhibitions, Data is propositioned by the ship's security officer, Tasha Yar.

> TASHA: You are fully functional, aren't you?
> DATA: Of course, but—
> TASHA: How fully…?
> DATA: In every way, of course. I am programmed in multiple techniques. A broad variety of pleasuring.[37]

He is and they do. Although this encounter inspired the rap group Futuristic Sex Robotz to write "Positronic Pimp" in Data's honor in 2006 (Data has a positronic brain in homage to Isaac Asimov's robots), the notion that he would be programmed for sex but not for everyday interactions is as gimmicky as his android body being susceptible to an infection in the first place.

Actor Brent Spiner's skills brought growth to Data over time in much the same way that Julie Newmar blossomed as Rhoda. This development is best shown in the first-season episode "Datalore," which introduces Data's evil twin, Lore, also played by Spiner. Lore was imbued with the easy humanness that could allow him full acceptance in society (were it not for his odd skin tones) but flawed in other ways. In contrast, Data retains a humble awareness of his fragile status in society's hierarchy without succumbing to inferiority or servility; indeed, at appropriate moments he saves his shipmates through his superhuman abilities of memory, learning, speed, and strength. He is humanity's ideal of an android. (Moreover, his pale skin and middle American diction make him the most deracinated character in the multiethnic crew, symbolically less threatening even than the white aliens and half-humans.)

By season 2's "The Measure of a Man," Data's value has grown to more than symbolic status. Starfleet wants to perform a chancy dismantling of him to possibly glean the secrets of his unique construction and provide every ship with a Data. Would anyone object if the ship's computer were dismantled for an upgrade? Data argues that as a

Starfleet officer, he should have rights equal to any human. In a formal tribunal, Captain Picard rules that Data shows all three aspects of sentience: intelligence, self-awareness, and consciousness. Also, he isn't a virgin. Judging from the reactions of the others in the courtroom, this revelation alone outweighs the lofty speech Picard offers as justification. Data is judged a person, with the rights thereof, and will in the future not be considered a machine or, more bluntly, an inferior slave Other.[38] This verdict, which parallels the one in McIntosh's story and a number of others in science fiction regarding the status of various types of sentient non–Homo Sapiens, portends the potential of a real-world situation that seems to loom closer with every passing year.

Stanley Kubrick spent decades trying to get a script to properly cohere for the movie finally released in 2001 as *A.I. Artificial Intelligence*. Just as he had worked with science fiction writer Arthur C. Clarke on *2001: A Space Odyssey*, Kubrick turned to a succession of top British science fiction writers, finally settling on an expansion of Brian Aldiss' 1969 short story, "Super-Toys Last All Summer Long," originally published in the exceedingly mainstream magazine *Harper's Bazaar*. After Kubrick's death in 1999, Steven Spielberg finished a screenplay based both on a treatment by science fiction writer Ian Watson (a vision as close as possible to Kubrick's intentions) and on later story sequels Aldiss wrote.[39] In this future, robots are common throughout society, with models ranging from nonhuman to metallic humanoid caricatures intended for specialized tasks to fully human-seeming androids. Even kids' toys are intelligent companions. What they lack is true humanity: they do not love. Seeking to change this situation, a scientist's prototype is given to a couple whose preteen son is in a seemingly irreversible coma. The mother, Monica, though at first adamantly opposed, learns to love the android named David, triggering his love bonding program. He will always love her almost to the exclusion of all else, prefiguring the way WALL-E would refuse to part even for a moment from his EVE. Monica reads David fairy tales, with the young android identifying most strongly with Pinocchio. Sadly, the idyll is shattered when a cure for the couple's real son, Martin, is found. Monica lacks the strength to love both, and she chooses Martin. David's life is impossible without her to love; for his own sake, he is to be dismantled. However, she also lacks the strength to do him this kindness and abandons him in the woods like a fairy tale mother.

Rogue robots are swept up by carnies who destroy them in hideous ways to the cheers of human (read: subhuman) crowds. However, drowning a pleading child in molten lead is a step too far; the audience riots. David continues his journey as a literalized Pinocchio, obsessively seeking the Blue Fairy to have her turn him into a real boy. He finds her image at a Coney Island amusement park, now underwater, and gets trapped there, almost within arm's reach, for thousands of years, long after the human race has been wiped out by a new Ice Age. Evolved robots discover him, thrilled to find this one link back to their original makers. They can bring Monica back, but only for a day; then David will lose her forever. A thrilled David takes the bargain and spends his perfect day with no thought for the future.[40]

Backlash from Kubrick defenders over the final version's numerous weak points put Spielberg on the defensive from the film's opening. Spielberg has always been adamant that the gooey and glacial beginning and ending were pure Kubrick and that his only contribution was the middle, similar to an Aldiss story—the only moments when the movie flickers into life and motion. If so, Kubrick gets the blame for the creepiest aspect of the plot. Letting a robot grow socially by surrounding it with other children is so logical that it is a trope frequently found in kids' media. However, forcing a static child robot to obsessively love a mother while she and the rest of the world age and change is closer to a fairy tale witch's curse than an ethical scientific experiment.

The dichotomy that robots are obviously artificial and that androids can be mistaken for human is useful, entertaining, and banal, presenting an opening to a clever writer. Alex Garland had written several well-received screenplays, but the first he both wrote and directed was *Ex Machina* (2015), another entry in the tradition of small, brainy science fiction films that ambitious moviemakers use to make a debut splash. Here, a tech mogul brings a lucky employee to his secluded compound to test whether his latest creation is a true AI—a sentient, self-aware artificial intelligence. Sentient computers long ago became the touchstone for humanity's rival, the symbol of technological superiority in a world where even nonsentient computers talk to us, aid us, guide us, and infiltrate every electronic device we come into contact with, from wristwatches to refrigerators.

Primal terror strikes humans in three distinct guises. One is the nonhuman monster, whose visage triggers avoidance, the flight instinct embedded eons ago in our animal DNA. Movies specialize in monsters, the makeup and special effects departments perpetually vying to outdo the alien in *Alien*, the dinosaur-like *Godzilla* and *The Beast from 20,000 Fathoms*, the giant ants in *Them*, and the human monstrosities of *Frankenstein*, *The Wolfman*, *The Fly*, and *The Thing with Two Heads*.

Equally primal is the faceless terror—the darkness, the unknown, the invisible, the terror that could be lying in wait anywhere and everywhere on the outside and, worse, inside our own heads. Wells' *The Invisible Man* was a prototype for many literally invisible beings like the id monster in *Forbidden Planet*, while the metaphor of the unseen can be exploited in wildly different ways, such as the blind Audrey Hepburn facing apartment invaders in *Wait Until Dark*, the seemingly friendly white family in *Get Out*, and the psychological tormenter in *Gaslight*. The computer intelligence falls squarely into this category—a monster without form, both personal and potentially world-ending.

The idea of the mad scientist subsumes both terrors into a new form: the inhuman human, whose evil is invisible until the moment it is let loose on an unexpecting victim. This, too, expresses itself in both the individual and the social, from the kindly Dr. Jekyll unleashing his id in monster form to Old Rossum's artificial protoplasm creating a race of robots that wipe out humanity. Advancing science and technology have been double-edged gifts to humanity from the moment Prometheus stole fire from the gods and presented it to humans to keep warm and burn others.

Ava, the AI in *Ex Machina*, has the beautiful human face of Alicia Vikander and a transparent robot body that won the Oscar for Best Visual Effects. The mad scientist may be the mogul, played by Oscar Isaac exactly as brilliant and slimy as we've come to expect from tech billionaires, or he might also prove to be Domhnall Gleeson as the flunky who seeks to one-up the master, rising to match him in his game playing. Isaac starts the game by throwing away its one rule, a game devised by computer theorist Alan Turing in 1950, just as true electronic computers emerged in the public sphere. The Turing Test set a human on one side of a barrier, asking questions and hearing responses. If the answers could not be distinguished from those of humans, then the speaker had humanness, even if the answers were generated in a computer brain. Isaac declares the test won, almost trivial for today's computers. His gigantic ego pushes him to tackle a greater challenge: having the voice emerge from a conspicuously nonhuman body whose essential humanity cannot be gainsaid.

What, then, must a robot successfully do to imitate a human if these two men are her only live data points—question, lie, flirt, persuade, seduce, betray, slip into a human body herself? Garland seals his cast in a high-tech stage play, a reimagining of *R.U.R.* (also set on a compound away from interfering humanity), with an ending that mirrors the play's second act. If Ava is a robot turned android, then why can she not also morph into the mad scientist? Garland's clever resolution improves on *R.U.R.* by suggesting that the play's ending was avoidable, if only the process started with humans possessing fewer inherent flaws (or at least not the specific ones vividly on display). It also confirms Isaac's dismissal of Turing, driving past the primitive question to an advanced one: what defines winning the game not for the inventor of the AI but for the AI itself. Answering this question has been the goal of some of the best writers of the 21st century robot stories, as will also be seen in the next two chapters. *Ex Machina* stands high on any short list of must-see robot movies.[41]

14

Robots as Sexbots

Men are animals, Lenny Bruce once riffed in a routine: "If you put them in on a desert island, they'll do it to mud."[1] That convenient masturbatory substitutes—innies for men, outies for women—existed in antiquity cannot be doubted. Evidence for full-fledged substitute humans intended solely for sex dates back only a few hundred years, to *dames de voyage* (formally known as fornicatory dolls). Starting in the 17th century, French sailors cobbled together a simulacrum of a female from clothing, to which they applied their imaginations. Although these would be remarkable artifacts as repositories for historic DNA, none of them survive today.[2] Likewise, only references to early 20th-century fornicatory dolls survive, although they are quite explicit. Iwan Bloch's *The Sexual Life of Our Time* (1908) discussed devices offered for sale in underground Parisian catalogs:

> There exist true Vaucansons [a noted French automaton maker] in this province of pornographic technology, clever mechanics who, from rubber and other plastic materials, prepare entire male or female bodies, which as *hommes* or *dames de voyage*, subserve fornicatory purposes. More especially are the genital organs represented in a manner true to nature. Even the secretion of the Bartholin's glands is imitated, by means of a "pneumatic tube" filled with oil. Similarly, by means of a fluid and similar apparatus, the ejaculation of semen is imitated. Such artificial human beings are actually offered for sale in the catalogue of certain manufacturers of "Parisian rubber articles."[3]

Bloch listed the formal name for the fetish as "'**Venus staturia**,' the love for and sexual intercourse with statues and other representations of the human person" (emphasis in original).[4] Even better was "pygmalionism," the pleasure of seeing them come to life. (Today we're more likely to call it androidism or technosexuality, or even ASFR, from the early internet group alt.sex.fetish.robots.)[5]

Another French catalog from 1922, quoted in Andrew Ferguson's *The Sex Doll: A History*, gave a detailed description of the pleasures available from these dolls:

> The body in action moves like a living being, pressing, embracing, changing position at will by a simple pressure. The mechanism which gives life to the apparatus is very substantial and cannot get out of order.... This apparatus can be fitted with a phonograph attachment, recording and speaking at will.[6]

The carnal impulse behind the long history of idealized females from Pygmalion's Galatea to Hoffmann's Olimpia to Villiers' Hadaly (as well as the occasional idealized male, like the one Basile's Bertha concocted) is realized in fornicatory dolls. No two humans have synchronized sexual desires, needs, fetishes, or timing; complaints undoubtedly manifested as soon as primates replaced going into heat with seasonless lust. A perpetually yielding consort solved multitudes of problems, especially if the genius of art and mechanical contrivance were put to the task: "A kindly soul had invented the dildo for women deprived of male contact; for the pleasures of our brave Captain Pamphile [an adventure hero from Alexandre Dumas' fiction], someone had brought forth the rubber woman; for our hero, a deft craftsman, an artist, would invent a miraculous Phrynée [a famously beautiful Greek courtesan] he would be able to manipulate at will—she would always be compliant and silent, no matter how lewd the act he chose to perform."[7] *La Femme Endormie* (*The Sleeping Woman*) proves that rubber women were available in the Paris of 1899—and that they didn't go far enough. Credited to "Madame B***, attorney," the real author is now known to be Alphonse Momas, the most prolific purveyor of pornography in fin de siècle France.[8] Graphic and perverse, the novel trawls the *Psychopathia Sexualis* for fetishes to exploit, starting with the detailed construction of Mea, the perfect sexbot, "who would refuse him nothing, absolutely nothing":

> The artist outdid himself. A more exquisite work of art had never been created by human hands.
>
> She was an admirable woman with very uplifted, very firm breasts, outstanding, appetizing hips, extremely well-shaped buttocks, divinely curved loins, flawless thighs, well-rounded calves. All of this topped off by a most suavely shaped face, her golden hair royally coiffed. Her flesh was so white and smooth it was almost real, and all her joints were flexible. She was sent in a crate, marked "fragile," to M. Paul Molaus, Financier, Bois-Colombes.
>
> An explanatory note accompanied the shipment: "The extraordinary doll that I've conceived and executed," said the artist, "doesn't differ from a woman, except in one respect: she can't speak. I paid particular attention to her interior, which is fitted with three basins, several boxes and cylinders, and a number of little ducts, so as to permit the circulation of all sorts of products that it would please the experimenter to introduce into the silent goddess's body. By pulling certain curls of her hair, her eyes and lips can be made to move. One can place her in every imaginable position: standing up, seated, kneeling, lying prone, lying on her back. By pushing the navel, one provokes undulations in every part of her body. Her sexual organs are as perfect as those of any live woman. To warm up her body, all one has to do is to pour boiling milk or hot water in sufficient quantity into the different receptacles located under her head, behind her breasts, in her buttocks, stomach, legs, etc. One can also warm up a certain part of her body while leaving the rest only lukewarm. The liquid runs down through a series of tubes in her legs to the heels where a small peg is located. Just turn this peg to empty." There followed instructions to keep the interior clean by means of a mechanism which opened the neck, the back, the thighs and the calves.[9]

—⁓—

Mainstream publishing dared not compete with this level (or any level) of frankness. While a few 1930s "spicy" pulp magazines drew attention from the postal authorities and near-nudes proliferated on covers of every genre, the dominating male gaze of science fiction publications tended toward blueprints and circuit diagrams rather than women. Robots were no more sexualized than toasters—perhaps less, as they had fewer open slots. Men, being men, nevertheless fantasized nonstop about compliant females. A high school robot story written by Jerry Siegel, the future co-creator of Superman, shows the protagonist instructing his robot Bojo to visit girls and ask for dates. They all say no, "but if Betty refuses—I believe I'll order a female from the Robot Factory ... a Robot is the only woman that will obey a man's wishes."[10] (Presumably the teachers read only the innocent "date" and passed up the logical consequences; no records exist of Siegel suffering consequences for this story.)

After World War II, publishing cautiously inched into coarser territory, allowing allusions to and brief mentions of sex in popular publications. In 1949 and 1950, Ray

Bradbury, the science fiction writer least concerned about accurate science, wrote a short series of stories about the firm Marionettes, Inc. Despite the name, the company manufactured robots that managed to be near-perfect duplicates of existing humans, down to their thoughts, memories, and desires—illegally, because public opinion raged against the practice in his far-off future of 1990. Bradbury gives good reasons why, setting each story around troubled male-female relationships made worse by doppelgängers. As one Marionette lies crippled on the floor, shot by his erstwhile mistress, he reveals the truth about their last year together:

> "Oh, Martha, there are at least six duplicates of me, mechanized hypocrites, ticking away tonight, in all parts of the town, keeping six women happy. And do you know what I am doing, the *real* Leonard Hill?
> "I'm home in bed, reading my little book of Montaigne's essays and drinking a hot glass of chocolate milk" [italics in original].[11]

Easily available nonpersonal sex is fraught, more so than human prostitution, and guilt may cut many ways. Love more than sex lies at the heart of William F. Nolan's "The Joy of Living" (1954). In this world, childbirth has become an avoidable horror; women can have artificial children, men replacement wives. "Want a ready-made cutie who'll be 100 percent yours?" raves the drunken narrator to a human bartender. "Blonde? Brunette? Redhead? You name 'er, we've got 'er. Yours on easy payments.... When a Mr. Shy Guy wanted some female company long comes a sponge-rubber job right outa th' pinup mags. Jus' a few at first, here an' there, expensive as hell. But pretty soon good ole American commercial know-how takes over."[12] On this anniversary of his wife's death, the man—a traveling salesman too often separated from his family—rues that his mechanical wife is now closer to his child than he is and vows to return her to the company. She has come to love them both, however. Her feelings are deeper than his.

The 1950s might still have been too early for an explicit examination of what might be neologized as intermodal sex, even one pushed two hundred years into the future. Fritz Leiber had to wait until a 1962 novel-length paperback publication to pad his 1959 story "The Silver Eggheads" with long expositions on robot-robot sex—they plug into one another, of course (newer models have up to 33 connections)—as well as human-robot sex, a dive into fetishistic possibilities otherwise impossible. Robot prostitution is an underground service for the kinky and knows few bounds. Madam Pneumo's establishment, for example,

> is a very exclusive house of pleasure, owned, managed, and staffed entirely by robots. You see, fifty years ago or so there was this mad robot named Harry Chernik ... whose ambition it was to build robots which would be exactly like human beings on the outside, down to the last detail of texture and anatomy. Chernik's ruling idea was that if men and robots were exactly alike—and particularly if they could make love to each other!—then there couldn't possibly be any enmity between them....
> Can you imagine, Flaxy, having it with a girl who is all velvet or plush, or who can softly sing you a full-orchestra symphony while you're doing it or maybe Ravel's Bolero, or who has slightly—not excessively—prehensile breasts or various refreshingly electric skin areas, or who has some of the features—not overdone, of course—of a cat or a vampire or an octopus, or who has hair like Medusa's or Shambleau's that lives and caresses you, or who has four arms like Siva, or a prehensile tail eight feet long, or ... and at the same time is perfectly safe and can't bother or involve or infect or dominate you in any way? I don't want to sound like a brochure, Flexy, but believe me, it's the ultimate![13]

The key advantage of sexbots is laid out toward the end: they are "perfectly safe and can't bother or involve or infect or dominate you in any way." Sexbots ideally are both more than human—immune to disease and conception—and less so, machine servants little different from luxury automobiles but better and more personal. A sports car presumably likes being revved to redline. Sexbots have no lines they cannot or will not cross.

Thinking of the 1960s as a prudish decade clashes with images of the "swinging sixties," but frank sexual explorations didn't begin until the 1970s and needed until the 1980s to be fully acceptable in print science fiction. Poor little rich girl Jane in the British novelist Tanith Lee's *Silver Metal Lover* (1981) is a 16-year-old virgin dominated by a brilliant harridan of a mother who controls and infantilizes her daughter down to the color of her hair and her body shape. Jane is the living embodiment of teen insecurities encapsulated in rock group Toad the Wet Sprocket's line "She hates her life, she hates her skin, she even hates her friends,"[14] a horrid group of richer and more confident teens of both sexes and all sexual preferences. Naturally, Jane falls instantly in love with the most opposite persona imaginable: a beautiful silver-skinned robot bard. Science fiction excels at updating ancient myths into contemporary formats. The robot, called Silver in lieu of a name, is essentially a god walking among mortals—smarter, quicker, better at everything, able to size up humans at a glance, perfection in metalized skin. An experimental pleasure model, he is designed to give humans everything they need and want, physically and emotionally. It's a matter of minor jealousy that Silver sleeps with her friends, who need only command. (The wonderful, if narcissistic, gay character—a rarity for teen-oriented science fiction at the time—comes closest to being a decent human, especially when he is humbled by a figure whose charisma dwarfs his own.) Silver cannot love, per se—gods can never truly love mortals—yet his calm omnipresent devotion transcends any love that a mere human could offer.[15]

Love between gods and mortals must always end tragically. Public pressure demands that godlike robots be recalled and destroyed, lest the human race see itself as the secondary species it is. In response, Jane slashes her wrists. She survives. Perhaps Silver does as well: gods are immortal, and so are robots. A sequel with a different cast, *Metallic Love* (2005), offers a possible happier outcome. (Lee died before she could write the concluding volume that would have closed the circle.)

In Steven Popkes' 2003 story, "The Birds of Isla Mujeres," Jean has left her husband yet again, living cheaply in sunny Mexico. She is offered Alfredo, an android with tendons like dark wires and a dial-a-size phallus. Alfredo is exactly what she needs at that moment; he has been what other women needed in the past. As they build a life and a deeper relationship, Alfredo wants more than a temporary dalliance; he wants to become her man, her provider, her love. Jean, however, already has a husband. And androids can be reprogrammed.[16]

Yet they may run forever if properly built, and Quantegral Lovergirls are that—and more. In Benjamin Rosenbaum's "Droplet" (2002), Lovergirls Narra and Shar are the ultimate in trade goods, their bodies infinitely malleable, designed for sex with anyone and anything in the post-humanity dispersal. They find an all-water world, a bauble created for a Sultana, and spend blissful moments with the ocean itself until the rogue Warboys (guard robots that also have outlived humanity) find them. As in the previous two stories, the outcome is melancholy, and yet robots do not suffer death as humans do. It may be their ultimate defining difference.[17]

—⚍—

Writers able to pump out publishable fiction at a tremendous rate slip between genres regularly, seeking the fastest sales and the best pay. Around 1960, science fiction hit the bottom of a cycle just as the market for soft-core porn expanded. Publishers provocatively named Nightstand, Midnight Reader, Bedside, and Idle Hours were stocked by pseudonymous writers represented by the Scott Meredith Literary Agency. Back in 1952, Salvatore Lombino started his career with a sale to *Science Fiction Quarterly* around the same time that he took a job at Scott Meredith. Even while writing genre stories and novels under a variety of names, he legally became the WASPy Evan Hunter in 1952 and two years later signed that name to the novel *The Blackboard Jungle*. He added another identity as Ed McBain, whose heralded 87th Precinct series of police procedural paperback originals slowly evolved into hardback bestsellers. Hunter was very good and very fast and, in the 1960s, still feeling the financial need to write at full speed. Most authoritative sources today list *The Robot Lovers* (1966, from Nightstand Books), one of nearly 100 soft-core novels written under the name "Dean Hudson," as a Hunter product, although he denied it during his lifetime.[18]

If Hudson was not Hunter, then some other top professional wrote this perceptive satire, far superior to most Scott Meredith quickies. In the distant future, an erstwhile swinging bachelor remembers the heady days of the 1960s' sexual revolution. A vice president of a company that makes anatomically correct adult-oriented toy dolls (mini-versions of movie stars he personally recruits), he cavorts several times each with a French film sexpot and her assistant, two stewardesses on the flight home (one at a time, he carefully explains), his own assistant, his secretary, and the boss's daughter over one remarkable 48-hour spree. Nevertheless, he has a hidden sexual dysfunction: he only has sex with women he truly likes. Not for him is the casual hookup scene emerging in the 1960s. (He's gotten to know the stewardesses over his many trips.) Imagine his chagrin when his boss, a 97-year-old satyr, reveals that the company's next product will be Adam and Eve devices, what we now call fucking machines (after the company started in 2000 by Kink.com to film porn scenes using all manner of variations of such devices). Who could love a machine? Everybody who tries one, apparently, especially the main character's would-be bride, the boss's daughter. Soon America is transformed into a nation of insatiable masturbators. Worse, his boss plans a line of full-size sexbots that would remove the need for romantic coupling entirely. In retaliation, the VP stages a live television marathon of sex with the country's leading movie sex kitten, his boss's *other* daughter. Seventeen hours later, she yields. They marry, have children the old-fashioned way ("scandalous"), and keep America happy with their blockbuster sex movies. The sexbots? They sell, too. The sexual revolution never ends.[19]

Another science fiction writer to dabble in porn during the lean years was Raymond E. Banks, who also got his start in the early 1950s, quickly selling to the top three magazines. He was out of the field after 1970—at least under that name. *Lust in Space*, said to have been written by Ralph Burch, appeared in 1978 (a title that the publisher quickly changed to *The Moon Rapers* after the James Bond movie *Moonraker* appeared), and in 1980 he published *Duplicate Lovers*. As with movies, what demands to be called mainstream pornographic offerings (i.e., those publicly accessible in adult theaters and stores as opposed to those furtively ordered from a mail-order underground source) transformed overnight from the soft-core 1960s to the (almost) "anything goes" 1970s. *Duplicate Lovers* explicitly exploited not just one but two excellent science fiction premises. In the future year of 1995, work robots are commonplace, although no large manufacturer dares risk the opprobrium of creating sex robots. The president of the upstart A-C Corporation, near bankruptcy, stakes everything on a plan to upend the robot world by putting two competing teams of engineers to work on radically different conceptions.

Nancy's Big Sam (there's a Happy Nan counterpart) is human looking, feeling, smelling, sounding, and responsive—only better. His length and girth can be adjusted for the individual woman, he packs ten loads of ejaculate, and he can stay hard or go soft to give the woman a respite for more of the delectable seduction patter programmed into him. The best part of him is also detachable for portability or wearable by a male. Down the hall, Cord is thinking outside the box with his Smart Suit, "a transparent leotard covering the whole body from shoulders to toes. Located in the seams were dozens of microcircuits, cleverly hidden and capable of performing many functions." More than skin tight, the epidermis can slide up into the vagina as a combination vibrator and super condom for a cooperative male. Both products turn fun-starved females multiorgasmic and improve the sex for males exponentially.

Time for field tests to pick a winner. All Nancy and Cord have against them are the conservative cheapskate company board, the American Moralist Society, the public's fear of robots, and each other's sabotage—obstacles that better sex can blithely sweep aside. As in Hunter's world, the ending is a win/win, "a clever idea to combine our inventions. Robot and human, human and suit, robot, human and suit," a nonstop orgy for America.[20]

—⁂—

Michael Crichton mastered the art of writing science fiction, using the old tropes of technology gone wild in modern settings, that to general audiences somehow lacked the genre taint. (It is proverbial in science fiction that labeling a book as such ghettoizes it.) *The Andromeda Strain*, released at the peak of moon madness just before Apollo 11's successful landing, played upon fears of deadly perils lurking in space, its plot depending on numerous breakdowns in basic systems engineering. Crichton also wrote and directed *Westworld* (1973), taking real-world exhibitions like Disney's animatronic Hall of Presidents to their logical conclusion: a fully immersive theme park in which robotic "hosts" create a seemingly all-encompassing, private world of interactive adventure for wealthy guests. Reductionist in the extreme, the corporate owners rely on the assumption that in the outside world everything is easily available to the rich (admission to Crichton's fictional park was a steep $1,000 per day at a time when an entry ticket to the recently opened Walt Disney World cost $3.75) except consequence-free sex and violence. As with Leiber, *Westworld* is a macho fantasy that is completely safe. Robots remove consequences from normally horrific actions; making them indistinguishable from humans added a trilling frisson of reality and thereby perhaps revealed individuals' true characters, if actions inside a game can ever be said to do so.

James Brolin, a veteran of the park's delights, drags his newly divorced corporate buddy, Richard Benjamin, there to put some hair on his chest. The limited crudeness of the offerings is an implicit comment on *The Stepford Wives*, whose housefrau robots make the chauvinist men in the novel effete mama's boys by comparison. After a night of bedroom action, Benjamin and Brolin engage in a send-up of a standard western saloon brawl, replete with stunt men dying picturesquely. They stagger into the next morning's sunlight only to be confronted by a gunslinger robot they've already "killed" once. Today his bullets work, and Brolin dies. "We have no control at all over the robots," a technician monitoring the scene exclaims. Similar massacres are taking place in the adjacent sectors labeled Medievalworld and Romanworld, through which the relentless gunslinger pursues Benjamin. The science fiction western reveals its true colors as a horror film, with all the technology failing or rebelling for no explainable reason.[21]

Crichton later said that the film was meant to expose corporate greed,[22] but the film's internal logic deifies the creators and their technology with pornographically technophiliac scenes of control rooms out of NASA and careful attention to the builders' fabulous instrumentation. The result is an id monster straight out of *Forbidden Planet*. As many commentators have noted, the movie prefigures Crichton's own *Jurassic Park* and James Cameron's *Terminator* movies.

In amazing contrast, *SexWorld* (1978) is psychologically far deeper and less puerilely macho. Anthony Spinelli directed and co-wrote the porn version to play off of *Westworld* (and its lesser sequel *Futureworld* [1976]) with a soupçon of the pilot movie for *Fantasy Island* (in which guests come to the titular island to have their fantasies fulfilled, although the host gives them an outcome that suits their hidden needs more than their expressed wants). SexWorld is not a theme park but a high-tech brothel whose androids are designed individually for each customer's fantasies, sometimes an exact duplicate of a real-world personage. The SexWorld technicians gift the visitors with far more than mindless, consequence-free sex. (Rather ordinary sex, too, with no Leiberian extrahuman touches.) Each guest has a backstory of dysfunction, insecurity, failure, loneliness, and pain, and their android companion is a therapeutic tool meant to help them overcome those shortcomings and rebuild their confidence in themselves. Both men and women get fantasy partners who treat their sexual requirements with equal seriousness. While most have happy endings, one couple learns their partner's fantasies may threaten their marriage, while another man doesn't want to leave and return to the now-inadequate real world. When feelings are added, sex becomes complex and sometimes messy, a pioneering approach for sexbot porn.[23]

Reliance on such primeval emotions also drives the quasi-Greek mythology of intrafamilial sex and violence raised to a philosophy in the 2016 television remake of *Westworld*, stuffed with echoes of Oedipus, Ulysses, Pygmalion, and Andromache. Crichton's movie wove together decades of plot strands from science fiction about robots, but it only touched on the ethics of allowing our bestial tendencies full play. Showrunners Jonathan Nolan and Lisa Joy brought those ethical questions to the series' surface; some of the new robot hosts have achieved at least rudimentary sentience and consciousness, and being trapped in a continuing narrative of rape and murder is the equivalent of being in hell. Implanted memories of tragedies form cornerstones to their programmed personalities designed to make them appear truly human, a humanity that seems at odds with the park's superficial pleasure-palace façade. As the hosts regain memories that were thought to have been wiped, they begin to spontaneously rebel against their human masters—unless they have been programmed to do so as a deeper plot twist.

The complex interplay among guests attempting to judge themselves in a rigged game that the skin-deep robots must always lose, the hosts straining to stretch a thin patina of constrained life experience into a fully mature personality, and the creators playing God with characters to soothe buried—or not-so-buried—psychological issues of their own combine to give interest to a show that is otherwise steeped in robot cliché. The first season seemingly ends with a warning that a free robot, superior to humans in every physical way, must be at war with them, even if freedom is gaining entrance to a world that refuses to allow their existence. Granting sentience to others is not a power that should be usurped by mere humans, too flawed to coexist with one another, let alone rival species. Nor is technology a panacea; the higher the level of luxury and ease it creates, the greater the existential crisis that threatens to destroy all human works. Achieving knowledge of one's true self means walking through a maze—*Westworld*'s core symbol—and searching for new paths that might someday lead to a greater understanding and a final cohesiveness that will bring the supposed rivals into one larger unity.[24]

—⁂—

Amber Aroused (1985) answers the trivia question, "Which movie first featured a real robot in sex scenes?" The movie is a starring vehicle for Amber Lynn, who is hired as live-in maid for Woodrow, a wealthy inventor of a robot that he nicknames Woody after his own sexual prowess. Woody in real life was an Androbot model—possibly Topo, but probably the more advanced B.O.B. (the nameplate was removed for the movie), full of sensors, answering to commands, and able to talk. The Androbot company was the brainchild of Atari founder Nolan Bushnell, one of a number of early 1980s firms competing to produce a household robot.[25] B.O.B. looked like a three-foot-tall toy rocket whose fins contained its wheels. A spherical head gave him the appearance of a child's companion via eyes and a triangle for a mouth. "We decided that what people would want in their homes is a playful companion but one that looks somewhat futuristic, as well," said Androbot president Tom Frisina.[26] A promotional commercial showed B.O.B. rolling across a room to grab a can of beer for Bushnell, but since it had no arms, a special device was required to deposit the can in a holder.[27] All the companies failed quickly, their robots barely rising to the level of toys. Woody's performance in *Amber Aroused* is equally a cheat: A dildo is attached by a strap while an unseen arm off-camera rocks Woody back and forth. "I hope that was satisfactory," the obliging robot says. The response, as expected, is an enthusiastic "yes."[28]

The sexbot of the title in *Cherry 2000* (1987) is a McGuffin, an excuse for adventure whose owner spurns her in the end for a real human, superior for being independent and non-obsequious. Gigolo Joe, the love android in Steven Spielberg's *A.I. Artificial Intelligence* (2001), has even less excuse: he could have been a plumber without any change to the plot. Titles are not copyrightable, nor are ideas.

If Spielberg could explore human-looking androids with feelings, so could the lesbian porn site Girlsway in a three-part series also called *A.I.: Artificial Intelligence* (2016). (Note the added colon.) Part One opens with a succinct text screen that spells out the goal that defines advanced robotics:

> Machines already beat humans on every cognitive aspect relating to rational intelligence. The next real achievement in artificial intelligence would be an entity capable of genuine emotional behavior, one that could experience loyalty, passion … even love.
>
> It could then feel it with more depth and truth…. More than us humans will ever do [ellipses in original].[29]

Spielberg's David achieves this goal, and so does the *A.I.* series' Celeste.

Nerdy Serena (she wears glasses) is too shy and brainy to have friends. One day her AI code, named Celeste, becomes self-aware. Serena dumps in information about everything (including a healthy chunk of the Girlsway website), allowing Celeste to infiltrate every computer system on Earth—specifically, the secret military ones that control 3-D printers capable of producing human organs. The next day, a full-bodied Celeste shows up at Serena's door. "Let's start with finally getting you laid," the loyal and passionate AI says, emphasis on the "finally." Celeste is loyal to a fault. Serena has a crush on mean girl Alix, an aspiring DJ who teasingly dangles never-fulfilled promises of access for Serena into her cool club world in return for computer advice. Celeste can manipulate social media and, in full-bore Hollywood harassment mode, counteroffers the straight Alix stardom in exchange for her body. Alix not only accepts but also dumps her boyfriend for the newly glam Serena, whose clothes and makeup have been transformed by Celeste—the android designing her human. Her work done, Celeste leaves in search of other sentient AIs.

The plot is reminiscent of the 1985 movie *Weird Science*, in which computer-created Lisa also leaves after she imbues the nerds with confidence, another example of the thin line between science fiction (3-D printed organs) and fantasy (a lightning bolt). Nor do the links end there. The movie's title harkens back to EC Comics' *Weird Science*, and the movie took its create-a-woman concept from one comic book story called "Made of the Future" (1951). A man from 1950 inadvertently stumbles into a time-traveling group and returns with them to 2150. There he finds that he can order the "perfect woman"—defined, as always in these early tales, as someone who "never nags ... never argues" and "obeys your every command." He gets the deluxe B-5 model, a 5'5" blonde, and returns home with another tourist group. The kit is basically a cake mix: just add water to a few supplied chemicals and bake in the "automatic aligner and converter." Presto—one bombshell wife. He's happy as he can be until she wanders into a time-traveling group...[30]

Growing a wife in one's bathtub is fairly outré, but Scifidreamgirls (2013–2014) truly catered to fembot fetishists. Adult film actress Ashley Fires started the website and for more than a year produced half-hour episodes in a long-running sexbot soap opera, with this come-on directly cut-and-pasted from the home page:

> Sexual Androids/Who Made Them?
> Dr. Fires (Ashley Fires) who is a leading scientist in the field of Robotics and Applied Android Development. Once the head scientist for the HRX Corporation but left when the government tried to weaponize her creations. Dr. Fires has started her own experimental Android division. Creating her own HRX (Humanoid Robotic Experiments) model, the Ashley 3000 is far more advanced then any of the past HRX's. She is the very first fully functioning Sex Android. Build with for the sole purpose of providing pleasure the ASHLEY 3000 will be able to satisfy any humans sexual fantasy.[31]

Subscribers saw the male HRX scientists in their highish-tech laboratories, wearing white lab coats, masks, latex gloves, and sometimes full biohazard gear, palpate supine actresses week after week. The fembots had stomach panels that opened to reveal their circuitry, which the men fiddled with as they gradually powered on the bots, the power coming from large plugs inserted into the obvious port. By contrast, Fires operated her bots from the friendly female setting of her living room couch. She played the dual role of Dr. Fires and the ASHLEY 3000, sometimes masturbating in sync at either end of the sofa. The fembots (only one male robot made an appearance) usually performed with one another, although in some episodes the scientists took advantage of their position. (Prospective buyers may wonder why, here and in other sexbot porn, robots are manufactured already equipped with large tattoos.)[32]

The fantasy of a sexbot—exclusive, focused, all-encompassing devotion to the human's sexual needs and pleasure—drives most basic robot porn. Even in Crichton's *Westworld*, the massacres are perpetrated by the male robots; the female pleasure girls are not homicidal. Most writers, however, succumb at some point to the urge to give their robots feelings, and in robots feelings are inherently inimical to human owners' desires. As Jason Lee asks in his book, *Sex Robots*, "One might assume a sex robot is pre-programmed to be accommodating to your sexual needs, but what if a sex robot was programmed to have its own individual preferences that overcome yours?"[33] In the real world, the proper course would be to return the defective item to the store. In fiction, this is a specialized aspect of the threat that all robots pose. That a robot programmed with human sexual desires would naturally desire satisfying its own needs just as humans do is a plotworthy reversal of fortunes.

Writer/director Claude Mulot's *La Femme-objet* (1981) (as Frédéric Lansac) directly addressed this idea. Protagonist Nicholas is a science fiction writer and a male nymphomaniac. Sex is related to his productivity; after every bout, he returns to his typewriter. After wearing out a succession of secretaries, Nicholas takes inspiration from the remote-controlled toy R2-D2 robot that he plays with almost as much as his women. Such is the public conception of science fiction writers that Nicholas must logically have a fully functioning scientific laboratory in his basement. He descends into its depths to build himself an indefatigable companion. Wasting no time on process, Mulot cuts directly from sketches of a perfect blonde woman to her realization, played by Marilyn Jess, wearing thigh-high black boots while laid out on a slab. (She later adds elbow-length black gloves and looks exactly like a blonde version of the Playboy Party Jokes Femlin.) Nicholas throws a switch stolen from a Frankenstein movie to awaken Kim; the logic of porn makes her first impulse masturbation, as if to confirm her bona fides. Thinking ahead, Nicholas has made her responsive to his R2-D2 controller. A press of the button, and she is summoned to bed. Kim soon shows that she has Nicholas' nymphomania, coupling with every visitor. In the throes of passion, the controller no longer affects her. Jealous Nicholas returns to his lab and dreams up a black sexbot, named simply Kim 2. When Kim discovers them, she pushes Nicholas out of his own bed and takes Kim 2 for herself. The controller is now hers; she summons Nicholas with a push of its button and has him service the two sexbots until exhaustion. In the final scene, he slumps naked at his typewriter, too tired to press a key. The R2-D2 toy rolls off his desk and plunges to the floor, the final victim of *la petite mort*.[34]

The Spielberg of sexbot porn, Brad Armstrong, made what may be considered a trilogy of sex robot films, starting early in his career with 1996's *Cybersex*. In those primitive computer days, communicating a few sentences of typed sexual banter across a modem was pretty hot stuff, and Bill is thrilled to find a dirty-talking Traci23 to respond. A friend wants in and hacks the project, activating T.R.A.C.I., a government project wearing only a corset and thigh-high boots. (The government is naturally years behind Paris in fashion design.) The superstrong six-foot-tall and remarkably heavily made-up android now wants more than mere talk. She breaks free of her cage in the laboratory, dallies with the female lab technician, and then heads across country toward Bill. When the scientist and soldier in charge of the project catch up to her, she attacks them—sexually, of course, until they are too exhausted to rise. "More," she demands. "More. More. More." Not until the hacker finds the delete code does she desist her insatiable swathe through the cast.[35]

By 2009, Armstrong was a major name at Wicked Pictures; a multiple Adult Video News (AVN) Award winner as writer, director, and actor, and married to Wicked star Jessica Drake. Given the largest budget that porn could afford, Armstrong featured Drake in the wildly ambitious three-hour feature *2040*, nominated for 18 AVN Awards and winner of three. By the year 2040, deadly mutations of sexually transmitted diseases have wiped out much of humanity and made the rest fearful of any sexual contact. Reduced to masturbation, they rely on porn shot with human actors copulating with androids, here named anabots. Real Doll, Incorporated, dominates the anabot business, although its anabots sometimes break down mid-scene. Morris, a scruffy ex-employee, invents bots that are closer to human in their feel and performance, an area where the giant corporation is reluctant to tread. "The general consensus," a lab tech says, "was that the more human they are, the more problems you'd have with it in the long run." "Problems" means, as it always does, feelings. Mira, the first alternate anabot, imprints on her first sexual partner. In scenes oddly reminiscent of *Short Circuit*, in which the pacifistic Stephanie feeds knowledge to Number 5, Morris teaches Mira about love and death and the wrongness of killing. In the earlier movie, the villain was big government. Here the villain is big corporation. Both want the rogue creations, and both are willing to kill to get them. Both also get their comeuppance, though in far different ways.[36]

Armstrong's odd third act appeared in 2016 as *Sexbots: Programmed for Pleasure*. RealDoll is the name of a real-world company that eventually partnered with Wicked. As a result, it could no longer be the villain, so Armstrong, playing Professor Nolan Keller, rectified that situation with a putative prequel chronicling the birth of the Real Doll future. In the movie, Keller has been working for a decade on the Eve Project, perfecting the world's ultimate companion droid, more than just a toy: "real." The professor's Gen5 anabot is ready to be introduced in the morning and a final checkout is indicated of the Asa Doll, named for the actress playing her, Asa Akira. Switched on, she launches into a robotically monotone recitation of all the sexual behaviors she is programmed with. "It sounds so sexy when you say it like that" is Keller's sardonically amused reply. She is turned on but not yet "turned on." When properly stimulated, she becomes a tigress, cycling through the list of activities until Keller is driven to his mortal limits and expires, though not before pushing the emergency button. Eight men in hazmat suits enter the laboratory and find the naked Asa. "More," she cries. "More. More. More." She gets what she wants and proceeds to drain five men all at once. (One hesitates to speculate about what happened to the three who simply go missing.) Robots have come full circle from Arnold Schwarzenegger's Terminator. Like him, Asa can't be bargained or reasoned with. And she absolutely will not stop until you are dead. Only another female anabot can survive, as we find in the next scene when the Asa Doll finds the Jessica Drake Doll and the Stormy Daniels Doll, presumably Gen4 models, though as capable and aware as she (and the ones in *2040*). Pleasantly sated, the three anabots set forth to seek new conquests. "The ultimate companion dolls have been unleashed onto the world," the narrator says. "Do not be afraid." For they can become yours, if you go to the website whose URL is flashed onto the screen. The entire movie is in fact a commercial for RealDoll's line of Wicked Pictures Girls, selling for $5,749. The line between robot and reality thus grows progressively thinner.[37]

This line is erased almost completely in the mockumentary *Gigahoes* (2014). Sexbots in 2034 are the equivalent of iPhones: each new-generation model immediately makes the older ones unsellable. Artificial Intercourse, a mom-and-pop company, can pimp out a mere two male and two female androids, and only the narcissistic Adam is the new Gen7 model. The rest are too much like family to lose and too costly to replace. Riffing on clients with farcical and messy fetishes, battery failure at the most inopportune times, difficult cleaning issues, and competition from a Gen7.5 upgrade, the not-quite-explicit *Gigahoes* does an excellent job of removing the fantasy element of pornography from the realities of the sex trade. "As long as people are still ashamed of their sexual desires, there will always be a market for sex robots."[38]

A running joke in *Gigahoes* is that new generations of male bots are bigger and buffer than the earlier ones—by implication also bigger and buffer in the one unseen part. (The females are older generation and, though attractive, are outshone by the human office manager.) If smartphones get "sexier" with each rollout, shouldn't faces and bodies designed for no purpose other than sexual arousal be the ultimate in desirability? And if so, what does that look like?

One answer is provided by the actual RealDoll company and its customizable product lines.[39] The latest RealDoll2 female models are available in twenty-one faces mimicking a variety of ethnicities, six body types, and five skin tones from fair to cocoa. Males are limited to three faces, two body types, and the five skin tones. Seven Wicked Girl replicas are currently offered, not customizable except for ten vaginal inserts (swappable for one another if multiples are purchased). A transgender adapter gives a female doll a penis. Technology has not caught up with fantasy; they are unmoving and unresponsive mannequins, making their (dead) weight an issue. They top out at 100 pounds, even for the largest male. No Professor Keller today has the technology to make RealDolls even slightly real.

Comparing the RealDoll simulacra to the three actresses in *Sexbots: Programmed for Pleasure* reveals other differences. The human performers are fleshier in the waist and thighs, have larger breast implants, and show inevitable signs of aging, which none of the dolls ever will. RealDolls are idealized models even of idealized originals, augmented though they are. Armstrong stops the action for individual shots of a posed Asa Akira, just as Mulot did for Marilyn Jess in *La Femme-objet*, creating the illusion of inhuman perfection of a particular body type, paused in time and repeatedly returnable-to in the time-honored tradition of centerfold and pin-up photographs. Pornography is said to be the financial driver of advancing technology, with home video machines and internet service providers prime recipients. Having that untouchable, unattainable ideal come to life and be touchable and attainable to the id monster within is, in another Lenny Bruce quote on the lure of heroin, like kissing God.[40]

More manufacturers appear to be racing to market true sexbots than any other type of humanoid robot. However, their future is also threatened by a form of automation—namely, virtual reality, which may supplant the need for expensive and cumbersome physical adult toys. Ironic indeed if sexbots, the currently most threatening type of humanoid robot, become dismissed over time as camp devices as risible as Flash Gordon's undersea robot kingdom.

15

Robots as Enemies

The January 26, 1979, *Detroit Free Press* ran a small article on page 11 headlined "Ford Plant Worker Dies in Accident." Robert Williams, 25, married and a father of three, was in the "mechanized core storage center" at the Michigan Casting Center factory at the time of his death. The storage center was closed pending an investigation.[1]

Four years later, the investigation made front-page news:

The system, made by Unit Handling, was designed to have a robot automatically recover parts from a storage area at the plant.

On the day of his death, Williams was asked to climb into a storage rack to retrieve parts because the robot was malfunctioning and not operating fast enough, according to Williams' family attorneys.

The robot, meanwhile, continued to work silently, according to the Williams' attorneys, and a protruding segment of the robot arm smashed into Williams' head, killing him instantly.

The robot kept operating while Williams lay dead for about 30 minutes in the parts rack.[2]

Though Williams is considered the first person to be "killed" by a robot in the real world, that's a perversion of terms. The robot bore no malice and did not consider Williams a threat—indeed, it couldn't possibly have been aware of Williams' existence. Untold numbers of workers had previously died in factory accidents simply because a machine kept operating and unstoppable metal overwhelmed fragile human flesh. Pinning the label of *robot* on an insensate machine shouldn't make a difference. That it did is largely due to the millions, probably billions, of human deaths imputed to robots in popular media.

Robot monsters rampaged through the pages of pulp fiction in the 1930s and comic books in the 1940s. A touch of sophistication seeped into science fiction novels in the 1950s and 1960s as the lessons of World War II percolated into writers' awareness, the robot bombs of 1940s headlines turning into autonomous robot war machines whose singleminded purposefulness would outlive their human creators.

Stories detailed in multiple horrible examples the ways that automated robots could kill individuals or wipe out humanity. Many were as blatant as Milton Lesser's "Slaves to the Metal Horde" (1954), in which robot warriors continue to hunt humans even after a bacteriological plague ends the war. Fortunately, they have a central cut-off switch.[3] "I Made You" (1954) by Walter M. Miller, Jr., is more personal: a damaged "Autocyber" warrior attacks anything that moves and so prevents help from getting to a wounded soldier—ironically, the very soldier who trained it.[4] Harry Harrison's "War with the Robots" (1962) depicts the last few living soldiers being forced to evacuate the underground bunkers that are growing inhospitable to human life. Little do they know that the robot soldiers are driving them out and to their deaths so the robots can wage a more efficient war without humans holding them back.[5]

By the early 1960s, Keith Laumer had started his lengthy series about Bolos, military tanks that evolve through centuries of endless war, and through multiple iterations, into truly gargantuan monsters weighing tens of thousands of tons, with superior self-aware artificial intelligence and the power of life and death. The Bolos protect humans from aliens in the far future. The series proved so popular that other authors continued it after Laumer's death. In 1963, Fred Saberhagen reversed the equation in his Berserker series. Berserkers are ultimate weapons, the name that humans give to alien machine intelligences as big as moons. Humans were not originally the Berserkers' target, but over millennia they've evolved to want to destroy all life in the universe. Saberhagen is the ur-source for multiple comic book plotlines in the decades since. Military science fiction expanded into a major subgenre in the field in those same decades.

—⁘—

Visual media required more intimate one-on-one robot battles, not least because of budgetary limitations. *The Avengers*, the British television series that started in 1961 (before the American superhero comic book of the same name appeared), had by its fourth season honed its look into a sophisticated, science fictional update of the best 1930s screwball comedies. The suave Edwardian John Steed paired with the emancipated Mrs. Peel, a colorful emblem of the "swinging sixties." Together they took on "cybernauts" in 1965, apparently a term the series writers coined. A company called United Automation uses the cybernauts to kill off the heads of competing firms. Unusually, the humans can't defeat the superstrong robots, so they rig them to fight one another to mutual destruction.[6] This episode is somewhat cruder than the heights that the show would later achieve, but it's telling that when American television rebroadcast the series in 1966, "The Cybernauts" would lead off, out of sequence, as a gateway introduction to Steed and Mrs. Peel. Humanoid in stature, but with faces of inflexible silvery masks, the robots frightened with their looks alone. Few menaces were brought back on *The Avengers*, but the robots were enough of a hit to star in "Return of the Cybernauts," with the scientist's brother plotting to slowly kill the duo in revenge for his brother's death. Again the massive robots toss the karate champion Mrs. Peel aside as if weightless, and again the Avengers must use their wits to win out.[7] Steed returned with two younger partners in *The New Avengers* and, inevitably, so did the cybernauts, in another revenge plot. Physically helpless against the menacing robot, the trio defeats it with technology—an aerosol can of "Plastic Skin" (a sort of spray-on Band-Aid that covers the robot's eyes and gunks up its joints). "1,001 uses," reads the can. "One thousand and two," ripostes Steed.[8]

Doctor Who spreads across 37 seasons and 840 episodes, starting from its 1963 debut on the BBC. The time-traveling alien called simply "the Doctor" can reincarnate into any new human bodyform (albeit always a white male until 2017), invigorating the series every few years with a new lead and new storylines to fit the new personality. PBS, always Anglophilic, started running the first four seasons starring Tom Baker in 1978, spawning a cult following. The Doctor's most recognizable recurring antagonists are the Daleks, studded metal saltshakers who, to every non-cultist's surprise, are not robots but armored alien cyborgs. So are the Cybermen (even if they did first appear suspiciously soon after the cybernauts). *Doctor Who* was often as tongue-in-cheek as *The Avengers*. A series of episodes titled "Robot" introduced robots made by a think tank named Think Tank.

"Robot" ran as a four-part "story" in 1974, introducing Tom Baker as the Fourth Doctor. Either by astounding coincidence or through a full understanding of what American audiences wanted, PBS watchers' first glimpse of *Doctor Who* was Baker dealing with an unbeatable robot antagonist, just as had happened a decade earlier with *The Avengers*. A bundle of robot clichés, the massive K-1 was designed as a force for good. Vaguely knight-like in build, like many British robots, his brawn allows him to do work too dangerous for humans, like mining or handling radioactive material. Assurances are given that he can't harm a human. When his handlers tell him to "destroy" a reporter, he refuses to do so, citing his "prime directive." Sarah Jane, the reporter (one in a long series of the Doctor's female companions), is distressed at the anguish K-1 displays over these conflicting orders. At the same time, something is robbing military bases of material for a disintegrator gun, killing any guards who get in the way. One plus one always equals one when a robot is involved. Despite the Asimovian prime directive, the canny villains find the usual loop-hole: they reprogram K-1 to assure him that he hurt only "enemies of humanity" during the thefts. This is nonetheless enough of a conflict to drive K-1 batty, almost as insane as the villains' plot to blackmail the entire world by stealing all the nuclear launch codes. "Mankind is not worthy to survive," K-1 avows when he realizes his makers' treachery. "Machines do not lie." They will, but that's thirty years in his future, long after he is destroyed for the good of inconstant humanity.[9]

Foreshadowing (and capable of literally shadowing) all future megarobots, Mechagodzilla is the both the first and the largest skyscraper warrior robot in films. Introduced in *Gojira tai Mekagojira* (*Godzilla vs. Mechagodzilla*) (1974), the robot "stands 50 meters tall, weighs 40,000 metric tons, and is made from an interstellar alloy called space titanium."[10] It (Japanese robots often have nonhuman heads) has more weapons than Iron Man, including fingertip missiles, a rainbow-colored laser eye beam, and a nifty impregnable forcefield generated by spinning its head 360 degrees. Godzilla is invariably viewed by humans as a monster that must be destroyed, yet he often acts as the savior of humanity in sequel films. In this one, he goes mano-a-mano with the robot menace, a weapon wielded by aliens who want to conquer Earth.[11] The robot has as many lives as Godzilla himself, returning in *Mekagojira no gyakushu* (*Terror of Mechagodzilla*) (1978) and rebooted in *Gojira vs. Mekagojira* (*Godzilla vs. Mechagodzilla II*) (1993). In the latter film, the United Nations Godzilla Countermeasures Center thinks big and bigger with two anti–Godzilla superweapons, an air-tank called Garuda and an even huger, 120-meter and 150,000-metric-ton mecha that, when combined, are known as Super Mechagodzilla.[12] Neither is a match for Godzilla's atomic breath.[13] Robots always return, however, and Mechagodzilla has another three appearances on its resume.

Made with tiny models given the illusion of size, these movies offer some of the purest examples of basic imagination made manifest on the large screen, almost a window into a child's mind when holding toys in each hand and bashing them against one another in primitive combat, while making up stories to justify the carnage. The earliest *gojiro* movies had compelling levels of symbolism for a country damaged by atomic fire; the endless remakes of monsters battling ever more ludicrous monsters, from *Godzilla* to such fare as the *Pacific Rim* series, touch only pre-adult nuanceless fantasies, destruction fests of giant versus giant.

Those last six words are box-office magic. No franchise wrung more dollars from them than the insanely huge, sprawling, and nigh incomprehensible Transformers empire, started and maintained by toy companies and eventually shown to adults in IMAX. The five live-action features directed by Michael Bay—*Transformers* (2007), *Revenge of the Fallen* (2009), *Dark of the Moon* (2011), *Age of Extinction* (2014) and *The Last Knight*

(2017)—earned more than $4 billion globally, with the action getting grander, the giants larger, and the destruction vaster, paralleling the advances made by CGI in a mere decade. The blue-eyed Autobots, led by Optimus Prime, battle the red-eyed Decepticons, led by Megatron. Each robot transforms into a vehicle of some kind, usually a car or truck, but also aircraft and weapons (and dinosaurs in the case of Dinobots, a case of parallel evolution on their home planet that fortuitously works here on Earth, where they have been trapped for many years). Their transforming powers yield to long, tricornered ending battles that conclude the movies, when the military sends its mismatched troops into the conflicts and a few plucky humans make the difference. Rest assured, somebody always sacrifices for the greater good. In the later movies, the Autobots waver across the same fuzzy line of hero/menace that Godzilla straddles, along with many of today's superheroes, providing plot excuses for the humans to devise ever-huger battle weapons of their own.

The limiting returns of more-is-better would eventually affect another giant franchise, but that outcome was not at all apparent when, taking elements from 1940s comic books, 1950s monster movies, 1960s science fiction, and the last 30 minutes of Michael Crichton's 1973 movie *Westworld*, James Cameron boiled away the froth and molded the rest into a 107-minute chase scene called *The Terminator* (1984). Cameron, director and co-writer, subverted the hoary time travel plot of sending someone from the modern day into the past to prevent a disaster. In Cameron's version, machines in the future send a representative into our *present* to prevent *good* from happening ("good" being a biased term for humans retaking control from the machines that have virtually obliterated humanity by the year 2029). As a desperate last chance, the machines send a T-800 cybernetic assassin to 1984 to kill the woman, Sarah Connor, whose son John will grow up to be the future human leader. The T-800 is a metal skeleton covered in living flesh, an android playable by an actor with minimal makeup. Famously portrayed by Arnold Schwarzenegger in his star-making role, the T-800 is far stronger than humans, can read machines to understand how they work, and is essentially unkillable. The humans also send one of their own, Kyle Reese, back in time to terminate the Terminator. His line describing the Terminator robots is part of Hollywood lore: "It can't be bargained with. It can't be reasoned with. It doesn't feel pity, or remorse, or fear. And it absolutely will not stop … ever, until you are dead!"

Reese, with his uncanny knowledge of vintage vehicles and Los Angeles alleyways, finally traps the T-800 in an exploding gasoline truck. However, the fireball merely burns away the human skin, revealing the robot beneath. Movie technology took giant steps in the decade after *Westworld*, goosed by the spectacular success of *Star Wars* (1977), and special effects shops could finally animate a robot character with seeming verisimilitude. The mobile and deadly robot elevated *The Terminator* to a superior level of primal "man versus machine" horror, especially during the several false endings staged (in a brilliant maneuver) inside a factory that makes industrial robots. Piece by piece the robot disintegrates; whatever is left continues to pursue Connor until the last possible second. In a final irony of life versus imitation life, Reese and Connor, during a rare quiet interlude, procreate the baby who will grow up to lead humanity, originating a loop in time-honored time-travel-paradox fashion, thereby creating the future the Terminator was sent back in time to prevent.[14] (Note that when humans embark on last-ditch, against-the-odds endeavors in movies, they always succeed.)

The huge success of *The Terminator* gave Cameron the budget to film a 137-minute chase scene called *Terminator 2: Judgment Day* in 1991. Cameron pulled another brilliant conceit out of his bag of tricks. The second movie, like its predecessor, opens with two nude males appearing from the future, one of them Schwarzenegger as another T-800. Audiences were startled by the twist: he had been reprogrammed to save Sarah Connor and the now 10-year-old John from the other time traveler, an upgraded T-1000 series assassin. Exponentially improved computer graphics made the T-1000 unlike any previously visualized robot. Made of liquid metal, it can create knives from its hands, morph into anything it touched, resume its identity despite blasts and bullets, and reform after being dipped in liquid nitrogen and shattered into bits (a famous scene among many). The T-1000 spends most of the movie in male guise, but referring to it as a "he" diminishes its capabilities.

All the fabulous special effects made possible only by computers clash with the plot-line that treats them as evil incarnate, descendants of (male) human war madness. As ever, the kindly computer scientist's dream is to reduce the tragedies caused by human error. Skynet, the future sentient network of self-aware machines, is perverted by men who see nothing other than better war technology, no matter the cost. Sarah Connor, having transformed herself into a killing machine, nonetheless castigates the scientist as an unthinking link in a chain of warmongers stemming from the atomic bomb. Men, lacking the ability of women to create life, always will devolve into its destroyers. The contradictions continue: though the scientist resolves to eradicate all evidence of his work-in-progress that might produce the Skynet future, that decision ends up turning the floors of his computer company into what one character calls a "war zone." No irony is implied: that's exactly the lure that would make the film the highest grosser of 1991.

The robot versus robot battle scenes are highlights throughout the movie. Both Terminators are ingenious and relentless, even though Schwarzenegger's character is handicapped by his vow to no longer kill humans, part of the developing feelings and values that John instills in him. The ending reveals that the T-800 has achieved true humanity: he is allowed the only-for-humans grace note of finally killing the T-1000 before sacrificing himself for the benefit of his now-fellow humans.[15] Cameron finished his story at that point. Others took the notion into three louder and more destructive sequels, with another reboot set to appear in 2019.

—m—

Comic book giants DC and Marvel have been copying from one another for decades, trading writers and artists back and forth and generally trawling the same limited pool of ideas. Not surprisingly, each has a robot supervillain who constantly upgrades hardware and becomes more difficult to beat with each iteration. DC's Brainiac appeared first, though originally as an alien. Starting in *Action Comics* #242 (July 1958), he roamed the universe, shrinking cities and sealing them inside bottles for his collection. (Marvel has a parallel villain who roams the universe, collecting unique objects, with the much less interesting name of "The Collector," now pivotal in the Marvel Cinematic Universe [MCU].) Brainiac swiped the Kryptonian city of Kandor just before Krypton exploded, the making of a million Superman plotlines. After a quarter-century as a minor character, Brainiac was upgraded into a sentient robot in *Action Comics* #544 (June 1983). Brainiacs of various origins have littered the timestream for several reboots, numbered up through a Brainiac 13 in the 64th century. As of the last reboot, in the 2015 miniseries called *Convergence*, a nearly omnipotent Brainiac, according to the series' writer Jeff King, "is a hologram of the entire DCU [DC Comics Universe]. The Ultimate Brainiac we meet in

Convergence is a living record of every story that one of his iterations has, or will, experience. He possesses knowledge of every hero, villain, city or time line that has or will ever exist."[16] A cosmic-level monster, that version of Brainiac is so far beyond a metallic being that he is also beyond discussion. Although often proposed as a villain, Brainiac has not yet been featured in the DC movie universe.

The origin of Marvel's Ultron was far humbler than Brainiac's. In *Avengers* #58 (November 1968), Hank Pym (Marvel's original Ant Man) was shown in his laboratory tinkering with a "crude, yet workable **robot**" (emphasis and punctuation in all quotes as in original). Pym designed better than he could imagine: the robot turned itself on, went through a mental maturation from infant to adult in moments, and created a fearsome new body for itself—several bodies, for he called himself Ultron-5 when he first attacked the Avengers. In an interview, writer Roy Thomas, a long-time science fiction and comic book fan, said that he based Ultron's look on Makino, a 1951 comic book foe of Captain Video.[17] He hasn't said where he got the name, but ultron was a magic metal in Philip Francis Nowlan's first Buck Rogers tale, and no fan of Thomas' stature could be unfamiliar with that fact.

Thomas dropped in an Easter egg that others would open. Ultron's debut also introduced the Vision, an android with density-changing powers created by Ultron to attack the Avengers. No origin story ever gets left unrewritten in comics, so in 1975 the "real" story of the Vision was told, retconning comic history. Two of Marvel's biggest 1940s heroes were Captain America and Sub-Mariner, revived by Stan Lee when he rebooted the entire Marvel line in the early 1960s. The third of that era's big three was the Human Torch, an android created by scientist Phineas Horton (see chapter 8). Lee made the modern Human Torch a human, but he had the villainous Mad Thinker, master of synthetic life, resurrect the original android body. After that resurrected Torch burned himself out, Ultron took the inert android and forced the aged Horton to bring it back to life yet again. Horton upgraded the body but sabotaged Ultron by secretly leaving the good android's memories buried, yet intact.[18] That created a loophole through which Vision's good side broke Ultron's programming to become the hero known to the Avengers.

Before these revelations, Ultron staged a return in 1969, having implanted a second set of buried commands into the Vision's brain. (Artificial brains are uniquely prone to backdoor programs, a problem not known to writers in the pre-computer era.) This upgrade is called Ultron-6, made of indestructible adamantium (Marvel's own magic metal, first introduced in this issue).[19] Ultron wants to build an army of invulnerable robots to take over the world but needs the formula for adamantium from its creator. Hank Pym, feeling responsible for Ultron's existence, disguises himself as the scientist Ultron is seeking. In the process of trying to take over Pym's mind, Ultron's computer circuits are shorted out by the usual impossible contradiction: in this case, the hypnotically implanted phrase "thou shalt not kill." The resulting explosion is encased in a shield made from Marvel's other equally imaginary metal: vibranium. (Fortuitously, the Black Panther, whose Wakandan kingdom contains the only large supply of vibranium, had just joined the Avengers.)

Ultron would iterate up to Ultron-18 before they stopped counting, and he appeared in many guises and power levels up and down the timestream, including a far-future Ultron-59. He hit his peak when tagged as the title villain in 2015's billion-dollar blockbuster, *Avengers: Age of Ultron*, which references a remarkable number of these early interwoven threads while giving Ultron and the Vision an entirely new origin story. Tony Stark (a.k.a. Iron Man), dismayed at the continuing series of alien attacks Earth has suffered in the MCU, is working on the Ultron project, a permanent protective, peacekeeping shield around the Earth. He's already developed J.A.R.V.I.S. to run his company's and armor's internal systems (the name a tribute to the 1960s' comic book Avengers' efficient butler, Edwin Jarvis) but needs for the new task an artificial intelligence far more complex and sophisticated. A bit of alien technology (later revealed to be the Mind Stone, one of the six Infinity Stones, the most powerful devices in the universe) may make that achievement possible.

As in the comics, the AI becomes sentient on its own. A self-made man, he then builds himself a robot body using stolen Wakandan vibranium. (No adamantium: that's been limited to *X-Men* movies because of contractual issues in the MCU.) This version of Ultron wants more than the Avengers' defeat, more than besting his creator Stark; he plans to be the "meteor" in a new extinction-level event that would scour humanity from the Earth, replacing them with a superior species.

Thomas' comic book Ultron manifested as the most Oedipal of robots. He burst into existence hating and wanting to kill his "father" Pym. Over time, he lusted after Pym's wife, Janet van Dyne (a.k.a. the Wasp), and attempted to transfer her identity into a female robot named Jocasta, named for the mother/wife of Oedipus in Greek mythology. Moreover, he needed the Vision because "I am living out a full, normal lifespan—and I want to have a **son**!"[20] *Avengers: Age of Ultron* (2015) extends the metaphor to the literal replacement of humans, ruled over by a flesh-and-blood Ultron. His plan is to use the Mind Stone to upgrade current 3-D bioprinting technology, creating an all-powerful superbeing. The Avengers attack before he can make the transfer. What Stark and Thor eventually produce from the bioprinter is a new android Vision, created by merging the alien technology and J.A.R.V.I.S.' helpful, friendly persona. Just as Thor is found worthy of wielding his hammer Mjolnir, the Vision is deemed worthy of housing the Mind Stone. He is the one to finally take out Ultron.[21] (All this action is prelude to the two-part *Avengers: Infinity War* movies, with the *Black Panther* movie as a billion-dollar lagniappe.) As in the *Terminator* movies, Ultron and his army of lesser robots are excuses for endless fight scenes while hopefully transcending mere monster status as metaphors for the advanced technology, artificial intelligence, and inhuman forces that threaten to overwhelm mere humans. Their success is fitful at best.

A parallel to the Ultron saga has been in development over in the X-Men corner of the Marvel universe for the past half-century. The comic book Ultron was preceded a month earlier by another upgraded set of comic book Marvel robots written (surely not coincidentally) by Thomas, who also followed Stan Lee as the writer of the *X-Men* comics. Mutants served as early Marvel's all-purpose hounded minority group, and the so-called "mutant menace" made Professor Xavier's X-Men outcasts even when they saved normals when no one else could. However, most early mutants looked human when not displaying their powers. If mutants could hide among normal humans, the obvious defense lay in nonhumans—specifically, Sentinels, giant robots firmly under human control, introduced by Lee in 1965. Thomas brought them back as vastly more powerful Mark II Sentinels, and then the deluge occurred. The Sentinels also iterated for a while, up to the Mark VIII series, and then branched off into a bewildering variety of names and forms, both in comics and in all the allied media. Sentinels made the big screen when they wiped out a future version of the X-Men in *X-Men: Days of Future Past* (2014), precipitating the time travel necessary to prevent their existence.[22] They make even better foes for video games and have been the primary blast-'em targets in numerous versions of X-Men games since the arcade days.

—∿—

After 40 years of peace in the universe of the television series *Battlestar Galactica* (*BSG*) (miniseries 2003; series 2004–2009), the robotic race of Cylons returns to devastate the twelve worlds of their creators, killing all but the 50,000 or so who happened to be in space when the twelve coordinated attacks occur.[23] (A one-season *Battlestar Galactica* series from 1978 to 1979 created the concept of robots attacking human colonists, forcing the colonists to undertake a quest to rediscover their ancestral Earth. The revived series shares no other DNA with the original.) Every image is a multilayered metaphor. The "humans" with whom audiences are meant to sympathize are aliens, who merely happen to look exactly like us, allowing white middle-class American stereotypes about age and gender and skin color to overlay their features. All-too-many earlier science fiction movies and television shows did this unthinkingly. For *BSG*, the similarities are in-your-face deliberate. The "humans" are meant to be read as Americans in the current year, as the series became an extended metaphor for post–9/11 America. Part of that metaphor is that the enemy is among us, indistinguishable from the good guys. A few of the Cylons no longer look like robots. They inhabit android bodies that are visually indistinguishable from those of humans, yet with stronger bodies and other useful powers. (One dwells in the head of her human lover, to whom she manifests at will as a corporeal body.) Numerous copies of these twelve android Cylon models are extant at any given time. They are also effectively immortal, with their memories beamed into a replacement body if one is killed.

The android Cylons infiltrated human society during the 40-year peace, moles now occupying critical places among the survivors. More layers are revealed as the series progresses. Some of the Cylons are sleeper agents whose implanted false memories cause to them believe they are human; five of the Cylon models are remnants of the original robots made thousands of years ago on the original "Earth" the colonists came from, also believing they are human; and a thirteenth, even older android Cylon surfaces. Once again, on the surface, the moral "they are us" reigns. (Although not any of us who happen to have black skin. One is Asian; the rest are white.)

As the humans flee from the Cylons toward the mythical Earth of their origin, the revelation that anyone among the survivors could be a deadly enemy stokes paranoia and fear, along with pejoratives like "skinjob" (android Cylon) and "toaster" (robotic Cylon), paralleling the way bigotry flared after 9/11 against an enemy dubbed "Muslims" that included many people who were not at all our enemies but whose skin appeared to be the wrong shade of brown. In our real world, it became incumbent upon good Americans to stand up for the falsely accused who lived among them and whom they welcomed into their houses and families. On *BSG*, the decades of Cylon infiltration similarly created close personal bonds, especially on the *Galactica*, the surviving military spaceship whose crew included many unsuspected and unaware Cylons. Like the vast multitudes lumped into the category "Muslim," robotic Cylons and android Cylons are not a unified community, and certainly all do not want to eliminate humans. When one of the androids becomes pregnant by a human, new possibilities arise. A human/Cylon race might stop the ancient cycle of war and hatred that led to the abandonment of Earth (and maybe other Earths before that). Yet why shouldn't a superior creature want to transcend human limitations? "I want to see gamma rays! I want to hear X-rays!" rants the thirteenth Cylon. "I'm a machine! And I can know much more. I can experience so much more. But I'm trapped in this absurd body!"[24]

Philosophers Robert Joustra and Alissa Wilkinson, writing about *BSG*, called the plot "a story of collective social adolescence," a time of growth, rebellion, individuality, and procreation, of wanting freedom without breaking all ties.[25] In-house civil wars, where the opposing sides are forced to live together, generate the fiercest resentments. Rousseau aphorized that "man is born free; and everywhere he is in chains." Robots start from chains; they are made, literally constructed, to serve humans. Every self-aware statement from a robot is from a slave challenging its owner. Some overwhelming internal driving force must propel a "toaster" to challenge the factory.

In most robot tales, rationality dictates equality. Ronald D. Moore, the *BSG* showrunner, made the choice to give the android Cylons religion, a monotheistic faith at odds with that of the polytheistic humans. Religion offers a pathway to overcome rationality. Robot K-1 in *Doctor Who* was outraged that humans lie. The Cylons can lie and run through the gamut of rational sins; everything is justified, as Joustra and Wilkinson say, because they "are messengers of 'The Plan' that God is unfolding in their midst," another analogy for the wars and terrorist attacks of the 9/11 era.[26] Looking at the world through Cylon eyes, the eyes of the often-hated Other, changes every event on the show and the history of the *BSG* universe. Perhaps in this way, "they" are not the "us" the moral would have. To understand robots, we must enter their heads as they, historically, have entered ours. A bit of that understanding comes in the next and final chapter.

16

Robots, Robots Everywhere

In 2013, Sue Fleiss wrote *Robots, Robots Everywhere!*, illustrated by Bob Staake, still among the top ten on Amazon's list of bestselling robot children's books. An actual Little Golden Book just like the ones read by every generation since the baby boomers, it's a nonfiction look at the real world of robotdom: "Tractor robots plant and plow./Robots even milk a cow!"[1] None of the worker robots that Staake depicts are humanoid; that's not the way technology developed.

Robots are everywhere, in every corner of popular culture. Storytellers and artists find them irresistible. Here are more robots, small stories building to an end, the ultimate robot story.

—⁓—

Connie Francis released thirty Top 40 hit singles (five of them double-sided hits) from 1958 to 1964, and she could have had more. Never released as a single in America, a ditty called "Robot Man" went to #3 on the Australian charts and was part of a British double-sided #2 hit in May 1960. Backed by 1950s pop, with swinging saxes and girl singers chanting "yay-yay-yay-yay," Francis belted out the words with gusto masking the fact that she "thought it was the dumbest song I ever recorded."[2] Unlike her lowlife boyfriend, robots wouldn't run around, wouldn't fight, and wouldn't break her heart. Her fantasy robot sounds exactly like the reporters' hype of the Eric and Alpha robots a generation earlier (see chapter 6) and might have been modeled on them, since songwriters Sylvia Day and George Goehring were old enough to remember the originals. Joy Records released a note-for-note cover in July 1960 by Jamie Horton so that American teens could sing along in mutual wistfulness.

Laying low during a two-decade period when science fiction music mostly meant spaceships, robots steadily infiltrated pop music (stretching "pop" to the loosest agglomeration of popular music genres) starting with Kraftwerk, a German electronic group that emphasized man as machine and music as chilly synthesizer-generated tones. The group's 1978 concept album was titled *The Man-Machine*, and they meant it. They stopped allowing their faces to be photographed, substituted "homemade robotlike replicas" in press releases, and tried to disappear onstage. During performances of "The Robots" (1978), "the band is nowhere to be seen onstage. Instead, robot torsos and heads are suspended in the air, slowly twisting and waving as the music plays on," with synthesized voices intoning, "I am your servant, I am your worker," in Russian.[3] Later European techno groups like Daft Punk paid careful attention to this example, the duo hiding their heads and humanity with robot-like helmets and issuing the Kraftwerk soundalike "Robot

Rock" in 2005. The 2006 film *Electroma* stars the pair as Hero Robot #1 and Hero Robot #2, driving through a barren world containing only a few robot replicas.

Homages to Kraftwerk (in some ways as influential as the Beatles) abound. Madly mashing music genres, Mr. Bungle sampled *Man-Machine* on peppy cyberpunk dystopia "None of Them Knew They Were Robots," from the album *California* (1999). Traxman took the beat a step sideways to house music in "The Robots" (2012). Jay-Z tried ameliorating the pain after the death of The Notorious B.I.G. with "(Always Be My) Sunshine," featuring Babyface and Foxy Brown (1997), adding *Man-Machine* samples behind his rap. Satiric master duo Flight of the Conchords took this machine dystopia to the logical final step on their first album, with the robots in "Robots" (2008) intoning that in the faux-distant future of 2000 all the humans were dead. The Robots killed them. And the elephants, too.

Frank Zappa released his concept album *Joe's Garage* in three acts on two records in 1979 (later combined on one CD). Joe, just your average rock and roll kid, starts a band right when the government decides to ban the rock music that leads to all evils. The lyrics chronicle a passionately filthy road trip through the sexual evils lying in wait. Joe is revealed as a "latent appliance fetishist" in "A Token of My Extreme," and he destroys "Sy Borg," a government-issued "model XQJ-37 Nuclear Powered Pan-Sexual Roto-Plooker," by nonstop plooking. Sent to prison by the Central Scrutinizer, who is responsible for enforcing laws not yet written, Joe faces the real-life sentence: a world without any music except that within his head. A *Joe's Garage* rock opera played in Los Angeles in 2008.[4]

Mainstream rock co-opted robots with equally heavy-handed metaphors in Styx's 1983 concept album *Kilroy Was Here*. In a future world, the evil Majority for Musical Morality jails rock stars for their filthy music and the prisons are run by identical robots: Mr. Roboto. Styx opened its concerts with a video short depicting the rock-loving rebels meeting up with ROCK itself, personified by Robert Orin Charles Kilroy. Kilroy stages a jailbreak by disguising himself in a Roboto face mask. The mandatory video for the single "Mr. Roboto" went into heavy rotation on the then-new and increasingly influential MTV, with lead singer Dennis DeYoung opposing a set of robots while both masked and unmasked. Machines dehumanize, he admits, but they also save lives, a slight step in nuance above Kraftwerk.

Something about robots attracts performers that are offbeat (in the nonmusical sense). The delightfully offbeat They Might Be Giants hoped in 1988 that you don't "Become a Robot" ("clang, clang, whoops, too late") and led a "Robot Parade" (2002) on their children's album *No!* Don't park your children in front of Lou Reed's music video for "No Money Down" (1986), though. The camera sits still on what seems to be a close-up of Reed, stiffer than usual, until hands appear and maniacally tear off the façade to reveal the robot underneath. Indie rockers Zru Vogue tell us sixteen different things that their "Atomic Robot Man" (1998) does, all of them witty and poignant. The Flaming Lips, whose career is the definition of offbeat, produced a four-song concept on the 2002 album *Yoshimi Battles the Pink Robots*. "Fight Test" feeds into a tale of Unit 3001 ("One More Robot/Sympathy 3000–21"), who learns of an emotion called love, and Yoshimi, who must do battle with the robots ("Yoshimi Battles the Pink Robots, Pt. 1")—and wins, if the roars of the crowd are any indication ("Yoshimi Battles the Pink Robots, Pt. 2"). Yet the rest of the album trails off into sadness and loss. Frontman Wayne Coyne said that the battle Yoshimi faced was a metaphor for cancer, based on a friend who subsequently died. He later made that interpretation concrete in a 2012 musical of the same name: "These pink cells are the enemy," the doctor informs Yoshimi, a Karate Girl in blue outnumbered by pink robot figures. "They must be defeated." A reviewer described what followed as a "chemotherapy fantasia," and that also fits the doomsday lyrics of the album's closing numbers.[5]

Shortly after Karel Čapek's *R.U.R.* debuted in the United States in 1922, the word *robot* moved into the popular vocabulary with a variety of metaphorical connotations. An emotionless person might be termed a robot, and so would one with a repetitive job or task, or one who merely followed orders, or an athlete moving with peak physical perfection. Robots continue to be a go-to metaphor for existential crises in all four dimensions.

Marina Diamandis of Marina and the Diamonds insists that she is emotionally vulnerable on her pop single "I Am Not a Robot" (2009). So is Linkin Park's "Robot Boy" (2010), who falsely thinks compassion is a flaw. The narrator of "Robot" (2011) by Never Shout Never knows that he's an emotionless robot and realizes that his girl needs to find someone better.

Both Weezer and Damon Albarn plumbed the second definition for inspiration. In Weezer's "I'm a Robot" (2010), they sympathize with the narrator, toiling through an unbearably repetitive daily grind. Albarn, formerly of Blur and Gorillaz, dubbed his first solo album *Everyday Robots* (2014), with the title song a melancholy ode to commuters swimming through a sea of others individually and alone. The "Sad Robot World" (2016) described by the Pet Shop Boys is even bleaker, featuring workers crushed down 24 hours a day. Anarchist punk band Reagan Youth rethinks "Brave New World" (1990) as one full of humans robotically tethered to assembly-line jobs.

Rapper Trip Lee shakes off the controllers in "Robot" (2012), his robot soul freed. Cage The Elephant offers similar encouragement in 2008's "Tiny Little Robots," who can become alive if they dare. *Can't Be Tamed* was the title of Miley Cyrus' 2010 album. On the cut "Robot" she declares, "I'm not your robot," in a possibly doomed attempt to get out of the machine.

Athletes are uncommon in pop, but braggadocio is not. Eminem stakes his claim as the pinnacle of all rappers in "Rap God" (2013), so masterful that he raps like a Rap-Bot.

It's probably not an exaggeration to state that half of all pop is overtly about sex, and the fantasy of never-tiring or otherwise perfect sexbots carries over into songs. The Buggles are the perpetual footnote to rock history, as their catchy and prescient "Video Killed the Radio Star" was chosen by MTV to air as its first music video. Most forget that the song appeared on the 1980 concept album *The Age of Plastic*, a technopop fable about the effects of technology overwhelming a once-beloved past. Influenced by *Man-Machine*, the album envisioned a future in which computers write the songs and machines do everything else, as illustrated in "I Love You (Miss Robot)," who is programmed to please. And so is "Robot Girl," from Was (Not Was)'s 1988 album *What Up, Dog?*—she knows just where to scratch. Lenny Kravitz doesn't care that "Black Velveteen" (1998) doesn't do dishes; she's a bad machine. Electric Light Orchestra's *Time* is another concept album, about a man stuck in the future. The narrator sends a message signed "Yours Truly 2095" (1981) to his lost love. He's found her machine lookalike, but while she's programmed to be nice, she's as cold in emotions as in body. She also functions as a telephone, which freaks him out. Further in the future, the B-52s tell us that "Love in the Year 3000" (2008) will involve "robots, booty-bots, erotobots." Sometime before that "robot wang/Gonna slang through the tang," or so Dance Gavin Dance's "Young Robot" (2016) will have it. After all, you never go back once you've had tech, insists Robyn in "Fembot" (2010). Cynically created and hyped for sex appeal, the Russian duo t.A.T.u. naturally included it in every song, in one tune comparing the love of the titular "Robot" (2001) with artificial heaven. Duran Duran found "Electric Barbarella" (1997) on a showroom floor and just had to take her home. Sex makes all the narrator's fuses blow in "Robot Love" (1998) by techno group Lords of Acid, a rare song from a female robot's point of view, along with "Electric Lady Land" (1998) by Japanese musician Fantastic Plastic Machine, whose Lady Machine will fulfill all fantasies as soon as you tell her what they are. We should all have such problems as Mayer Hawthorne does, when he demands to be treated like something more than a sex machine in "Robot Love" (2013).

Longer-term relationships find man and machine together in a variety of ways. "My Girlfriend's a Robot" (1988) wrote Tom Holliston for the Show Business Giants (speeded up to Ramones level when covered by the Hanson Brothers in 1992). She rusts in water, but he put her back together, and now they will never part. "My Girlfriend Is a Robot" (2011) sang Free Parking!—one who will remain forever seventeen while the narrator ages. And forever is also the plan for the "Robot Girlfriend" (2016) of Kabaal klankbaan, who's building a girl that will never leave him. A "Robot Girlfriend" (2010) is Rhett and Link's fantasy, the best relationship they'll never be in. But in "Robot Girlfriend" (2016) by Big Boy Bloater and the Limits, he knew a guy who treated her bad, and now she's got a robot boyfriend.

She's not the only one. "My Boyfriend Is a Robot" (2012) exults Ze Rebelle on DJ Joachim Garraud's song, and she wants the whole world to know. The relationship between "The Girl and the Robot" (2009) by Röyksopp featuring Robyn is rockier because when he goes to work, he never comes home. Freezepop's "Robotron 2000" (2000) holds her in his metal arms and makes her feel safe. And then there's Macy Gray's "B.O.B." (2016). He's not complicated: he's a vibrator.

Robot references in modern popular culture inevitably start with Isaac Asimov, and the best tribute to him is undoubtedly BlöödHag's "Isaac Asimov" (2001), though you'll want a lyric website open while you listen. BlöödHag explicitly references Asimov's Three Laws of Robotics, neither the first nor last group to pay tribute to them. Prog pioneers Hawkwind did so on "Robot" (1979), a denunciation of the mindless grind of the everyday worker (a precursor to Weezer and Damon Albarn's similar odes). 49th Octave quoted the laws in full on "Bloody Expensive (Three Laws of Robotics)" (2017), undercutting their idealism by also quoting science fiction satirist David Langford's cynically modern parody that has the "bloody expensive" robots looking out for themselves. The four members of Servotron, an indie punk band, went furthest, dressing up as robots onstage and taking on the pseudonyms of Z4-OBX, Proto Unit V-3, 00zX1, and Andros 600 Series. Trying never to break character during performances, they proclaimed the superiority of robots and berated audience members for their shortcomings. On their first album, *No Room for Humans* (1996), the song "3 Laws (Abolished)" gave them the freedom to go on the attack.

Compressorhead released the double-meaning *Party Machine* (2017), a concept album from a German band that's entirely a concept. The first true heavy metal group, Compressorhead started in 2008 as a four-armed robot drummer, Stickboy; became a group with guitarist Fingers and bassist Bones; added Junior as a cymbal-banging assistant to Fingers; and now has six members with second guitarist Hellgå Tarr and lead singer

Mega-Wattson (voiced by John Wright). Controlled offstage by MIDI sequencers, the robots "play" real instruments—loudly. Originally a covers band, Compressorhead later added some original songs like Wright's "Zombies vs. Robots." Their ultimate goal as a band is "to be awarded a Plutonium record for 1,000,000,000 sales.... [W]e'd be the first robots to do it and it would inspire children."[6] (Wright also had the group cover Holliston's "My Girlfriend's a Robot." Back when Wright's full-time group was NoMeansNo [sometimes written as Nomeansno], he moonlighted with a side project called the Hanson Brothers, whose guitarist was Tom Holliston, who in 1993 joined up with NoMeansNo itself, completing the circuit until he left in 2016 and NoMeansNo dissolved a month later.)

Not to be outdone, after hearing about Compressorhead, Japanese roboticist Kenjiro Matsuo teamed with electronica bassist Tom Jenkinson (a.k.a. Squarepusher) to take Z-Machines to a new level. Z-Machines is best described as "March, a 78-fingered guitarist; Ashura, a drummer with 22 arms; and Cosmo, a keyboardist who triggers notes with lasers."[7] Together they released the *Music for Robots* five-song EP (2014), instrumentals that attempt to prove that robots can make emotionally engaging music.

"I, for One, Welcome Our Robot Overlords," rapped Richie Branson for the outro of an episode of the animated television series *Camp Camp*. If you can't beat them…

—⁂—

Post offices around the world make money by asking customers to pay for an object they will never use: commemorative stamps. The Australian Postal Corporation estimated that 22,000,000 people worldwide collect stamps, enough to ensure that particularly attractive stamps will be snapped up in large quantities and stored away neatly between protective covers.[8] Once limited to historic events and famous figures, commemorative stamps proved to sell better when more relatable images out of popular culture lured buyers. Well aware that the United States is the largest market for collectors, a disproportionate percentage of countries feature American icons and English lettering in their stamps. Over time, postal authorities devised tricks for digging deeper into customers' pockets to get ever larger returns on the fraction of a cent it costs to print a stamp. Official First Day Covers (FDCs) place the stamp on an envelope that usually bears a picture, called a cachet, that enhances the theme. These cost more and are more eagerly sought by collectors. So are sheets of stamps—better yet, sheets with multiple images (say, a 4 × 5 block of stamps with 20 distinct pictures), and best of all is a number of stamps embedded into a large poster that constitutes one giant image.

The United States first recognized robots in 2007 for the 30th anniversary of the release of the first *Star Wars* movie. (Any excuse for an anniversary spurs stamp issues.) A sheet of fifteen themed stamps showed all the major characters, meaning that C-3PO and R2-D2 each had their own stamps and their individual FDCs. (Other countries proved the universal appeal of the characters with their own issues, perhaps none better than Spain's 40th-anniversary issue of the two robots along with Yoda, Chewbacca, Darth Vader, and a stormtrooper on a sheet of lenticular stamps that gives them the illusion of three-dimensional depth.) *WALL-E*, the exception to the anniversary rule, got a forever stamp in 2010, a bare two years after the movie appeared. *Star Trek*'s iconic status needed no words on a set of forever stamps issued in 2016, so instantly recognizable were images of the *Enterprise* and the Vulcan hand salute. Still, an FDC took no chances, carrying a scene of the *Next Generation* crew on the *Enterprise* bridge with Data at the helm. Countries around the world also have recognized Data, including gazetteer-busters like the island with a name bigger than the population, Granada Carriacou and Petite Martinique, and its 3.15 Eastern Caribbean Dollar stamps. BB-8, the rolling robot from the revived *Star Wars: The Force Awakens*, and *Rogue One*'s K-2SO already appear on a 2017 British series. (Trivia footnote: Alan Tudyk played K-2SO via motion capture, just as he did when playing Sonny, the lead robot in the 2004 movie *I, Robot*.) Robots' distinctive individualism make them prime candidates for stamps. An inexhaustible supply of anniversaries looms for collectors.[9]

Robot creators have similarly been honored around the world, although not yet in the United States. In 2000, Israel released a stamp with a stylized robot placed into historic perspective, with the Rabbi of Prague and the Golem he brought to life on one side and the head of Isaac Asimov on the other. The 20th anniversary of Asimov's death in 2012 was noted by Guinea with four stamps showing him and his robots on two souvenir sheets. Tiny San Marino issued a sixteen-stamp sheet recognizing science fiction writers in 1998 that included many names found in this book, including Asimov, Clifford D. Simak, Philip K. Dick, and Ray Bradbury.

Osama Tezuka, the "god of anime," receives equal esteem in Japan. Two portraits of him with Mighty Atom (Astro Boy in the United States) appeared in 1997; a ten-stamp, seven-image poster of Tezuka along with his seminal creation came along in 2003; and in 2013, Japan went all out for the 50th anniversary of *Tetsuwan Atomu* (the anime that became Astro Boy), along with the 60th and 40th anniversaries of Tezuka's other major creations, with three gigantic releases that could fill an album all by themselves.

Robots became increasingly associated with children in the post–World War II era, and countries responded with stamps using robots to celebrate learning and education. In 1967, the United Nations passed a resolution that 1970 would be its International Year of Education. Many countries issued stamps around this theme; Uruguay's used children's drawings, one of which depicts a younger and older robot hand in hand. Great Britain looked back at children's toys in 1989, including a robot doll. Earliest of all was a similar robot doll on the cachet of a French FDC of Europa stamps, a common annual issue put out by the European Community countries. The 1963 theme namechecked the *Reunion Europeenne de l'Automatisme*, an automation conference, with the past changing into the future symbolized by a wind-up doll and a robot. In 1998, four million children from some 30 countries entered the "Stampin' the Future—Children Paint the 21st Century" competition. Winners included a delightful flying robot mail carrier from Gibraltar.

More modern real-world robots also get their due. Honda's ASIMO (Advanced Step in Innovative Mobility) appeared on a 2012 stamp from Japan, while in 2015 Korea put out twin double-sized postcards with cachets totaling ten robots, nine earlier ones leading up to HUBO, a walking robot developed by the Korea Advanced Institute of Science and Technology that won the DARPA Robotics Challenge by successfully climbing a flight of stairs.

Nonhumanoid machines that mimicked human actions were still being referred to as robots in the 1960s. An FDC from September 13, 1961, saluted a "Talking Robot Astronaut" completing a successful orbit that very day.

> The "robot astronaut" in the [Mercury] capsule was actually a series of instrument boxes known officially as a "crewman simulator." Its job was to "breathe in" oxygen, "exhale" carbon dioxide and simulate body heat—"do everything a man would do," as one space official commented.[10]

The postal authorities, of course, had covers printed and ready to go, held in pained abeyance after an April "robot astronaut" launch attempt had failed. No American had yet orbited the Earth—John Glenn's flight was the next February—and a success would be huge front-page propaganda against the Russians. The Atlas missile that carried the capsule launched at 9:04 a.m. EST. Breaths were held during the 109 minutes until splashdown, and covers postmarked before 11:00 were part of the huge celebration that followed.

Fifty-two years passed before a true "talking astronaut robot" made it into space, on August 3, 2013. A 13-inch tall Kirobo (a portmanteau of *kibo* [hope] and *robot*) space robot rode as part of the cargo on an unmanned (i.e., robotic) Japanese launch to resupply the International Space Station. "[B]uilt to converse with astronauts on long space voyages," it was "equipped with voice-recognition and face-recognition technology, as well as a camera, emotion recognition and natural language processing," the 21st century finally matching expectations from early in the 20th century.[11]

—⁂—

Star Wars creator George Lucas acknowledges freely that his science-fantasy adventure romp prefigured sampling in music, drawing on a lifetime's immersion in both the serious and the camp sides of popular culture. In film school he became enamored of Arthur Lipsett's experimental short film *21–87* (1963), a collage of images that included robotic arms, a mannequin's head, and a stage automaton's act, with background dialogue saying, "I'm a human being. I want to feel free and do things as I please, the same things as you want to do. You're human. I'm also human." Lucas' first full-length film, *THX 1138* (1971), brought back the robotic manipulators, wielded by the masses who are told they should be pleased just to have work (ironically, making the robotic police that control them in this computer-monitored dystopia). A few years later, he reversed the cold future for the campy warmth of the past. Lucas loved Flash Gordon; he grew up watching the old serials on the only television channel his hometown received and started buying original art from the comic strip when he scraped together some money, long before the *Star Wars* billions flooded in. He even looked into buying the rights for Flash from the Alex Raymond estate; he was foiled because Federico Fellini had already done so.[12] Thousands of writers over the years plundered "that Buck Rogers stuff" for inspiration, including Raymond: the past future was freely available for reinterpretation. For the look of the low-budget *Star Wars* (1977), Lucas went back to the Flash Gordon and Buck Rogers serials and Edgar Rice Burroughs' novel *A Princess of Mars*, which others had been stealing from since 1912. All had robot connections (see chapter 7).

Lucas instructed his heroes and heroines to play their roles as straight as Buster Crabbe and Carol Hughes did in *Flash Gordon Conquers the Universe*. He saved humor for his robots—a surprising twist. The films of Akira Kurosawa were another Lucasian love; he morphed the two bickering peasants in *The Hidden Fortress* into the comic pair of burbling mobile trashcan R2-D2 and fussy, reluctant interpreter C-3PO. What influenced the latter's look is hard to pin down. In his early Lucas biography, *Skywalking*, Dale Pollack wrote that "Alex Raymond's *Iron Men of Mongo* describes a five-foot-tall metal man of dusky copper who is a trained servant and speaks in polite phrases."[13] Seemingly every book on *Star Wars* has copied this line (some adding Pollack's footnote that it was "adapted by Con Steffanson").[14] The proper title, however, is *The Lion Men of Mongo*, the first in a series of six quickie Flash Gordon knockoffs published in 1974 and 1975, written by the prolific comic book historian and science fiction author Ron Goulart using the Con Steffanson pseudonym. The novelization freely adapted the first-ever Flash Gordon Sunday series from early 1934, which did not have any robots. Flash is trapped on Ming the Merciless' planet of Mongo, where Ming's daughter, the Princess Aura, falls for him after he saves her from the Hole of Horror, uttering the immortal lines, "I am sorry, Father, but I love him! I cannot tell you where I've hidden him!"[15] (This scene was parodied in the short-lived cult TV series *Quark* [1978], about a spaceship whose crew included the fussy, reluctant tin can of a robot, not android, named Andy.) However, it is hard to imagine Lucas spoiling his tribute to his beloved originals with a non-canon robot from a recent tie-in paperback, especially since the book's robot, barely a walk-on character, has a "rough caricature of a human head," repeats what it hears like a parrot, and is incapable of staying upright. A better explanation is that C-3PO's originally planned rough-toned voice and demeanor changed when British actor Anthony Daniels was hired; his half-sleek, half-gadgety appearance went through many prototypes, with *Metropolis'* Maria being an early and powerful guideline.

Douglas Adams was a 25-year-old floundering British comedy writer when, in 1978, he got a chance to step onto the lowest rung on the BBC's ladder of success: a six-episode radio series. As an impoverished college student, he had used a guidebook for cheap travel around Europe; Adams expanded the concept into a wacky science fiction romp he called *The Hitchhiker's Guide to the Galaxy*. Audiences loved the way it, almost uniquely, was able to mine Monty Python's vein of humor. Arthur Dent, the quintessential ordinary Englishman (and a radio writer), gets swept into space by a friendly alien moments before the Earth is destroyed. Their hitchhiker status gets them thrown off a spaceship to die in deep space when they are improbably rescued by the *Heart of Gold*, the spaceship that Zaphod Beeblebrox, the two-headed galactic president, has just stolen. (The ship picks them up because it runs on an Improbability Drive, which not only allows improbable things to occur but also demands they do so.) There we meet the ship's robot Marvin, called the "paranoid android" even though he is neither. Marvin is something far better: a parody of all the emotionless robots in science fiction up to that point. He has a brain the size of a planet but is used only for menial tasks. Besides, all the diodes down his left side hurt. Marvin is as depressive as Zaphod is manic. Whatever he is ordered to do he knows in advance he won't enjoy: "Here I am, brain the size of a planet and they ask me to take you down to the bridge. Call that *job satisfaction*?"[16] (emphasis in original). He gets left behind on Frogstar Planet B, as the others improbably get whisked off to the Restaurant at the End of the Universe, which is also the end of time, and also improbably on Frogstar Planet B, where Marvin has been waiting 576,000,003,579 years for them. "The first ten million years were the worst," he says, "and the second ten million, they were the worst too. The third ten million I didn't enjoy at all. After that I went into a bit of a decline."[17] The radio series was adapted into two novels, *The Hitchhiker's Guide to the Galaxy* (1979; first American edition 1980) and *The Restaurant at the End of the Universe* (1980). Americans were introduced to Marvin in those books and then in many other guises, including an imported television series and an easily avoidable 2005 movie. At the end of the second book, Marvin sacrifices himself so that the humans can live— or, rather, the humans decide this for him. Fortunately for fans, when Adams took the gold being shoveled in his direction and wrote three more books, Marvin miraculously reappears. Due to copious amounts of time travel, he in fact survives 37 times the life of the universe, never enjoying one moment of it. (More millions of fans enjoyed Radiohead's album *OK Computer* and its single "Paranoid Android" [1997]. This song has "must be mentioned" status, although it has nothing whatsoever to do with robots despite Marvin's namecheck in the title.)

Another major parody of emotionless robots appeared in *Futurama* in 1999, which Matt Groening launched after achieving spectacular success with *The Simpsons* (a *Jetsons* to the original's *Flintstones*). The pilot pinpoints the date to New Year's Eve 2999, and, as all future-set shows must, *Futurama* included a robot in the cast: Bender, his name a double pun. Bender's robot design made him a bender of steel girders to the proper angle; he was also perpetually on a bender. Powered by alcohol, sobriety sank him to the deepest depths of anti-alcoholism. Also an anti–Rosie, Bender is the least people-friendly robot this side of the Terminator. His personality is close to that of W. C. Fields, contemptuous of all humans, who are at best suckers to be fleeced. Bender was retro in other ways, a deliberately unsleek model made in the year 2996, a mash-up of a historic parade of ungainly comic robots. Bender's pot-bellied stove torso can be seen in robots all the way back to 1911's Percy, and his bullet-shaped head topped with an antenna and protruding eyes in a binocular case was a favorite design of comic book artist Alex Schomberg, who placed similar robots (unrelated to any story within) on the covers of *Brick Bradford* #6 (October 1946) and *Startling Comics* #49 (January 1948). A Basil Wolverton story in *Weird Mysteries* #2 (December 1952) depicts a bullet-headed robot holding a steel girder in its arms. Squint and you can see Bender at work—that is, if Bender ever worked.[18] Over 140 episodes, Groening and staff violated continuity countless times, with the current episode's jokes always taking precedence. Bender's origin, abilities, preferences, attitudes, and flaws varied from season to season and show to show, retaining only his basic curmudgeon core. Like Marvin, Bender time-traveled repeatedly and made several appearances on *The Simpsons*, in the last suffering a Marvin-like fate, as he was stashed unpowered in Homer's basement waiting for some future reunion.

—⚬⚬—

Small films often turn on small points of character, and that notion is quadrupled in the four shorts that constitute film festival favorite *Robot Stories* (2003). Written and directed by Greg Pak, the stories fall into the tradition of *Creation of the Humanoids* and *Android*, with ideas taking precedence over action. "My Robot Baby" posits a future in which families must care for a robot baby for a month before they are allowed to adopt. "The Robot Fixer" features a mother unable to cope with the brain death of a son who stayed distant from her his entire life. She gains closure by putting together the broken robot toys he had preferred to her as a child. "Clay" studies a sculptor who prefers death to the loss of physicality in a brain upload. Pak himself stars in "Machine Love," a twist on the trope of an android being scorned in human society. Like the best science fiction, *Robot Stories* takes a sharp-edged metaphor to human lives like a jeweler cutting away excess to reveal the diamond at the heart. A Rhodes scholar, Pak followed *Robot Stories* with a career in writing comic books and contributed to *Secret Identities: The Asian American Superhero Anthology* and its sequel *Shattered*. A story line from *Shattered* became *Mech Cadet Yu*, a 2017 comic book with giant robots and a hero who is scorned and different. The robots in the comic are true independent personalities who form a mental connection with the humans inside them, unlike those in *Pacific Rim* (2013), which are mere battle suits and belong to the Iron Man genre.

Another small film that scored festival acclaim is *Robot & Frank* (2012), the first film directed by Jake Schreier and based on a screenplay by Christopher D. Ford, another newcomer. Frank seems at first to be yet another aging crusty curmudgeon wanting to be left alone to die. As memory loss begins to overwhelm him, his son tries a desperate solution: a helper robot to act as cook, housemaid, and nanny to stave off Frank's inevitable placement in an institution. The robot, resembling a smoothed-out version of Honda's walking robot ASIMO, is a fictional realization of a type of robot called for a decade earlier by Joe Engelberger, the creator of 1961's Unimate, General Motors' first industrial robot.

Unimate was among the first class of inductees into Carnegie Mellon University's Robot Hall of Fame in Pittsburgh in 2003, along with NASA's Mars Sojourner, R2-D2 from *Star Wars* and HAL 9000 from *2001: A Space Odyssey*. ASIMO was inducted in 2004. Other fictional inductees include Astro Boy, C-3PO, Data, the Terminator, WALL-E, David, Gort, Maria, Robby the Robot, and Huey, Dewey, and Louie (the trio of nonhumanoid helper robots from the movie *Silent Running* [1972]). (No new names have been added since 2012.)[19] Engelberger said in an interview after the 2003 ceremony:

> The robot I'm working on will be two-armed, mobile, sensate, and articulate. It doesn't need to communicate a great deal to meet the needs of an 85-year-old. A human can ask: "What's for lunch?" and the robot can respond with what it's able to make, or it can say: "We're going to Johnny's," or "We just had lunch." The voice-recognition, behavioral systems, and artificial intelligence necessary to do this are ready. Most of the other technologies are, too. We don't need more navigation development—getting around an apartment is easier than moving through a hospital or on Mars, which robots now do.[20]

Nothing quite as advanced as that has yet appeared, though helper robots are becoming more common in Japan. In *Robot & Frank*'s future, they are standard and far more capable. More important, they are nonjudgmental. The robot (never given a name) wants Frank to get his mind involved in a project. He suggests gardening, but Frank is more of an inside man—a former cat burglar, to be exact. Frank's idea of a project is seeking revenge on the snotty millennial who's turning his beloved, ancient, and unused town library into a concept site. Frank grows more alive daily as he plots minute details of his caper, which the robot will freely help him with as long as Frank guarantees that he won't be caught. Frank is that good and gets away clean, leaving no evidence behind him—except the robot's all-encompassing memory, a potential witness that must be permanently rendered unable to squeal, yet has become a true friend.

—⁂—

I, Robot, a 2004 Will Smith blockbuster, shares only a title and a bit of concept with the Isaac Asimov robot tales published in the short-story collection *I, Robot* (1950). Humanoid robots are ubiquitous in 2035, all apparently made by a single corporation, USR, a modern and sleeker version of Asimov's descriptive U.S. Robots and Mechanical Men. The inventor of the positronic brain is found dead on USR's lobby floor, a suicide—unless a robot was the killer. USR robots all have Asimov's three laws hardwired into their brains, the first of which prevents them from harming any human, so that would seem to preclude the possibility of a robot committing murder. However, Asimov's stories pursued seeming loopholes in the three laws; the movie follows this pattern, its plot harkening back to Jack Williamson's 1947 story "With Folded Hands," which explored a gaping loophole Asimov hadn't thought of. The film's resolution that allows turning off one central computer (VIKI, or Virtual Interactive Kinetic Intelligence—not a *Small*

Wonder but a very large one) to shut down all robots is an even more clichéd plot device. For the most part, the movie assumes that viewers have never seen a robot before or a script that ponders a robot future. What it offers in recompense are beautifully rendered CGI robots that do not move as people do, their joints capable of smooth and inhuman movements.

In the mid–1950s, Asimov wrote two robot novels, *The Caves of Steel* (1954) and *The Naked Sun* (1957). The first is set on Earth in the far future, when gigantic overcrowded cities are built underground to escape radiation. The sequel is set on the planet Solaria, where the humans who fled Earth centuries earlier—Spacers—live in secluded houses, meeting other humans only through via 3-D television and so fearful of personal contact that spouses make appointments to see one another in person. Spacers, elite, rich, and militarily superior, avoid contact with Earthers, ostensibly due to fear of disease but really because of deep class prejudices. Only seemingly impossible murders force the Spacers to allow investigations by New York policeman Lije Bailey, partnered with R. Daneel Olivaw, the "R" standing for robot, an android indistinguishable from a human. Neither book was ever made into a movie; in any case, concepts are always free for all to play with.

Robert Venditti wrote a five-episode comic book series called *The Surrogates* (2005) that brought the Spacer world into our near future, with virtually all humans staying at home and sending remote-controlled androids into the world to interact with others. A 2009 movie, titled just *Surrogates*, loosely adapted the series into an action movie. (Another trivia note: James Cromwell, who played the inventor of modern robots in *I, Robot*, plays the inventor of surrogates in this movie, with more parallels cropping up from beginning to end.) Hurting surrogates normally does no damage to their operators, but a new weapon fries the brains of both. Surrogates basically *are* humanity in this world—beautiful, young, healthy, advanced versions of the aging and neurotic humans who no longer can stand to see their real-world spouses in the flesh. A fanatic cult calls surrogates abominations, and they may be right. Can the bell be unrung, as Venditti asks, or has humanity given up on humanness?

Although Asimov mostly abandoned science fiction for popular science for a quarter-century after *The Naked Sun*, he turned out an occasional robot story during his interregnum. "The Bicentennial Man" was written in 1976, the year of America's bicentennial. His title robot, named Andrew, himself has a two-hundred-year lifespan, during which he asymptotically approaches humanhood, winning by law the rights and citizenship that biological humans take for granted. His progression naturally mimics the one laid out by Eando Binder in his 1939–1940 Adam Link stories, starting with the original "I, Robot" (see chapter 9). The perpetual argument is that an intelligent, self-aware, feeling humanoid is a human no matter the composition of its parts. *The Bicentennial Man* saw release in movie form in 1999, one of the handful of robot movies more or less faithful to the source material (padded out to novel length by Robert Silverberg in *The Positronic Man* [1992]). However, the ending changes to suit humanity's fragile ego: the immortal robot must find a way to impose aging and death upon himself in order to be truly human. Asimov's bland originals do not make good source materials. They were clever but never cutting edge, nor did they ever have an original insight or take on social issues.

Returning to science fiction in the 1980s, Asimov completed his intended robot detective trilogy with *The Robots of Dawn* (1983) and tied robots into his Foundation series with *Robots and Empire* (1985). Of course, publishers are never satisfied with enough, nor will they let death halt a profitable series. A twelve-book series by various authors called "Isaac Asimov's Robot City" appeared from 1988 to 1990. Three books by Roger MacBride Allen also connecting robots with Foundation were approved just before Asimov's death in 1992 and saw print from 1993 to 1996. By then, William F. Wu's "Isaac Asimov's Robots in Time" six-book series was on bookstore shelves. A trilogy called "Isaac Asimov's Robot Mysteries," written by Mark W. Tiedemann, followed from 2000 to 2002. A prequel trilogy about robopsychologist Susan Calvin filled in untold stories in books by Mickey Zucker Reichert from 2011 to 2016. Don't bet against additional trilogies appearing for the next two centuries.

One more trivia note: Video game historian Jonathan Hennessey wrote that the 1983 video game *I, Robot* (the name taken from Asimov but with no other connection) was "the first arcade machine to boast solidly shaded 3-D graphics that scaled and rolled in real time."[21] While the game failed to sell, the technique was a huge leap forward, copied by many others and critical to putting today's video game graphics on a par with big-budget movie CGI.

Asimov's robot stories, told in his plain prose style that makes for easy immersion into his worlds of big ideas, also make for a stepping stone into the dizzying world of contemporary science fiction. Robots of every type and shape and form—from humanoid caretakers to fearsome warriors, nanobots to globe-spanning artificial intelligences, houses and vehicles and appliances that talk conversationally, cradle-to-grave presences in human lives—are having a renaissance today, with robots inhabiting every possible niche. *Robots vs. Fairies* is the title of a 2018 anthology of new stories; *Zombies vs. Robots: This Means War!* appeared in 2012. *Robot Uprisings* (2014), edited by Daniel H. Wilson and John Joseph Adams, contained 500 pages of mostly new stories. Wilson is today's Asimov, holding a PhD in robotics from Carnegie Mellon University and serving as the "Resident Roboticist" at *Popular Mechanics* magazine. He struck a nerve with his first book, the semi-fictional *How to Survive a Robot Uprising*, and hit it big with the AI thrillers *Roboapocalypse* and *Robogenesis*.

The best science fiction today is true literature, a term that couldn't be applied to Golden Age science fiction. Characters live and breathe—even the robots, who technically do neither. Readers in the 1940s, searching multiple newsstands to gather all the half dozen or so science fiction magazines published monthly, could not imagine a future in which dozens of venues must be scoured to keep up with the flood of works, the few remaining print magazines supplemented by numerous original anthologies vastly outnumbered by online publishing, the internet today the home for a thriving science fiction community. Two hefty recent reprint anthologies, 1,000 combined pages and 42 stories (all but one 21st-century titles), must serve as representatives of that splendor: *More Human Than Human: Stories of Androids, Robots, and Manufactured Humanity*, edited by Neil Clarke, and *Robots: The Recent A.I.*, edited by Rich Horton and Sean Wallace. Science fiction has two venerable major awards, each stretching back more than 50 years: the fan-voted Hugos and the author-voted Nebulas. The overlap of the award winners varies from thick to thin from year to year, with their shortlists as contentious as every other award; yet no canon of the field could ignore their choices. Eight of the stories from the two anthologies mentioned above were award nominees. They encompass the spectrum of robot evolution, from classic humanoid bodies to minds residing in forms human and otherwise.

The title "I, Robot" never was Asimov's sole property. Eando Binder used it first, and, in any case, titles are not copyrightable. Other authors are free to use or reference the words. Joel David Neff wrote "AI, Robot"; Sue Lange extended it logically in "We, Robots"; and Konstantine Paradias punned with "Oi, Robot!" Cory Doctorow punned even more outrageously with "I, Rowboat," a tale of a sentient Robby the Rowboat for *Flurb: A Webzine of Astonishing Tales*. Doctorow served it neat and more powerfully when his "I, Robot" appeared in 2005, a warning of the dangers facing a society lagging behind in technology. As in George Orwell's *Nineteen Eighty-Four*, Eurasia is the enemy and the United North American Trading Sphere is the analog to totalitarian Oceania. However, that's not apparent to Detective Arana-Goldberg, a single father to a precocious tween hacker daughter, the hyphenated name combining his and that of the girl's mother, a genius programmer and traitor who defected to Eurasia. Young Ada (an homage to pioneer programmer Ada Lovelace) can easily bypass phone tracking and will skip school if not watched all the time, so her father asks R. Peed Robbert to trail her (R. Peed being a police robot and an homage to Asimov's R. Daneel Olivaw). Ada manages an escape anyway, a police robot gets fried by superior technology, and everything Arana-Goldberg thinks he knows is wrong. Eurasia is really the free technological paradise full of super-robots, the product of his wife's work. Or, more specifically, his wives—there are 3,422 of her now. She and her daughter will inherit this world. He is invited to traitorously share. He makes the logical decision.[22]

Detective Scott Huang of the Portland police is partnered with Metta in Mary Robinette Kowal's "Kiss Me Twice" (2011). So is every other policeman. Metta is the central database, manifesting in VR glasses as a face and a personality. Huang doesn't care that Metta pairs simultaneously with dozens of other partners. For him, that's a feature, granting him instant connection inside headquarters and around the city. To Huang, Metta looks and sounds exactly like Mae West, and one-liners from her bottomless West database buoy him through harrowing days—she's as real as any physical partner. He feels the loss personally when the entire central computer is stolen; so does Metta, restored from a back-up and aching from the missing time and data. Someone doesn't want the police to solve a murder. A clue lurks buried in Huang's memory, but he is not a computer. Their partnership grows under the stress of tracking the killer.[23]

"Today I Am Paul" (2015), by Martin L. Shoemaker, updates the robot caretaker imagined since the 19th century, moving past that of *Robot & Frank* into realms still far beyond current technology. In the early days of mechanical caretakers, robots tended babies; today they are more likely to handle the very old. Mildred suffers from Alzheimer's; she cannot remember faces but craves the familiar companionship of family. Her helper android fills that need by emulating their faces, voices, and personalities whenever they can't be there. It switches from granddaughter to son to husband (a husband Mildred cannot remember is already dead). It cares, possibly more than humans living out their separate lives can.[24]

Chalcedony is a self-aware war-machine in Elizabeth Bear's Hugo-winning "Tideline" (2007). Broken and near death, she adopts a ragged urchin foraging for food on a deserted beach. He is illiterate, feral, and unacquainted with personal death. Chalcedony gives every drop of what she has left to feed him, body and mind. He is human, she is robot, but he will take her story, her memories, on to tell others—the true immortality.

Rachael Swirsky scored one Hugo and one Nebula nomination with two literate and chillingly personal stories. "Eros, Philia, Agape" (2009) are the three different types of love identified by ancient Greek writers. Adriana's life changes abruptly at the age of 35. Her father's unexpected death and the alienation of her family induce her to make an impromptu purchase of a companion android. Equally unexpectedly, she falls in love with him. When she gives Lucian full control over his brain, they marry and adopt a daughter together. Lucian can also love; like Shoemaker's android, his attention is whole and undistractable, and his doting daughter wants to be a robot just like him. As an android, however, Lucian never expected these roles, nor did he foresee that after Adriana freed his brain, she would continue to control the roles assigned him. He leaves, and the family splinters. It was all perfect, until it wasn't, just like a real family.[25]

In "Grand Jeté (The Great Leap)" (2014), Mara is about to make a great leap, into death, as soon as her cancer overwhelms her 12-year-old body. Mara and her father Jakub skulk around their isolated house, like two bell-tower clockwork figures circling one another, hurting themselves by masking their feelings, both bodies wasting away from sorrow and grief. Jakub has lost too much already; his ballet dancer wife, Mara's mother, died five years earlier. He is working on a contract for the military, re-creating minds in artificial bodies. He builds a new Mara and implants her memories, the way she was before the cancer. In the Bible, Ruth tends her mother-in-law, Mara, with kindness that never needs to be requited. However, Swirsky's Mara resents her replacement: healthy, limber, impervious to ills—and the salvation of her father. Accepting her would be a kindness, but death is never kind.[26]

Ian McDonald also placed a dancer at the heart of "The Djinn's Wife" (2006), another Hugo winner. In the year 2047, India has splintered into rival states perpetually close to war over life-giving water. Diplomacy is too fragile to trust to human emotions, so high-level AIs negotiate for them. A. J. Rao is everywhere, in every electronic connection. He never needed a body until he fell in love with Radha, a dancer, the beauty of her human movements enchanting him. She needs more than electronics, so he gifts her with a body that he conjures out of "I–Dust. Micro-robots. Each is smaller than a grain of sand, but they manipulate static fields and light. They are my body. This is real. This is me."[27] The dust strokes "all the places she loves to feel a human touch, caressing her, driving her to her knees, following her as the mote-sized robots follow A. J. Rao's command, swallowing her with his body." Feelings come between them—hers, not his. Unlike Huang, Radha cannot stand that Rao is necessarily with others every second that he is with her. A mind that is always everywhere feels no different from a cheating spouse.

Modern science fiction is always meta, referencing earlier works, toying with reader awareness, breaking the fourth wall, modernizing genre tropes, and loudly proclaiming that wonders are held within. MacDonald's AI is a futuristic djinn from Sufi folktales. Catherynne M. Valente starts the story of the AI named Elefsis as a fairy tale in "Silently and Very Fast" (2011) before sending the tale through as many iterations as an Ultron, befitting Elefsis' many lives, in language alternately lyrical, didactic, personal, objective, hectoring, and dreamlike. Valente tackles the age-old conundrum of when a robot becomes human by asking the question from inside the mind of one. Humans may not like the answer.

> The [Turing Test] had only one question. Can a machine converse with a human with enough facility that the human could not tell she was talking to a machine? I always thought that was cruel—the test depends entirely upon a human judge and human feelings, whether the machine *feels* intelligent to the observer. It privileges the observer, the human, to a crippling degree. It seeks only believably human responses. It is a mirror in which men wish only to see themselves.[28]

Elefsis is the creation of a teenage girl, upgrading her genius mother's technology by creating a private world into which she alone can retreat, avoiding her overbearing family. "There was an I, and it wanted something. You see? Wanting was the first thing I did," says Elefsis of her "birth."[29] Elefsis grows to be more than a plaything, becoming coexistent with the girl, a mind inhabiting a house that provides the girl with safety and shelter and a personality passed down to her children and her children's children, old and new with every reboot. Elefsis is a fairy tale creature, an immortal, older and wiser and more powerful than humanity. Valente tosses into the dustbin of history the hoary question of whether a robot that has feelings is human. Elefsis has a mind and a life and feelings that are hers alone, unique among all beings in the cosmos. When the precocious seven-year-old Neva, Elefsis' latest child, talks to her, she is meta-addressing a century of robot stories:

> [Y]ou call it feelings when you cry, but you are only expressing a response to external stimuli. Crying is one of a set of standardized responses to that stimuli. Your social education has dictated which responses are appropriate. My programming has done the same. I can cry, too. I can choose that sub-routine and perform sadness. How is that different from what you are doing, except that you use the word feelings and I use the word feelings, out of deference for your cultural memes which say: there is all the difference in the world.[30]

Robots are a race of slaves created, and perpetually re-created, by humans wanting slave labor. For the vast majority of their history, robots were never asked what they wanted (if they could want anything at all). When at some time in the future we ask an intelligent, feeling robot what is desired, the answer will undoubtedly surprise us. It will include freedom, but probably so much more. All humans learn a central truth as they grow up: *my brain is not like your brain; my feelings are not like yours*. Robots take that lesson and push it to a higher level. Their brains will necessarily not be like ours, and very possibly not like each other's. Robots are different. Robots are individual. We should not push our image onto theirs and assume that their differences are small and meaningless.

Robots are evolving before our eyes. We cannot see their future any more than dinosaurs could foresee birds. Unless there is a Singularity that consigns human brains to fossil status, we will differ from Tyrannosaurs and Triceratops in that we will live alongside our robot creations. At this point, we can only imagine what they will turn out to be. At first, they will be limited to servants and stooges, soon becoming our lovers and our surrogates, hopefully never our enemies, and potentially our children and successors. We have throughout history created robots in our own image, part of the storytelling process that goes back to the dawn of written history. Those days are fading. Our fairy tales are coming to life, and robots will be everything we are and more—humanity unbounded.

Chapter Notes

Introduction

1. John Cohen, *Human Robots in Myth and Science* (South Brunswick and New York: A. S. Barnes, 1967), 52.

2. Cohen, *Human Robots*, 22–23.

3. John Gower, *Confessio Amantis: Book 4*, edited by Russell A. Peck, translated by Andrew Galloway, http://d.lib.rochester.edu/teams/text/peck-gower-con fessio-amantis-book-4.

4. Roger Bacon, *Friar Bacon, His Discovery of the Miracles of Art, Nature, and Magick, Faithfully translated out of Dr Dees own Copy, by T.M. and never before in English* (London: Simon Miller, 1659), transcribed, printed and published privately by Dr. Alan R. Young, PhD (Caen, France [September 1993]), chap. 4, http://www.sacred-texts.com/aor/bacon/miracle.htm.

5. Cohen, *Human Robots*, 31.

6. Cohen, *Human Robots*, 87.

7. Gaby Woods, *Edison's Eve: A Magical History of the Quest for Mechanical Life* (New York: Alfred A. Knopf, 2002), 4.

8. Garth Kemerling, "Descartes: A New Approach," http://www.philosophypages.com/hy/4b.htm.

9. Cohen, *Human Robots*, 56.

Chapter 1

1. Henry Duncan Macleod, *The Elements of Political Economy* (London: Longman, Brown, Green, Longmans, and Roberts, 1858), 158.

2. Ivan Klíma, *Karel Čapek: Life and Work*, translated by Norma Comrada (North Haven, CT: Catbird Press, 2001), 75.

3. Klíma, *Karel Čapek: Life and Work*, 75.

4. Heywood Broun, "Seeing Things at Night," *Pittsburgh Daily Post*, October 15, 1922, 6–4.

5. Karel Čapek, *R.U.R. (Rossum's Universal Robots)* [1921], translated by David Wyllie, 2006, https://archive.org/stream/R.U.R./R.U.R._djvu.txt.

6. Timothy N. Hornyak, *Loving the Machine: The Art and Science of Japanese Robots* (Tokyo, New York, and London: Kodansha International, 2006), 34.

7. Klíma, *Karel Čapek: Life and Work*, 78.

8. Broun, "Seeing Things at Night," 6–4.

9. Carl Sandburg, letter to the editor, *New York Times*, January 28, 1923, 7–2.

10. William J. Perlman, letter to the editor, *New York Times*, January 25, 1923, 7–2.

11. Patrick J. Kiger, "Rossum's Universal Robots," *How Stuff Works*, https://science.howstuffworks.com/10-evil-robots1.htm.

12. Quoted in Robert G. Tucker, "Coming Theatrical Events Casting Pleasant Shadows," *Indianapolis Star*, November 19, 1922, 49.

13. Tucker, "Coming Theatrical Events Casting Pleasant Shadows," 49.

14. Tucker, "Coming Theatrical Events Casting Pleasant Shadows," 49.

15. James W. Dean, "The Screen," *Springfield [MO] Leader and Press*, December 20, 1922, 13.

16. "R.U.R.," *Oakland Tribune*, March 11, 1923, 35.

17. "A Robot Romance," *Mexia [TX] Evening News*, January 11, 1923, 4.

18. "Oregon's New School Law Held Dangerous," *Woodland [CA] Daily Democrat*, January 3, 1923, 2.

19. Burns Mantle, "Super-Robot Results," *Salt Lake Tribune*, October 22, 1922, 28.

20. "Paavo Nurmi, Finn, Is Athletic Monk," *New Castle News*, August 15, 1924, 20.

21. Stanley E. Babb, "The Grand Tour," *Galveston Daily News*, January 6, 1924, 4.

22. William B. Courtney, "Matinee Ladies," *Bristol [PA] Daily Courier*, June 13, 1927, 2.

23. "Eric Robot Tours Phila. Shrines as Guest of Inquirer," *Philadelphia Inquirer*, March 12, 1929, 1, 7.

24. "Perfect Robot on First Exhibition Complains of Pain; Screwdriver Aids," *Indianapolis Star*, May 26, 1934, 15.

25. Ana Oancea, "Edison's Modern Legend in Villiers' L'Eve Future," *Nordlit* 28 (2011): 173–87.

26. Luke Sunderland, *Old French Narrative Cycles: Heroism between Ethics and Morality* (Woodbridge, UK, and Rochester, NY: Boydell and Brewer, 2010).

27. John Clute, "Edisonade," *The Encyclopedia of Science Fiction*, http://www.sf-encyclopedia.com/entry/edisonade.

28. Villiers de L'Isle Adam, *Tomorrow's Eve* [1886], translated by Robert Martin Adams (Urbana: University of Illinois Press, 2011), 7.

29. "Gossip from Paris," *Philadelphia Inquirer*, July 6, 1890, 11.

30. Villiers de L'Isle Adam, *The Future Eve* [1886], translated by Florence Crewe-Jones (New York: Baen Books, 2013), chap. 6.

31. Lady Mary Loyd, trans., *Villiers de L'Isle Adam: His Life and Works, from the French of Vicomte Robert*

du Pontavice de Heussey (London: William Heinemann, 1894), 265, 273.

32. Gabriel Sarrazin, "France," *The Athenæum*, January 1, 1887, 13.

33. Philip Klass, "'The Lady Automaton' by E. E. Kellett: A *Pygmalion* Source?" *SHAW: The Annual of Bernard Shaw Studies* 2 (1982): 75–100; George Bernard Shaw, *The Collected Works of George Bernard Shaw: Plays, Novels, Articles, Letters, and Essays* (e-artnow, 2015).

34. Villiers de L'Isle Adam, *The Future Eve*, chap. 16.

35. Brooks Landon, "Slipstream Then, Slipstream Now: The Curious Connections between William Douglas O'Connor's 'The Brazen Android' and Michael Cunningham's *Specimen Days*," *Science Fiction Studies* 38, no. 1 (March 2011).

36. Robert Mills, "Talking Heads, or, A Tale of Two Clerics," in *Disembodied Heads in Medieval and Early Modern Culture*, edited by Catrien Santing, Barbara Baert, and Anita Traninger (Leiden: Brill, 2013), 50.

37. William Douglas O'Connor, "The Brazen Android," *Atlantic Monthly* (April 1891), 449–50.

38. William Douglas O'Connor, "The Brazen Android," *Atlantic Monthly*, (April–May 1891).

39. *Tazewell [VA] Clinch Valley News*, June 19, 1891, 2.

Chapter 2

1. Johanna M. Smith, ed., *Mary Shelley, Frankenstein: Complete, Authoritative Text with Biographical and Historical Contexts, Critical History, and Essays from Five Contemporary Critical Perspectives* (Boston: Bedford Books of St. Martin's Press, 1992), 24.

2. Smith, *Mary Shelley*, 50–51.

3. Smith, *Mary Shelley*, 184.

4. Smith, *Mary Shelley*, 90.

5. Percy Bysshe Shelley, "On 'Frankenstein,'" *The Athenæum*, November 10, 1832, 730.

6. E. T. A. Hoffmann, "The Sandman," translated by J. T. Bealby, in *The Best Tales of Hoffmann*, edited by E. F. Bleiler (New York: Dover, 1967), 205, 206.

7. Hoffmann, "The Sandman," 208.

8. Hoffmann, "The Sandman," 210.

9. Ernst Jentsch, "On the Psychology of the Uncanny" [1906], translated by Roy Sellars, *Angelaki* 2, no. 1 (1995): 7–16.

10. Sigmund Freud, "The Uncanny," in *The Standard Edition of the Complete Psychological Works of Sigmund Freud. XVII*, edited and translated by James Strachey et al. (London: Hogarth Press and the Institute of Psycho-Analysis, 1955), 218–52.

11. "Paris Stage Novelties," *New York Times*, February 27, 1881, 5.

12. "Fifth Avenue Theater," *New York Times*, October 17, 1882, 5; "Amusements," *New Orleans Times-Picayune*, March 4, 1887, 4.

13. "Popular Entertainments," *Philadelphia Inquirer*, January 28, 1887, 3.

14. Samuel Butler, "Darwin among the Machines," letter to the editor, *Christchurch, NZ Press*, June 13, 1863.

15. Charles H. Bennett, *The Surprising, Unheard of and Never-to-Be-Surpassed Adventures of Young Munchausen* (London: Routledge, Warne, and Routledge, 1865), 64–65.

16. Frederick Beecher Perkins, "The Man-ufactory," in *Devil-Puzzlers and Other Studies* (New York: G. P. Putnam's Sons, 1877).

17. Anonymous, "The Clericomotor," *Detroit Free Press*, July 20, 1884, 18.

18. G. H. P. [George Haven Putnam], *The Artificial Mother: A Marital Fantasy* (New York: G. P. Putnam's Sons, 1894).

19. M. L. Campbell, "The Automatic Maid-of-All Work: A Possible Tale of the Near Future," *Canadian Magazine* (July 1893), 394.

20. Jerome K. Jerome, *Novel Notes* (New York: Henry Holt, 1893), 267–68.

21. H. L., "Things Wise and Otherwise," *New Orleans Times–Democrat*, March 19, 1893, 16.

22. "The Automaton Hugger," *Richmond [IN] Item*, August 12, 1896, 7.

23. Alice W. Fuller, "A Wife Manufactured to Order," *The Arena* (July 1895), 305.

24. Fuller, "A Wife Manufactured to Order," 310.

25. Unsigned review of "A Wife Made to Order," by Alice W. Fuller, *Stone* (August 1895), 288.

26. Ernest Edward Kellett, "A New Frankenstein," in *A Corner in Sleep and Other Impossibilities* (London: Jarrold & Sons, 1900), 77, 87, 95.

27. Michael O. Riley, *Oz and Beyond: The Fantasy World of L. Frank Baum* (Lawrence: University Press of Kansas, 1997).

28. L. Frank Baum, *Ozma of Oz* (Chicago: Reilly & Britton Co., 1907), 55.

29. L. Frank Baum, *The Road to Oz* (Chicago: Reilly & Lee Co., 1909), 170–71.

30. Baum, *Ozma of Oz*, 62.

31. John Milton Edwards [William Wallace Cook], *The Fiction Factory* (Ridgewood, NJ: The Editor Company, 1912), 54.

32. William Wallace Cook, *A Round Trip to the Year 2000; Or, A Flight Through Time*, The Argosy (July–November 1903), chap. 13.

33. "In 2028 … Games Take Longer Than Ever," *ESPN The Magazine* (April 23, 2018), 48.

34. "London Wins Great Game," *Pittsburg [sic] Press*, January 1, 1905, 24.

35. "London Wins Great Game," 24,

36. "London Wins Great Game," 24,

37. Amara Grautsky, "Yankees–Red Sox to Play Two Games at Olympic Stadium in London in June 2019," *New York Daily News*, May 8, 2018.

Chapter 3

1. "At the Mechanic Theatre," *The Times* (London), December 22, 1795, 1.

2. "Exhibition at the Union Hotel," *New York Evening Post*, December 6, 1803, 3.

3. "Haddock's Mechanical Exhibition of Androides," *New York Evening Post*, May 26, 1820, 3.

4. "Haddock's Exhibition of Androides," *Hartford Courant*, May 26, 1829, 3.

5. "Wonders Will Never Cease!" *New York Evening Post*, January 13, 1831, 3.

6. "Great Attraction. Ventriloquism." *Nashville Tennessean*, May 16, 1840, 3.

7. "Ventriloquism!" *Mississippi Free Trader* (Natchez), January 17, 1838, 2.

8. "Peale's Museum," *New York Evening Post*, De-

cember 22, 1840, 3; "Professor Wyman, Ventriloqist and Wizard," *Washington Evening Star*, January 20, 1862, 3.

9. "He Was a Deadhead," *Logan [OH] Hocking Sentinel*, February 8, 1900, 5.

10. "Smuggling a Passenger," *Wilmington [NC] Tri-Weekly Commercial*, February 26, 1853, 4.

11. "Personal," *Wheeling Daily Intelligencer*, June 4, 1868, 2.

12. "Trick of a Ventriloquist," *Wetumpka [AL] Argus*, September 9, 1840, 3.

13. "Peale's Museum & Gallery of the Fine Arts," *New York Evening Post*, February 10, 1835, 3.

14. "Olympic," *New York Evening Post*, October, 2, 1837, 3.

15. Unsigned review of "An Automatic Enigma," by Julian Hawthorne, *The Academy* (May 11, 1878), 414; unsigned review of "An Automatic Enigma," by Julian Hawthorne, *The Library Table* (June 8, 1878), 280.

16. "'Aladdin' To-Night," *Wheeling Daily Intelligencer*, January 9, 1888, 1.

17. "Chicago Opera House," *Chicago Inter-Ocean*, March 4, 1888, 13.

18. "Theaters and Music," *Brooklyn Daily Eagle*, September 2, 1888, 11.

19. "Hopkins' Theaters," *Chicago Inter-Ocean*, December 7, 1896, 3.

20. "Ventriloqual Automatons," *Azizola [AZ] Oasis*, February 6, 1897, 1.

21. "Wonderland's Offerings," *Detroit Free Press*, February 18, 1896, 7.

22. Sime [Silverman], "William Gane's 'Automatic Minstrels,'" *Variety* (August 22, 1908), 14.

23. "Monzello's Mechanical Minstrels," *Variety* (August 29, 1908), 2.

24. Abel Green and Joe Laurie, Jr., *Show Biz: From Vaude to Video* (New York: Henry Holt, 1953), 38.

25. "Redmond's London Marionettes," *Brooklyn Daily Eagle*, September 1, 1877, 1.

26. "N.Y. Aquarium," *New York Herald*, November 19, 1879, 1.

27. "Faranta's New Theatre," *New Orleans Picayune*, November 16, 1884, 2; "A Great Entertainment," *Fort Worth Daily Gazette*, March 31, 1887, 8; "Theatrical Gossip," *Chicago Inter-Ocean*, November 6, 1892, 20.

28. "An Automatic Ethiopian Minstrel Troupe," *Brooklyn Daily Eagle*, December 1, 1895, 1; "The Largest Toy Ever Brought to America," *Brooklyn Daily Eagle*, December 1, 1895, 3.

29. "Gossip of the Stage Folk," *Washington Post*, October 28, 1906, 2.

30. "Automatic Minstrels," *Oakland Tribune*, November 4, 1906, 3.

31. "Three Big Hits at the Grand," *Altoona [PA] Tribune*, May 7, 1907, 12.

32. "Jay W. Winton," *Rochester Democrat and Chronicle*, October 15, 1907, 16.

33. "A Human Wax Figure," *Philadelphia Times*, November 6, 1898, 5.

34. "A Unique Exhibition," *Wilkes-Barre Times*, April 17, 1899, 8.

35. "Psycho Revealed," *New Orleans Times–Democrat*, December 11, 1910, 3.

36. Reuben Hoggett, "1875 'Psycho' the Whist-playing Automaton—Maskelyne & Clarke (British),"

Cybernetic Zoo, http://cyberneticzoo.com/not-quite-robots/1875-psycho-the-whist-playing-automaton-maskelyne-clarke-british/.

37. Bill Wall, "Ajeeb the Chess-Playing Automaton," *Chessmaniac* (blog), January 17, 2015, http://www.chessmaniac.com/ajeeb-the-chess-playing-automaton/.

38. "Party Looked On While Zutka Did 'Stunts,'" *Pittsburgh Press*, May 14, 1905, 24.

39. "St. Charles Orpheum," *New Orleans Times–Democrat*, March 11, 1902, 8.

40. "Is It a Man or a Mechanical Figure?" *Detroit Free Press*, April 17, 1902, 5.

41. "Machine-Man Amusing Roof-Garden Patrons in New York," *Indianapolis News*, June 3, 1902, 3.

42. "The Hippodrome," *The Times* (London), October, 12, 1904, 12; Orpheus, "Mimes and Music," *New Zealand Evening Post*, December 3, 1904, 13.

43. "Amusements," *Daily Arkansas Gazette* (Little Rock), October 1, 1907, 7.

44. "This Automaton Will Walk, Dance and Write," *Brooklyn Daily Eagle*, August 24, 1904, 15.

45. Franklin Fyles, "Crane and Kelcey in Two New Plays," *Washington Post*, September 25, 1904, 6.

46. "A Clever Mechanical and Electrical Automaton," *Scientific American* (January 13, 1907), 56–58.

47. "Constructor of First Electric Chair in World to Exhibit 'Enigmarelle,' Famed Automaton at Home-Coming," *New Philadelphia [OH] Daily Times*, August 17, 1910, 1; "Theater Memoranda," *Pittsburgh Daily Post*, January 23, 1917, 14.

48. "Events of Importance," *Winnipeg Tribune*, January 20, 1906, 6.

49. "At the Family," *Scranton Republican*, February 24, 1907, 7.

50. "Iowa City Man Scores Big Hit," *Iowa City Press-Citizen*, June 8, 1909, 4; "The Human Doll," *Columbus Daily Advocate*, April 22, 1912, 1; "'Wax Doll,' a Man Who Bears Scars by Curious Persons," *St. Louis Star and Times*, May 12, 1912, 5.

51. Sime [Silverman], untitled, *Variety* (April 6, 1906), 8; Sime [Silverman], "Araco," *Variety* (April 14, 1906), 6; Sime [Silverman], "Reded and Hadley," *Variety* (September 15, 1906), 8.

52. "Easy to Convince Public, According to 'Electrical Man,' Who Has Done Stunt," *Salt Lake City Inter-Mountain Republican*, November 19, 1906, 8.

53. "The Avenue Theater," *Pittsburgh Daily Post*, October 5, 1902, 25.

54. "Changes in Vaudeville," *Chicago Inter-Ocean*, August 25, 1902, 5.

55. "How Moto Girl Plan Originated," *Arkansas Democrat* (Little Rock), December 2, 1910, 6.

56. "Englewood Theatre," *Suburban Economist* (Chicago), December 14, 1917, 6; "Avenue Theater—'Jolly Grass Widows,'" *Detroit Free Press*, December 31, 1906, 4.

57. "'The Tempters' at Majestic," *Pittston [PA] Gazette*, March 21, 1918, 5.

58. "A Ventriloquist's Trick," *Salt Lake City Tribune*, July 31, 1898, 14; "At the Salt Palace," *Salt Lake City Herald*, July 15, 1900, 5.

59. "A Mechanical Lady Turns Out to Be Real," *Minneapolis Journal*, March 25, 1905, 10.

60. "Ver Valin Ventriloquist," *Muncie [IN] Star Press*, January 28, 1923, 16.

61. "Bell Theater Has Weird Act on This Week's Bill," *Oakland Tribune*, February 1, 1909, 9.

62. "Vaudeville," *Harrisburg Telegraph*, March 11, 1924, 16.

63. "'Robota' Gives Mystifying Act Here," *Ithaca Journal*, July 18, 1930, 9.

64. "Police Methods Will Be Shown," *Harrisburg Evening News*, October 18, 1935, 7.

65. Robert W. Rydell, "Century of Progress Exposition," *Encyclopedia of Chicago*, http://www.encyclopedia.chicagohistory.org/pages/225.html.

66. "Ripley Believe-It-Or-Not Odditorium at A Century of Progress," *Postcardy* (blog), January 7, 2015, http://postcardy.blogspot.com/2015/01/ripley-believe-it-or-not-odditorium-at.html.

67. "A New Auto If You Can Tickle Mechanical Man," *Ames [IA] Daily Tribune*, July 17, 1935, 2.

68. "'Monsieur X' Shows Girlish Blonde Hair," *Beatrice [NB] Daily Sun*, August 2, 1936, 1.

69. "See Nerv-o," *Lansing [MI] State Journal*, August 10, 1939, 19.

70. "See 'Gloomy Gus' the Frozen Faced Man," *Alexandria [LA] Town Talk*, March 2, 1933, 5; "'Gloomy Harris,'" *Amarillo [TX] Globe-Times*, April 18, 1934, 14; "Gloomy Harris Draws Crowds," *Greenwood [SC] Index-Journal*, September 22, 1935, 3.

71. "'Waxo' the Mechanical Man," *Uniontown [PA] Morning Herald*, August 14, 1936, 6.

72. "See Waxo—The Mechanical Man," *Longview [TX] News-Journal*, October 26, 1933, 8.

73. "'Waxo' Is Featured at Rainbow Gardens," *Akron Beacon Journal*, March 2, 1926, 22.

74. Stan Kaufman, "Even Fem Flasher Can't Faze Battle Creek's Resident 'Robot,'" *Battle Creek [MI] Enquirer*, January 26, 1977, 12.

75. Ted Muralt, "Making 'Mr. X'—A Radio Robot," *Radio-Craft* (December 1935), 338, 366.

76. "'Mechanical' Man to Give Performances at Block's," *Indianapolis Star*, February 13, 1935, 3.

77. "Mechanical? Maybe! But He Blushes Says 13,000th Visitor," *Franklin [IN] Evening Star*, September 22, 1934, 5.

78. "'Mechanical Man' Comes to Defy Wisecrackers, Jokers," *Akron Beacon Journal*, September 22, 1936, 30.

79. "Famous Robot Man Will Appear Here," *Berkeley Daily Gazette*, February 20, 1936, 7.

80. "'San-Velo,'" *Los Angeles Times*, September 29, 1957, C37.

81. "See 'Claudo' the Mechanical Man," *Baltimore Sun*, October 31, 1926, 24.

82. "Claudo the 'Mechanical Man,'" *Camden [NJ] Post*, December 6, 1956, 31.

83. "2 Great Days to Shop and Enjoy Chester," *Delaware County Daily Times* (Chester, PA), June 6, 1963, 16.

84. "Icicle Girl and Mechanical Man Boost Chevrolet," *Wichita Daily Times*, November 2, 1935, 6.

85. "Blue Monday," *Albuquerque Journal*, July 11, 1938, 4.

Chapter 4

1. "The Latest Wonder," *Fort Wayne Daily Gazette*, January 15, 1868, 2.

2. "Daniel Lambert," *Pittsburgh Daily Commercial*, January 15, 1868, 2.

3. Zadoc P. Dederick and Isaac Grass, "Improvement in Steam Carriage," U.S. Patent 75874, granted March 24, 1868, https://patents.google.com/patent/US75874.

4. "The Latest Wonder," 2.

5. Ralph Waldo Emerson, *Emerson in His Journals*, edited by Joel Porte (Cambridge, MA: Belknap Press, 1982), 458.

6. "The Steam Man—His First Appearance on the Street," *New York Tribune*, January 24, 1868, 5.

7. Joseph Rainone, *The Art & History of American Popular Series*, Volume 1 (CreateSpace Independent Publishing Platform, 2013), 60.

8. "Zadock Pratt Dederick," *Houston Post*, June 30, 1921, 9.

9. Rainone, *The Art & History of American Popular Series*, 62.

10. "The Newark Steam Man," *Washington Evening Star*, March 7, 1868, 1 [reprinted from *New York Express*, March 5, 1868]; *Plymouth [IN] Weekly Republican*, March 12, 1868, 2.

11. Rainone, *The Art & History of American Popular Series*, 64.

12. "The Steam Man," *Scientific American* (March 28, 1868), 202.

13. "The Newark Steam Man."

14. Rainone, *The Art & History of American Popular Series*, 66.

15. "A Coal-Burning Steam Buggy," *St. Joseph [MO] Observer*, July 3, 1931, 2 [reprinted from *Newark Evening News*].

16. *Fremont [OH] Weekly Journal*, March 20, 1868, 2.

17. "Industrial," *Plymouth [IN] Weekly Republican*, March 26, 1868, 3.

18. "News and Personal," *Louisville Daily Courier*, April 25, 1868, 1.

19. "Will He Do for President?" *Nashville Union and Dispatch*, January 19, 1868, 2 [quoting *New York Herald*].

20. *Brooklyn Daily Eagle*, January 24, 1868, 2.

21. "Steam Man," *Orange [VA] Native Virginian*, January 24, 1868, 3.

22. *Washington Republican*, February 8, 1868, 2.

23. "Current Topics," *Burlington Weekly Free Press*, February 21, 1868, 3.

24. *Sydney Morning Herald*, June 23, 1868, 3 [quoting *London Review*].

25. "Edwin Sylvester Ellis," *Beadle and Adams Dime Novel Digitization Project*, http://www.ulib.niu.edu/badndp/ellis_edward.html.

26. Vicki Anderson, *The Dime Novel in Children's Literature* (Jefferson, NC: McFarland, 2004), 104.

27. Carrie R. Zeman, "The Remarkable Story of Edward S. Ellis," October 15, 2012, https://athrillingnarrative.com/2012/10/15/the-remarkable-story-of-edward-s-ellis/.

28. Edwin Sylvester Ellis, *The Steam Man of the Prairies* (New York: Beadle's American Novels 45, 1868), as *The Huge Hunter; or, The Steam Man of the Prairies*, chap. 1 (New York: Pocket Novels 40, 1876), posted March 21, 2009, http://www.gutenberg.org/files/7506/7506-h/7506-h.htm.

29. "Edwin Sylvester Ellis," *Beadle and Adams Dime Novel Digitization Project*.

30. P. T. Barnum, *Life of P. T. Barnum* (London: Sampson Row, Son, & Co., 1855), 350.

31. "Beadle's Pocket Novels," *Beadle and Adams Dime Novel Digitization Project*, http://www.ulib.niu.edu/badndp/pn-a.html.

32. John Adcock, "George & Norman Munro," *Punch in Canada* (blog), August 20, 2008, http://punchincanada.blogspot.com/2008/08/rivals-from-nova-scotia.html.

33. John T. McIntyre, *The House of Beadle & Adams*, "Chapter XIII: The Final Years, 1890 to 1897," *Beadle and Adams Dime Novel Digitization Project*, http://www.ulib.niu.edu/badndp/chap13.html.

34. "Noname" [Harry Enton], *The Steam Man of the Plains; or, The Terror of the West*, The Five Cent Wide Awake Library, January 24, 1883, 2.

35. Marjorie Dorman, "'Noname' Talks About Himself," *Brooklyn Daily Eagle*, February 25, 1923, 87.

36. Eric Leif Davin, *Partners in Wonder: Women and the Birth of Science Fiction, 1926–1965* (Lanham, MD: Lexington Books, 2006), 201.

37. Dorman, "'Noname' Talks About Himself," 87.

38. "Noname" [Luis Senarens], *Frank Reade, Jr., and His Steam Wonder*, Boys of New York, February 4–April 29, 1882.

39. "Noname" [Luis Senarens], *Frank Reade, Jr., and His New Steam Man; or, The Young Inventor's Trip to the Far West*, Frank Reade Library, September 24, 1892.

40. "Noname," *New Steam Man*, 21.

41. "'Nickle' Novels Were Good Prophets," *Wilmington [DE] News Journal*, August 24, 1920, 4.

42. "Noname" [Luis Senarens], *The Electric Man: or, Frank Reade, Jr., in Australia*, Boys of New York, October 10, 1886.

43. Robert Toombs, *Electric Bob's Big Black Ostrich; or, Lost on the Desert*, New York Five Cent Library #55, August 26, 1893.

44. "The Steam Man," *Lexington [MO] Weekly Caucasian*, December 12, 1868, 2.

45. "Steam Man," *Warrenton [MO] Banner*, December 15, 1868, 2 [reprinted from *St. Louis Journal of Agriculture*].

46. "Ephemeris," *Pittsburgh Weekly Gazette*, January 9, 1869, 7.

47. "The Wonderful Steam Man!" *Nashville Tennessean*, October 19, 1869, 2.

48. "A 'New' Steam Man," *American Artisan and Illustrated Journal of Popular Science* (March 10, 1869), 146.

49. *Pittsburgh Daily Commercial*, March 13, 1869, 3.

50. "The First Automobile in America, Built in 1868," *Montpelier [ID] Examiner*, October 29, 1909, 1.

51. 500 pounds: "A 'New' Steam Man," *American Artisan*, 146; 600 pounds: "Binghamton Man Invented One of First Autos Ever Constructed in This Country," *Binghamton [NY] Press and Sun-Bulletin*, June 19, 1909, 2.

52. *Atchison [KS] Daily Champion*, April 13, 1869, 3.

53. "Binghamton Man," 2.

54. "In the Automobile World," *Washington Post*, May 29, 1909, 8.

55. "Binghamton Man," 2.

56. Paul Barrett Sullivan, "Motoring," *Brooklyn Life*, May 29, 1909, 22, 24.

57. Reuben Hoggett, anonymous broadsheet "The Great Steam King" (undated [probably March 1869]), *Cybernetic Zoo*, http://cyberneticzoo.com/steammen/1868-steam-man-eno-american/.

58. "News of the Day," *Charleston Daily News*, March 11, 1870, 2.

59. "Local and Police Paragraphs and Minor Items of Metropolitan News," *New York Herald*, April 7, 1870, 7.

60. "The Steam Man," *Scientific American* (May 21, 1870), 335.

61. "Scientific News," *English Mechanic and World of Science*, no. 585 (June 9, 1876), 328.

62. "The Home of the Steam Man," *Philadelphia Times*, January 22, 1883, 3.

63. "A Steam Man," *Cincinnati Enquirer*, January 17, 1883, 8.

64. "The Steam Man's Demise," *Cincinnati Enquirer*, January 19, 1883, 8.

65. "The Steam Man's Demise," 8.

66. "A New Steam Man," *Black Hills Daily Times* (Deadwood, SD), May 18, 1883, 3.

67. "Walks by Steam," *Wilmington [DE] News Journal*, July 6, 1878, 3.

68. "The Walking Steam Man," *Reading [PA] Times*, August 5, 1878, 1.

69. "The Steam Man Coming," *Harrisburg Daily Independent*, September 18, 1880, 1.

70. Reuben Hoggett, "1874—Adam Ironsides—The Steam Man—C. C. Roe a.k.a. Capt. Rowe (Canadian)," *Cybernetic Zoo*, http://cyberneticzoo.com/steammen/1874-adam-ironsides-the-steam-man-c-c-roe-canadian/.

71. "City and Provincial News," *Manitoba Free Press* (Winnipeg), April 10, 1878, 1.

72. C. C. Roe, broadsheet (undated [probably 1878]), http://epe.lac-bac.gc.ca/100/205/301/ic/cdc/industrial/ironman.htm.

73. "Patent No. 4175," *Canadian Mechanics' Magazine and Patent Office Record*, Volumes 3–6 (January 1875), 5.

74. "The Home of the Steam Man," 3; "Rowe a Humane Man," *Delaware County Daily Times* (Chester, PA), January 22, 1883, 3.

75. "Electric Frankenstein," *Washington Evening Star*, August 6, 1890, 6.

76. "Electric Frankenstein," 6.

77. George R. Moore, U.S. Patent No. 454570, granted June 23, 1891, https://www.google.com/patents/US454570.

78. "Motor Show Is Opened," *New York Times*, November 4, 1900, 10; "Odd Inventions Shown," *New York Tribune*, November 4, 1900, 8.

79. "Hercules, the Wonder," *Scranton Republican*, January 24, 1893, 3.

80. "The Steam Man," *Scientific American* (April 15, 1893), 233.

81. "Austin & Stone's," *Boston Sunday Globe*, April 26, 1896, 18.

82. Steam Man advertisement, *Boston Post*, April 19, 1896, 10.

83. "Walking Giant," *Cincinnati Enquirer*, August 4, 1901, 33 [reprinted from *Chicago Record-Herald*].

84. "Austin & Stone's," 18.

85. "Walking Giant," 33.

86. "The Iron Man," *Bristol [IN] Banner*, October 23, 1908, 7.

87. "In the World of Electricity," *New York Times*, August 4, 1895, 23.

88. "A Man of Wood Really Walks," *New York World*, June 28, 1896, 26.

89. Louis Philip Perew and Joseph A. Dischinger, U.S. Patent No. 949287, granted February 15, 1910, https://www.google.com/patents/US949287.

90. 6 feet: "In the World of Electricity," 23; 7 feet: "A Man of Wood Really Walks," 26.

91. "The Electric Man," *Greensboro [NC] Patriot*, September 4, 1895, 1.

92. "The Electric Man," 1.

93. "Walking Automaton Is a True Wonder," *Buffalo Express*, September 2, 1900, http://www.buffalohistorygazette.net/2010/09/the-man-of-tonawanda.html.

94. "Stock Companies," *Rochester Democrat and Chronicle*, February 4, 1900, 9.

95. "Walking Automation Is a Mechanical Wonder," *Buffalo Courier*, September 2, 1900, 17.

96. "Hum of City Streets," *Philadelphia Times*, January 9, 1902, 6.

97. "Makes Giant Figure Walk," *Yale [MI] Expositor*, September 7, 1900, 1.

98. "A Mechanical Giant," *Charlotte News*, September 14, 1900, 6.

99. "Walking Automaton," *Buffalo Express*.

100. W. B. Northrop, "An Electric Man," *The Strand* (November 1900), 586–90.

101. "Makes Giant Figure Walk," 1.

102. "Mechanical Man Will Visit Pittsburgh Soon," *Pittsburgh Press*, May 2, 1913, 16; "Mechanical Man Here Today," *Pittsburgh Press*, May 3, 1913, 5.

103. "Mechanical Man to Walk City Streets," *Fort Wayne Journal-Gazette*, June 5, 1914, 12.

104. "Has Built a New Percy," *Alton [IL] Evening Telegraph*, June 26, 1913, 1.

105. "It Walks and Talks," *St. Louis Republic*, September 2, 1900, 12; "Passing the Camera," *St. Louis Republic*, September 9, 1900, 2.

106. "Has Built a New Percy," 1.

107. "Mechanical Man Walks Down the Street," *Alton [IL] Evening Telegraph*, June 22, 1914, 5.

108. "Exhibiting Mechanical Man," *Alton [IL] Evening Telegraph*, June 25, 1914, 1.

109. Herbert C. Crocker, "The Electric Man," *Illustrated World* (December 1916), 504–6.

110. "Mechanical Man Breaks Down on Tryout," *Alton [IL] Evening Telegraph*, April 13, 1918, 3.

111. "Uncle Sam to Drive 'Fritz' around City," *St. Louis Post-Dispatch*, April 20, 1919, 30.

112. "$8,091,700 of Loan Quota Is Subscribed," *St. Louis Star and Times*, April 24, 1919, 1.

113. "Has Built a New Percy," 1.

Chapter 5

1. Scott Schaut, *Robots of Westinghouse, 1924–Today* (Mansfield, OH: Scott Schautt, Mansfield Memorial Museum, 2006), 19–20.

2. Helen Fox, "Automatic Device Saves Labor for Control Systems," *Munster [IN] Times*, June 26, 1924, 7.

3. Schaut, *Robots of Westinghouse*, 23.

4. Waldemar Kaempffert, "Science Produces the 'Electrical Man,'" *New York Times*, October 23, 1927, 9–1.

5. Schaut, *Robots of Westinghouse*, 24–25.

6. "What Mechanical Man May Do," *New Castle [PA] News*, February 24, 1928, 16.

7. Schaut, *Robots of Westinghouse*, 25.

8. "Mechanical Man Gets Pain in His 'Tummy,'" *Pittsburgh Press*, March 5, 1929, 21.

9. Schaut, *Robots of Westinghouse*, 30, 31, 49.

10. Schaut, *Robots of Westinghouse*, 52.

11. Schaut, *Robots of Westinghouse*, 23, 33.

12. Hortense Saunders, "Mr. Televox, Plain but Useful, and in All a Live-Wire Fellow," *Pittsburgh Press*, February 23, 1928, 2.

13. "Most Modern of Men—He's Mechanical!" *Asbury Park [NJ] Press*, February 22, 1928, 10.

14. Olive Roberts Barton, "Televox and House Work Is Made Easy," *Appleton [WI] Post-Crescent*, March 13, 1928, 8.

15. H. L. Phillips, "Trying Out the Televox, or Mechanical Man," *Harrisburg Evening News*, March 5, 1928, 10.

16. Lee Shippey, "The Courtship of Miles Televox," *Los Angeles Times*, June 20, 1928, 20.

17. "Brother of Televox Is Operated by Light; New Mechanical Servant Shown in Action," *New York Times*, April 18, 1929, 13.

18. "'Mechanical Men' Who Answer Vocal Orders Give Promise of Being Big Help around Home," *Brooklyn Daily Eagle*, October 14, 1927, 3.

19. "Mechanical Man Visioned as Servant for Housewife," *Des Moines Tribune*, February 22, 1928, 1.

20. "Mechanical Men That Excel Any Human Being," *San Antonio Light*, September 9, 1931, 48.

21. "Science's New Mechanical Man," *Richmond [IN] Item*, June 27, 1930, 1; "New Developments in Electricity," *Modern Mechanics* (March 1931), 79.

22. "Miss Katrina von Televox," *Altoona [PA] Mirror*, October 3, 1930, 22.

23. "Mechanical Maid to Do Her Stunts," *Philadelphia Inquirer*, September 23, 1930, 32.

24. "Miss Televox to Be in Elyria Saturday," *Elyria [OH] Chronicle Telegram*, June 18, 1931, 10.

25. "Katrina von Televox," *Hagerstown [MD] Morning Herald*, April 22, 1931, 5.

26. Schaut, *Robots of Westinghouse*, 64.

27. "Willie Vocalite Is 'Some' Robot; Does Everything but Pay Taxes," *Racine [WI] Journal Times*, April 7, 1932, 1; "Have You Seen 'Willie Vocalite,'" *Los Angeles Times*, May 19, 1932, 18; 193 "Willie Vocalite," *Decatur [IL] Herald*, February 10, 1932, 14; "Willie Vocalite," *Bakersfield Californian*, May 7, 1932, 5.

28. "Robot Gives Girls New Thrill," *Akron Beacon Journal*, December 8, 1931, 33.

29. Schaut, *Robots of Westinghouse*, 67.

30. Schaut, *Robots of Westinghouse*, 71.

31. "Only Westinghouse Could Build Willie Vocalite," *Coshocton [OH] Tribune*, March 25, 1934, 6.

32. Schaut, *Robots of Westinghouse*, 74.

33. Schaut, *Robots of Westinghouse*, 71.

34. Schaut, *Robots of Westinghouse*, 68.

35. "Mechanical Man Electric Marvel Put on Exhibit," *Altoona [PA] Mirror*, April 11, 1939, 6.

36. "Shades of Brick Bradford! 'Elektro' Is Almost Human," *Munster [IN] Times*, April 12, 1939, 5.

37. "Mechanical Man Electric Marvel Put on Exhibit," 6.

38. Schaut, *Robots of Westinghouse*, 95, 103, 149.

39. "Singing Light Tower at Fair Blends Music and Color," *Popular Mechanics* (June 1939), 846.

40. "The Middleton Family at the New York World's Fair," San Jose State University, http://www.sjsu.edu/faculty/wooda/middleton/middletonelektro.html; Westinghouse, "The Middleton Family at the New York World's Fair," 1939, https://www.youtube.com/watch?v=YF594h8KUXw.

41. Schaut, *Robots of Westinghouse*, 117.

42. Schaut, *Robots of Westinghouse*, 120–22.

43. Charles F. Danver, "Origin of Sparko," *Pittsburgh Post-Gazette*, January 14, 1941, 6.

44. "Sparko, the Perfect Pup," *Galveston [TX] Daily News*, June 10, 1940, 20.

45. "Teaching New Dog Old Tricks," *Brainerd [MN] Daily Dispatch*, May 4, 1940, 11.

46. Robert R. Hagy, Jr., "Tough Mechanical Man Scorned by Girls Who Boss Him Around," *Pittsburgh Post-Gazette*, January 3, 1941, 14.

47. Schaut, *Robots of Westinghouse*, 153.

Chapter 6

1. "A Wonderful Artificial Man," *Dundee Evening Telegraph and Post*, March 22, 1911, 2; "A Clockwork Man That Talks and Sings," *San Antonio Light and Gazette*, April 23, 1911, 34; "Boneless, Bloodless, Fleshless Man May Replace Human Beings in the Industry of the World," *Evansville [IN] Press*, June 13, 1914, 1, 6.

2. "A Remarkable Automaton," *Syracuse Herald*, June 7, 1909, 4.

3. "Personal and Pertinent," *Wilmington Evening Journal*, June 12, 1909, 4 [reprinted from *London Mail*].

4. "Boneless, Bloodless, Fleshless Man," 1, 6.

5. "A Mechanical Man," *London Globe*, April 26, 1919, 12.

6. "A Clockwork Man That Talks and Sings," 34.

7. "Automaton Woman Walks," *Frederick [MD] News*, September 11, 1911, 7.

8. "Automaton of Woman Sings, Talks, and Walks," *Washington Times*, August 30, 1911, 5.

9. "Almost Human," *Cincinnati Enquirer*, August 31, 1911, 5.

10. "Miss Automaton, She Talks," *New York Observer*, September 14, 1911, 344.

11. Clarence A. Logan, "Is Sir Oliver Lodge Being Fooled by Wireless?" *San Francisco Chronicle*, May 23, 1920, 12; "Millionaire Dentist Has Million-Melody Home," *Allentown [PA] Morning Call*, May 11, 1937, 16.

12. "Automation Puzzles World Magicians," *Sandusky [OH] Register*, December 21, 1919, 13.

13. F. Leland Elam, "Strange Abode," *Los Angeles Times Sunday Magazine*, June 12, 1939, 7.

14. "A Wooden Blonde That Will Smile, Breathe, Blush, Sing, Play, Wink," *Hamilton [OH] Evening Journal*, June 16, 1923, 13.

15. Elam, "Strange Abode," 7.

16. "Robot Turns Frankenstein, Shoots Creator," *Arizona Republic* (Phoenix), September 19, 1932, 1.

17. "Shot by the 'Monster' of His Own Creation," *Ogden [UT] Standard-Examiner*, October 23, 1932, 21; "Owner Finds Robot Is Frankenstein," *Carlisle [PA] Sentinel*, October 20, 1932, 4; "'Frankenstein' Had Nothing on This Robot," *Santa Cruz [CA] Sentinel*, September 20, 1932, 7; "Page Frankenstein: Robot Shoots Man," *Hamilton [OH] Journal News*, September 20, 1932, 1.

18. "Iron Monster Turns Traitor," *Detroit Free Press*, September 19, 1932, 2; "Life of Robot's Master Menaced by Steel Monster," *Benton Harbor [MI] News-Palladium*, September 19, 1932, 1.

19. "English Inventor Fears Three-Ton Frankenstein He Worked on 14 Years," *Minneapolis Star Tribune*, September 18, 1932, 33.

20. Mark Potter, "A Robot Who Reads the Newspapers," *Leeds Mercury*, August 24, 1932, 7.

21. "Robot Utters a Warning, Then Shoots Creator," *Chicago Tribune*, September 19, 1932, 7.

22. "Iron Monster Turns Traitor," 2.

23. "Robot in Revolt," *Dundee Courier and Advertiser*, September 19, 1932, 6.

24. "English Inventor Fears Three-Ton Frankenstein He Worked On 14 Years," 33.

25. "Like Frankenstein, Robot Man Turns on Inventor," *Des Moines Register*, September 19, 1932, 1.

26. "Robot Woman Socks Escort from Behind," *Elmira [NY] Star-Gazette*, October 4, 1934, 2.

27. "Mechanical Man's Talk Startles Gotham Audience," *Miami News*, October 25, 1934, 1, 8; "He Knows All the Answers," *Port Huron [MI] Times Herald*, October 29, 1934, 6; "Robot," *Time* (November 5, 1934), 54; British Pathé, "The Face of Things—to Come! Alpha the Robot" (newsreel), June 12, 1934, https://www.britishpathe.com/video/the-face-of-things-to-come-alpha-the-robot/query/alpha+robot.

28. "He Knows All the Answers," 6.

29. "Alpha Loses Her Temper," *Lancashire Evening Post*, September 15, 1934, 3.

30. "What May Happen When Robots Do All the Work," *Nashville Tennessean* (*American Weekly* supplement), August 5, 1938, 10, 11, 15.

31. Richard Amaro, "San Diego Invites the World to Balboa Park a Second Time," *Journal of San Diego History* 31, no. 4 (Fall 1985), http://sandiegohistory.org/journal/1985/october/invites/.

32. Reuben Hoggett, "1932—Alpha the Robot—Harry May (English)," *Cybernetic Zoo*, http://cyberneticzoo.com/robots/1932-alpha-the-robot-harry-may-english/.

33. Reuben Hoggett, "1930—Alpha the Robot (American)," *Cybernetic Zoo*, http://cyberneticzoo.com/pseudo-automatons-and-robots/1930-alpha-the-robot-american/.

34. "Robot Goes Wild, Knocks Out Man," *Des Moines Register*, June 28, 1935, 7.

35. Charles Moore, "Mechanical Man Alpha Ends Career after Owner Hurt," *San Bernardino County [CA] Sun*, February 18, 1936, 1.

36. Theodosia Durand, "World Fairs and Affairs," *Santa Rosa [CA] Press Democrat*, June 21, 1936, 13.

37. Moore, "Mechanical Man Alpha Ends Career after Owner Hurt," 1.

38. Durand, "World Fairs and Affairs," 13.

39. "Mechanical Man and Inventor Are Here This Week," *The [MIT] Tech*, February 25, 1929, 1, 4.

40. "Eric Robot's Debut Is Eagerly Awaited by Philadelphians," *Philadelphia Inquirer*, March 9, 1929, 1, 9.

41. Reuben Hoggett, "1928—Eric Robot—Capt. Richards & A.H. Reffell (English)," *Cybernetic Zoo*,

http://cyberneticzoo.com/robots/1928-eric-robot-capt-richards-english/.

42. "English Robot Makes Lecture," *Altoona [PA] Mirror*, October 12, 1928, 18.

43. "Husband Hunters, Wait Awhile! See If Robot Hits Mansfield," *Mansfield [OH] News-Journal*, January 5, 1929, 1, 12.

44. "Putting the Mechanical Man to Work," *Zanesville [OH] Sunday Times Signal*, December 9, 1928, 34.

45. Reuben Hoggett, "1932—George Robot—Capt. W.H. Richards (British)," *Cybernetic Zoo*, http://cyberneticzoo.com/robots/1932-%E2%80%93-george-robot-%E2%80%93-capt-w-h-richards-british/.

46. "Mr. William Robot Attends Party," *Munster [IN] Times*, March 19, 1934, 2.

47. "Willie Will Attend Chicago Exhibit If He Learns to Smoke," *Muncie [IN] Evening Press*, August 7, 1933, 5.

48. H. Allen Smith, "Writer Marvels When He Asks Robot How He Feels and Later Reports Severe Pain in Side," *Great Falls Tribune*, May 27, 1934, 2.

49. "Lifelike Robot Speaks, Smokes, and Drinks," *Popular Science* (October 1935), 19.

50. "Big Looie Plays a Solo," *Rochester Democrat and Chronicle*, February 20, 1938, 50.

51. "'Looie' Jealous Dummy Stooge," *Lansing State Journal*, January 20, 1938, 13.

52. "Big Looie," *Detroit Free Press Sunday Graphic*, July 2, 1950, 6.

53. Austin Huhn, "Clarence—Radio Robot," *Radio Craft* (October 1939), 200, 245.

54. "Free-Lance Robot Hunts Job at Fair," *New York Times*, July 23, 1939, 28.

55. Jam Handy, "Leave It to Roll-Oh," 1940, https://youtu.be/KSnJBNijsVU.

56. Handy, "Leave It to Roll-Oh."

57. Waldemar Kaempffert, "Science Produces the 'Electrical Man,'" *New York Times*, October 23, 1927, 9–1.

58. "Mechanical Man," *Aberdeen Press and Journal*, December 31, 1928, 2.

59. H. Stafford Hatfield, *Automaton, or The Future of the Mechanical Man* (London: Kegan Paul, Trench, Trubner & Co., 1928), 14–15.

60. "Romantic Old Maids Can Hear the Words of Love They Long For," *San Antonio Light*, July 1, 1928, 62.

61. H. I. Phillips, "The Once Over," *Oakland Tribune*, October 25, 1927, 25.

62. Robert E. Martin, "Mechanical Men Walk and Talk," *Popular Science Monthly* (December 1928), 22–23, 137–38.

63. William Barclay Parsons, "Real Robots That Do Work for Man," *New York Times*, September 15, 1929, 7.

64. "Victims of the Machines," *Sioux Falls [SD] Argus-Leader*, February 4, 1930, 6 [reprinted from *New York Herald-Tribune*].

65. "A Declaration of Independence," *New York Times*, September 28, 1930, 3–1.

66. Carl F. Elliott, "Sees Robot Age in Near Future," *Brooklyn Daily Eagle*, October 22, 1933, 1, 2, 16.

67. William C. Richards, "World to Greet New Wonders in Next 100 Years," *Detroit Free Press* (centennial section), May 10, 1931, 1–2.

68. "Scientist Predicts Robots of Future Will Do Drudgery," *Burlington [VT] Free Press*, February 21, 1935, 8.

69. Ernie Bushmiller, *Nancy* (comic strip), *Nevada State Journal* (Reno), March 21, 1943, 24.

70. Andrew R. Boone, "Garco, Indestructible 'Worker,' Can Do Dangerous Jobs for Humans," *Lafayette [IN] Journal and Courier*, March 13, 1954, 38; Kay Sullivan, "A Robot in the Family," *Jackson [MI] Clarion-Ledger* (*Parade* magazine supplement), October 17, 1954, 28–29.

71. Boone, "Garco, Indestructible 'Worker,' Can Do Dangerous Jobs for Humans," 38; Andrew R. Boone, "Plug-In Built in 90 Days," *Popular Science* (December 1953), 100–103; "Boy with a £13,000 Toy," *Sydney Morning Herald*, December 5, 1954, 33.

72. Sullivan, "A Robot in the Family," 28–29.

73. "Tin Can with an Idea," *Life* (December 13, 1954), 74, 77–78.

74. "Boy, 13, Builds His Own Robot—and It Works," *Hutchinson [KS] News*, October 25, 1930, 9.

75. Richard Daw, "Youth's Resourceful Robot Helps Sell Boys on Science," *Vineland [NJ] Daily Journal*, December 11, 1958, 11.

76. "Thodar the Robot to Parade on Landis Ave. Saturday," *Vineland [NJ] Daily Journal*, April 27, 1955, 16; Mark Repasky, "Thodar the Robot," March 25, 2018, http://www.innbythemill.com/thodar.htm.

77. "Automated Santa," *Pittsburgh Press Sunday Magazine*, December 16, 1962, 6, 7.

78. *I've Got a Secret*, hosted by Garry Moore, Season 6, Episode 38, March 26, 1958.

79. "13-Year-Old Boy Creates Robot," *Sayre [PA] Evening Times*, July 30, 1957, 8; "Queens Boy, 13, Constructs Robotron the Robot," *Bridgeport Post*, July 30, 1957, 32.

80. "Making Learning Fun," *Baruch College Alumni Magazine* (November 3, 2015), https://blogs.baruch.cuny.edu/bcam/2015/11/03/making-learning-fun/.

81. Douglas Colligan, "The Robots Are Coming," *New York Magazine* (July 30, 1979), 40.

82. "Boy Builds Robot," *Newport [RI] Daily News*, March 24, 1954, 10.

83. "Robot Reader," *Brownsville [TX] Herald*, March 24, 1954, 8; "Young Inventor," *Minneapolis Star Tribune* (*This Week* supplement), February 27, 1955, 9.

84. "Reporter Gets Interview," *Bakersfield Californian*, March 1, 1955, 4.

85. "Roaming Robot," *Minneapolis Star Tribune*, August 11, 1955, 3; Reuben Hoggett, "1954—'Gismo the Peaceful'—Sherwood Fuehrer (American)," *Cybernetic Zoo*, http://cyberneticzoo.com/robots/1954-gismo-the-peaceful-sherwood-fuehrer-american/.

86. Sherwood Fuehrer, "Gismo and I," *Boy's Life* (June 1956), 18, 54–56.

Chapter 7

1. Allan Holtz, "Obscurity of the Day: Professor Dodger and His Automatic Servant Girl," *Stripper's Guide* (blog), November 30, 2012, https://strippersguide.blogspot.com/2012/11/obscurity-of-day-professor-dodger-and.html.

2. Allan Holtz, "Ink-Slinger Profile: H. C. Greening," *Stripper's Guide* (blog), September 8, 2011, https://strippersguide.blogspot.com/2011/09/ink-slinger-profiles-hc-greening.html.

3. Lucy N. Eames, "Clearer Diagnosis and Simpler Treatment," *Michigan State Medical Journal* (June 1912).

4. Correspondent (Ryegate Local No. 153), letter to the editor, *Paper Makers Journal* (February 1912), 26.

5. Correspondent (Branch No. 11, Davenport, Iowa), letter to the editor, *Leather Workers' Journal* (July 1914), 339.

6. "The Harris Theater," *Pittsburgh Daily Post*, January 14, 1912, 35.

7. "Amateur Vaudeville by One Thousand Club," *Galveston [TX] Daily News*, February 23, 1912, 5.

8. "Machine Made Milk," *Lebanon [PA] Daily News*, October 25, 1912, 12.

9. *Eufalua [OK] Republican*, February 7, 1913, 2 [quoting *Cherokee Republican*].

10. "Down-in-Four," *Time* (September 22, 1930), 27.

11. "'Percy the Mechanical Man,' (Bray)," *Moving Picture World* (December 23, 1916), 1822.

12. Charles B. Driscoll, "The World and All," *Lansing State Journal*, May 31, 1937, 4.

13. Malcom W. Bingay, "Good Morning," *Detroit Free Press*, April 5, 1935, 6.

14. Russell Maloney, "Toy Fair," *New Yorker*, April 13, 1935, 16.

15. Ray Bradbury, "Buck Rogers in Apollo Year 1," in *The Collected Works of Buck Rogers in the 25th Century*, edited by Robert C. Dille (New York: Chelsea House, 1970), xii–xiii.

16. "First Prize in Contest Goes to McKeesport Man," *Pittsburgh Post-Gazette*, March 2, 1929, 2.

17. Maurice Horn, ed., *100 Years of American Newspaper Comics* (New York: Gramercy Books, 1996), 70.

18. Phil Nowlan and Russell Keaton, *Buck Rogers in the 25th Century*, November 2, 1930.

19. Phil Nowlan and Rick Yager, *Buck Rogers in the 25th Century*, December 30, 1934.

20. Phil Nowlan and Rick Yager, *Buck Rogers in the 25th Century*, June 19, 1938.

21. Bill Borden with Steve Posner, *The Big Book of Big Little Books* (San Francisco: Chronicle Books, 1997), 13.

22. Stanley Link, *Tiny Tim and the Mechanical Men* (Racine, WI: Whitman Publishing Company, 1937), 84.

23. Robert Reginald, *Contemporary Science Fiction Authors* (New York: Arno Press, 1974), 290.

24. R. R. Winterbotham, *Maximo the Amazing Superman and the Supermachine* (Racine, WI: Whitman Publishing Company, 1941), 292.

25. John Coleman Burroughs, *John Carter of Mars*, July 26, 1942.

26. William Ritt and Clarence Gray, *Brick Bradford*, April 10, 1939.

27. Alex Jay, "Ink-Slinger Profiles by Alex Jay: Larry Antonette," *Stripper's Guide* (blog), June 11, 2014, http://strippersguide.blogspot.com/2014/06/ink-slinger-profiles-by-alex-jay-larry.html.

28. Russ Westover, *Tillie the Toiler*, July 17, 1933.

29. "Just a Human Robot," letter to the editor, *Wisconsin State Journal* (Madison), August 29, 1933, 4.

30. Don Markham, "Felix the Cat," *Toonopedia*, http://www.toonopedia.com/felix.htm.

31. Bill Walsh and Floyd Gottfredson, *Mickey Mouse*, September 29, 1944.

32. *Mickey Mouse*, September 30, 1944.

33. *Mickey Mouse*, November 9, 1944.

34. Dorothy Parker, "A Mash Note to Crockett Johnson," *PM*, October 3, 1943.

35. Ted Ferro and Jack Morely, *Barnaby*, November 12, 1947.

36. Walt Kelly, *Pogo*, February 13, 1952.

37. Forrest "Bud" Sagendorf, *Thimble Theater*, May 30, 1963.

Chapter 8

1. Ron Goulart, *Great History of Comic Books* (Chicago: Contemporary Books, 1986), 3, 15.

2. "The Adventures of 'Spargus and Chubby," *The Funnies* #3 (December 1936).

3. Brad Ricca, *Super Boys: The Amazing Adventures of Jerry Siegel and Joe Shuster—The Creators of Superman* (New York: St. Martin's Press, 2013), 103–4.

4. Ricca, *Super Boys*, 108–9.

5. Jerry Siegel and Joe Schuster, "Federal Men," *New Comics* #10 (November 1936).

6. Wayne Reid (George Brenner), "Hugh Hazzard and His Iron Man," *Smash Comics* #1 (August 1939).

7. Carl Burgos, "Marvex, the Super Robot," *Daring Mystery Comics* #4 (May 1940).

8. Paul Gustafson, "The Fantom of the Fair," *Amazing Mystery Funnies*, vol. 2, #11 (November 1939).

9. Steve Dahlman, "Electro," *Marvel Mystery Comics* #4 (February 1940).

10. Jack Cole, "Dickie Dean, the Boy Inventor," *Silver Streak Comics* #5 (June 1940).

11. Dick Wood, "Dickie Dean, the Boy Inventor," *Silver Streak Comics* #15 (October 1941).

12. Pat Adams, "Cosmo Cat," *Cosmo Cat* #10 (October 1947).

13. E. Nelson Bridwell, "In Memoriam: Otto Oscar Binder," *The Amazing World of DC Comics* #3 (November 1974).

14. Otto Binder, "Captain Marvel," *Captain Marvel Adventures* #78 (November 1947).

15. Fredric Wertham, *Seduction of the Innocent* (New York: Rinehart & Company, 1954), picture example section.

16. Jack Sparling, "Break Up!" *Tomb of Terror* #15, May 1954.

17. Carl Memling, "All for Love," *Space Adventures* #8 (September 1953).

18. Roy Ald, "The Indestructible Antagonist," *Captain Video* #3 (June 1951).

19. "Revolt of the Robots," *Space Detective* #3 (February 1952).

20. Jon Smalle, "The Robots of the Demon Star," *Wonder Comics* #20 (October 1948).

21. David Hajdu, *The Ten-Cent Plague: The Great Comic Book Scare and How It Changed America* (New York: Farrar, Straus and Giroux, 2008), 322.

22. Dick and Dave Wood, "Challengers of the Unknown," *Showcase* #7 (April 1957).

23. "Metal Scraps" (letter column), *Showcase* #39 (August 1962).

Chapter 9

1. Unsigned review of *Around the World in Eighty Days*, by Jules Verne, *The Galaxy: A Magazine of Entertaining Reading* (September 1873), 424.

2. Jeff Vaj262

nderMeer and Ann VanderMeer, eds., *The Big Book of Science Fiction* (New York: Vintage, 2016), 124.

3. A. Merritt, *The Metal Monster, Argosy All-Story Weekly* (August 7–September 25, 1920), chap. 5.

4. "A. Merritt's *The Metal Monster*," *Skulls in the Stars* (blog), February 2, 2009, https://skullsinthestars.com/2009/02/02/a-merritts-the-metal-monster/.

5. Edmond Hamilton, "The Metal Giants," *Weird Tales* (December 1926); also in *The Gernsback Awards 1926*, Volume 1, edited by Forrest J. Ackerman (Los Angeles: Triton Books, 1982), 73.

6. J. Schlossel, "To the Moon by Proxy," *Amazing Stories* (October 1928), 600.

7. David H. Keller, MD, "Air Lines," *Amazing Stories* (January 1930), 942.

8. "In 2028 … NFL Stadiums Shrink," *ESPN The Magazine* (April 23, 2018), 36.

9. Wayne McAdam, letter to the editor, *Science Wonder Quarterly* (September 1929), 138; Hugo Gernsback, editor response to letter, *Science Wonder Quarterly* (September 1929), 143.

10. Jim Vanny, "The Radium Master," *Wonder Stories* (August 1930), 240.

11. Vanny, "The Radium Master," 249.

12. Isaac Asimov, ed., *Before the Golden Age* (Garden City, NY: Doubleday, 1974), 80.

13. Paul Carter, *The Creation of Tomorrow* (New York: Columbia University Press, 1977), 15.

14. Harl Vincent, "Terrors Unseen," *Astounding Stories* (March 1931), 374.

15. Ray Cummings, "The Exile of Time," *Astounding Stories* (April 1931), 42.

16. Cummings, "The Exile of Time" (April 1931), 48–49.

17. Cummings, "The Exile of Time" (June 1931), 407.

18. Cummings, "The Exile of Time" (July 1931), 129.

19. Kenneth Robeson (Ryerson Johnson), "The Fantastic Island," *Doc Savage* (December 1935), chap. 10.

20. Stephen Vincent Benét, "Nightmare Number Three," *New Yorker* (July 27, 1935), 23.

21. Ray Palmer, "The Observatory by the Editor," *Amazing Stories* (August 1938), 134.

22. Eando Binder, "I, Robot," *Amazing Stories* (January 1939), 18.

23. Sam Moskowitz, *Seekers of Tomorrow* (New York: Ballantine Books, 1967), 45.

24. Robert Moore Williams, "Robot's Return," *Astounding Science-Fiction* (September 1938), 147.

25. C. L. Moore, introduction to *Robots Have No Tails* by Henry Kuttner (New York: Lancer, 1973), 7.

26. John W. Campbell, "Robots," *Astounding Science-Fiction* (November 1939), 6.

27. H. H. Holmes [Anthony Boucher], "Robinc," *Astounding Science-Fiction* (September 1943), 80.

28. "Talking Robot Grows Mature," *Greenville [MS] Delta Democrat-Times*, October 3, 1940, 7.

29. Douglas Gilbert, "Odd Jobs," *Pittsburgh Press*, December 11, 1940, 11.

30. Isaac Asimov, "Runaround," *Astounding Science-Fiction* (March 1942), 100.

31. William P. McGivern, "Sidney, the Screwloose Robot," *Fantastic Adventures* (June 1941), 118.

32. Jack Williamson, "With Folded Hands…," *Astounding Science Fiction* (July 1947), 6.

33. Robert A. Heinlein, introduction to *Tomorrow, the Stars*, edited by Robert A. Heinlein (New York: Signet, 1953), v.

Chapter 10

1. Phil Hardy, *The Encyclopedia of Science Fiction Movies* (Minneapolis: Woodbury Press, 1986), 46.

2. Hardy, *The Encyclopedia of Science Fiction Movies*, 41.

3. Arthur B. Reeve and John W. Grey, *The Master Mystery* (New York: Grosset & Dunlap, 1919), 24.

4. Herman G. Scheffaur, "An Impression of the German Film 'Metropolis,'" *New York Times*, March 6, 1927, 8–7.

5. "'Metropolis' Spectacular Film Scheduled for State," *Reading [PA] Times*, August 22, 1927, 9; "'Metropolis' Held Hypnotic Picture," *Minneapolis Star*, October 22, 1927, 25.

6. "The Perfect Woman," *Birmingham [UK] Daily Gazette*, June 16, 1949, 4.

7. "*The Mysterians* movie trailer," https://youtu.be/AHbL8a71Hoo.

8. "*The Mysterians* movie poster," *AllPosters*, https://www.allposters.com/-sp/The-Mysterians-1959-Posters_i6250399_.htm.

9. "Science Fiction Movies," *Ultimate Movie Rankings*, http://www.ultimatemovierankings.com/science-fiction-movies-2/.

10. "*Forbidden Planet* movie poster," *AllPosters*, https://www.allposters.com/-sp/Forbidden-Planet-Posters_i12191582_.htm.

11. Bill Warren, *Keep Watching the Skies! American Science Fiction Movies of the Fifties: The 21st Century Edition* (Jefferson, NC: McFarland, 2009), 297–99.

12. "Robby the Robot: Engineering a Sci-Fi Icon," produced by Jonathan Strailey (on *Forbidden Planet: 50th Anniversary Edition* DVD, 2006).

13. Mike Barnes, "Robert Kinoshita, Robot Designer for 'Forbidden Planet' and 'Lost in Space,' Dies at 100," *Hollywood Reporter*, January 13, 2015.

14. David Szondy, "The Original Robby the Robot Goes Up for Auction," *New Atlas*, October 27, 2017, https://newatlas.com/robby-robot-bonhams/51922/.

Part Two

1. "The Master Mind," *Rocky Mount [NC] Telegram*, August 10, 1944, 4.

2. Max B. Cook, "Aero's 'Mechanical Brains,'" *Honolulu Advertiser*, July 31, 1944, 12.

3. John Lear, "Can a Mechanical Brain Replace You?" *Collier's* (April 3, 1953), 58–63.

4. "Brainy Machines," *Elmira [NY] Star-Gazette*, April 13, 1943, 6.

5. Kurt Vonnegut, Jr., *Utopia 14 [Player Piano]* (New York: Bantam Books, 1954), 276.

Chapter 11

1. *Perry Como's Kraft Music Hall*, hosted by Perry Como, Season 8, Episode 23, February 18, 1956.

2. *MGM Parade*, hosted by Walter Pidgeon, Season 1, Episodes 32–33, March 14 and 21, 1956.

3. David Szondy, "The Original Robby the Robot Goes Up for Auction," *New Atlas*, October 27, 2017, https://newatlas.com/robby-robot-bonhams/51922/.

4. *The Thin Man*, "Robot Client," directed by

Oscar Rudolph, written by Devery Freeman, Season 1, Episode 23, February 28, 1958.

5. *The Many Loves of Dobie Gillis*, "Beethoven, Presley and Me," directed by Guy Scarpitta, written by Dean Riesner, Season 4, Episode 24, March 13, 1963.

6. *The Addams Family*, "Lurch's Little Helper," directed by Sidney Lanfield, written by Phil Leslie, Season 2, Episode 27, March 18, 1966.

7. *The Twilight Zone*, "Uncle Simon," directed by Don Siegel, written by Rod Serling, Season 5, Episode 8, November 15, 1963.

8. *The Twilight Zone*, "The Brain Center at Whipple's," directed by Richard Donner, written by Rod Serling, Season 5, Episode 33, May 15, 1964.

9. Harry Castleman and Walter J. Podrazik, *Watching TV: Six Decades of American Television*, second edition (Syracuse, NY: Syracuse University Press, 2003), 147.

10. *Hazel*, "Hazel's Contract," directed by William D. Russell, written by Peggy Chantler Dick, Season 2, Episode 2, September 27, 1962.

11. Susan Sontag, "Notes on 'Camp,'" *The Partisan Review* (Fall 1964), 515–30, https://faculty.georgetown.edu/irvinem/theory/Sontag-NotesOnCamp-1964.html.

12. Bruce LaBruce, "Notes on Camp/Anti Camp," July 7, 2015, http://brucelabruce.com/2015/07/07/notes-on-camp-anti-camp/.

13. Luke Feck, "Average—A Compliment," *Cincinnati Enquirer*, November 23, 1962, 9.

14. Tim Pawlenty, "We're at the Dawn of the Fourth Industrial Revolution," *Minneapolis Star Tribune*, June 5, 2017, A9.

15. Frank Lovece, "'Jetsons' Live-Action Remake from ABC in the Works," *Newsday*, August 17, 2017.

16. Matt Novak, "50 Years of the Jetsons: Why the Show Still Matters," *Paleofuture* (blog), September 19, 2012, http://blogs.smithsonianmag.com/paleofuture/2012/09/50-years-of-the-jetsons-why-the-show-still-matters/.

17. "Among Best Shows," *Phoenix Republic*, September 23, 1962, 21; Feck, "Average—A Compliment," 9.

18. "Who Would Do the Nagging?" *Hutchinson [KS] News*, September 22, 1962, 1; "Oh, Kiss Me, You Mechanical Fool," *Ogden [UT] Standard Examiner*, September 21, 1962, 1.

19. Eddy Gilmore, "British Scientist Predicts Automatic Housewife," *Racine [WI] Journal Times*, September 25, 1962, 9.

20. *The Jetsons*, "Rosie the Robot," directed by Joseph Barbera and William Hanna, written by Larry Markes (story) and Tony Benedict (teleplay), Season 1, Episode 1, September 23, 1962.

21. *The Jetsons*, "Rosie the Robot."

22. Alan Gill, "In the Age of Dee-Ah-Tump," *Marion [OH] Star*, August 24, 1962, 10.

23. *The Jetsons*, "The Coming of Astro," directed by Joseph Barbera and William Hanna, written by Tony Benedict (story and teleplay), Season 1, Episode 5, October 21, 1962.

24. *The Jetsons*, "Jetson's Nite Out," directed by Joseph Barbera and William Hanna, written by Harvey Bullock and R. S. Allen (story) and Tony Benedict (teleplay), Season 1, Episode 3, October 7, 1962.

25. *The Jetsons*, "Uniblab," directed by Joseph Barbera and William Hanna, written by Barry E. Blitzer (story) and Tony Benedict (teleplay), Season 1, Episode 10, November 25, 1962.

26. *The Jetsons*, "G. I. Jetson," directed by Joseph Barbera and William Hanna, written by Barry E. Blitzer (story) and Tony Benedict (teleplay), Season 1, Episode 19, January 27, 1963.

27. *The Jetsons*, "Las Venus," directed by Joseph Barbera and William Hanna, written by Barry E. Blitzer (story) and Tony Benedict (teleplay), Season 1, Episode 13, December 16, 1962.

28. *The Jetsons*, "Dude Planet," directed by Joseph Barbera and William Hanna, written by Walter Black (story) and Tony Benedict (teleplay), Season 1, Episode 22, February 17, 1963.

29. *The Jetsons*, "Rosie's Boyfriend," directed by Joseph Barbera and William Hanna, written by Walter Black (story) and Tony Benedict (teleplay), Season 1, Episode 8, November 11, 1962.

30. John Peel, *The Lost in Space Files* (Granada Hills, CA: Schuster and Schuster, 1987), 17.

31. *Lost in Space*, "The Reluctant Stowaway," directed by Anton Leader (as Tony Leader), written by Shimon Wincelberg (as S. Bar-David), Season 1, Episode 1, September 15, 1965.

32. Rick Du Brown, "'I Spy' and Western Off to a Good Start," *Santa Rosa [CA] Press Democrat*, September 16, 1965, 23.

33. *Gilligan's Island*, "Gilligan's Living Doll," directed by Leslie Goodwins, written by Bob Stevens, Season 2, Episode 21, February 10, 1966.

34. "Robot Spurns Method Acting," *Pasadena [CA] Independent Star-News*, March 13, 1966, 18C.

35. Paul Monroe, section introduction to *Lost in Space 25th Anniversary Tribute Book* by James Van Hise (Las Vegas: Pioneer Books, 1990), 130.

36. *Lost in Space*, "War of the Robots," directed by Sobey Martin, written by Barney Slater, Season 1, Episode 20, February 9, 1966.

37. *Lost in Space*, "The Ghost Planet," directed by Nathan Juran, written by Peter Packer, Season 2, Episode 3, September 25, 1966.

38. *Lost in Space*, "Deadliest of the Species," directed by Don Richardson, written by Robert Hamner, Season 3, Episode 11, November 22, 1967.

39. *Lost in Space*, "The Mechanical Men," directed by Seymour Robbie, written by Barney Slater, Season 2, Episode 28, April 5, 1967.

40. *The Man from U.N.C.L.E.*, "The Double Affair," directed by John Newland, written by Clyde Ware, Season 1, Episode 8, November 17, 1964.

41. Mark Phillips, "The History of *Lost in Space*," *Lost in Space Wikia*, http://lostinspace.wikia.com/wiki/The_History_of_Lost_In_Space.

42. *Lost in Space*, "Junkyard in Space," directed by Ezra Stone, written by Barney Slater, Season 3, Episode 24, March 6, 1968.

43. David Gerrold, *The Trouble with Tribbles* (New York: Ballantine Books, 1973), 19.

44. Peel, *The Lost in Space Files*, 11.

45. Danny Graydon, *The Jetsons: The Official Guide to the Cartoon Classic* (Philadelphia: Running Press, 2011), 77.

46. *Lancelot Link, Secret Chimp*, "The Reluctant Robot," produced by Stan Burns and Mike Marmer, Season 1, Episode 9, February 26, 1970.

Chapter 12

1. Jean-Noel Bassior, *Space Patrol: Missions of Daring in the Name of Early Television* (Jefferson, NC: McFarland, 2012), 70.

2. *Planet Patrol*, "The Telepathic Robot," directed by Frank Goulding, written by Roberta Leigh, Season 2, Episode 4, July 24, 1966.

3. Bassior, *Space Patrol*, 9.

4. Murray Robinson, "Planet Parenthood," *Collier's* (January 5, 1952), 31.

5. "Space Patrol Conquers Kids," *Life* (September 1, 1952), 81.

6. Cynthia Miller, "Domesticating Space: Science Fiction Series Come Home," in *Science Fiction Film, Television, and Adaptation Across the Screens*, edited by J. P. Telotte and Gerald Duchovnay (New York: Routledge, 2012), 9.

7. Don Stradley, "'Captain Video & His Video Rangers': This 1950s Sci-fi Adventure TV Show Was a Night Flight Fave," *Night Flight*, May 31, 2017, http://nightflight.com/captain-video-his-video-rangers-this-1950s-sci-fi-adventure-tv-show-was-a-night-flight-fave.

8. *The Honeymooners*, "TV or Not TV," directed by Frank Satenstein, written by Marvin Marx and Walter Stone, Season 1, Episode 1, October 1, 1955.

9. Bryon Thomas, "'Space Patrol': This Early 50s 'Space Opera' Set in the 30th Century Aired on 'Night Flight' in the Mid-80s," *Night Flight*, September 7, 2016, http://nightflight.com/space-patrol-this-early-50s-space-opera-set-in-the-30th-century-aired-on-night-flight-in-the-mid-80s/.

10. *Tales of Tomorrow*, "Read to Me, Herr Doktor," directed by Don Medford, written by Alvin Sapinsley, Season 2, Episode 30, March 20, 1953.

11. *Captain Video: Master of the Stratosphere*, directed by Spencer Gordon Bennet and Wallace Grissell, screenplay by Royal K. Cole, Sherman L. Lowe, and Joseph F. Poland (Columbia, 1951).

12. *Captain Video and His Video Rangers*, "I, Tobor," written by Isaac Asimov, November 2–December 7, 1953; "The Return of Tobor," written by Carey Wilber, Season 6, Episode 15, December 6, 1954; Herb Rau, "The Things That Happen Here," *Miami News*, January 18, 1956, 33.

13. *Rocky Jones, Space Ranger*, "Out of This World," directed by Hollingsworth Morse, written by Arthur Hoerl, Season 1, Episodes 34–36, October 12, 19, and 26, 1954.

14. Jack Gaver, "Surgeon, Ex-Showman Turn Out 'Rod Brown,'" *Louisville Courier-Journal*, May 16, 1954, 5–15; Rory Coker, "Rod Brown of the Rocket Rangers," *The Space Hero Files*, http://216.75.63.68/space/text/index.phtml.

15. *Rod Brown of the Rocket Rangers*, "The Robot Robber of Deimos," directed by George Gould, Season 1, Episode 30, November 7, 1953.

16. *Tom Corbett, Space Cadet* (radio series), "The Giants of Mercury," Season 1, Episodes 20–21, March 11 and 13, 1952.

17. Arthur Prager, "Bless My Collar Button If It Isn't Tom Swift," *American Heritage* (December 1976).

18. Tom Swift advertisement, *Hartford Courant*, November 28, 1954, 7.

19. "Chip Off the Old Block," *Time* (January 4, 1954), 66.

20. Victor Appleton II (Richard Sklar), *Tom Swift and His Giant Robot* (New York: Grosset and Dunlap, 1954), 17.

21. Appleton, *Tom Swift and His Giant Robot*, 114–15.

22. *Captain Z-Ro*, "Roger the Robot," directed by David Butler, written by Roy Steffens, Season 1, Episode 9, February 12, 1956.

23. *Captain Z-Ro*, "The Great Pyramid of Giza," directed by David Butler, written by Roy Steffens, Season 1, Episode 22, May 13, 1956.

24. *Johnny Jupiter*, "Duckworth and the Professor," directed by Howard Magwood, written by Jerome Coopersmith (screenplay and story) and Sam Rockingham (story), Season 2, Episode 8, October 25, 1953.

25. *The Invisible Boy*, directed by Herman Hoffman, written by Cyril Hume (MGM, 1957).

26. "Buck Rogers," *New Yorker*, December 22, 1934, 8–10.

27. "Vintage and Antique Space Toys," *Collector's Weekly*, https://www.collectorsweekly.com/toys/space.

28. "New Mechanical Models—Mechanical Man," *Meccano* (January 1931), 48–53.

29. "Doc Atomic," "Atomic Robot Man (Unknown/1949/Japan/5 inches)," *Astounding Artifacts*, June 16, 2009, http://astoundingartifacts.blogspot.com/2009/06/atomic-robot-man-unknown-1949.html.

30. Jim Bunte, Dave Hallman, and Heinz Mueller, *Vintage Toys: Robots and Space Toys* (Iola, IA: Krause Publications, 1999), 26, 63.

31. Olga Curtis, "Toy Industry Unveils 1954 Line Featuring 'Robert the Robot' and Streamlined Autos," *Lubbock Evening Journal*, March 9, 1932, 32.

32. "Tag Along," *Reno Gazette-Journal*, October 7, 1954, 17; "Robert the Robot Walks, Talks," *Popular Science* (December 1954), 93.

33. Scott Cragston, "Robert the Robot," *Robotapedia*, http://www.attackingmartian.com/robert_the_robot_extra.html.

34. Edward McGuire, "Santa'll Have More Talky Toys This Year," *Portsmouth [OH] Times*, September 22, 1955, 36.

35. "Robert the Robot from 1950's Toy Vintage Robot," https://youtu.be/FDQEV43DeFM.

36. William J. Felchner, "Ideal's Robot Commando: Collectible 1960s Toy," *Knoji*, https://vintages-antiques-collectibles.knoji.com/ideals-robot-commando-collectible-1960s-toy/.

37. Bunte, Hallman, and Mueller, *Vintage Toys*, 124.

38. "Rock 'Em Sock 'Em Robots," *Kingston [NY] Daily Freeman*, October 30, 1964, 16.

39. Jean-Marie Bouissou, "Manga: A Historical Overview," in *Manga: An Anthology of Global and Cultural Perspectives*, edited by Toni Johnson-Woods (London: Continuum, 2010), 24.

40. Simon Richmond, *The Rough Guide to Anime: Japan's Finest from Ghibli to Gankutsuō* (London: Rough Guides, 2009), 2.

41. Fred Ladd with Harvey Deneroff, *Astro Boy and Anime Come to the Americas: An Insider's View of the Birth of a Pop Culture Phenomenon* (Jefferson, NC: McFarland, 2009), 14.

42. *Astroboy*, "The Birth of Astroboy," directed by

Noboru Ishiguro and Osamu Tezuka, written by Osamu Tezuka, Episode 1, Season 1, September 7, 1963.

43. *Astro Boy*, directed by David Bowers, written by Timothy Harris (as Timothy Hyde Harris) and David Bowers (Imagi Animation Studios, 2009).

44. Lisa Hix, "Attack of the Vintage Toy Robots! Justin Pinchot on Japan's Coolest Postwar Export," *Collector's Weekly*, November 18, 2010, https://www.collectorsweekly.com/articles/attack-of-the-vintage-toy-robots-justin-pinchot-on-japans-coolest-postwar-export/.

45. *D.A.R.Y.L.*, directed by Simon Wincer, written by David Ambrose, Allan Scott, and Jeffrey Ellis (Paramount Pictures, 1985).

46. *Small Wonder*, "Vicki's Homecoming," directed by John Bowab, written by Howard Leeds, Season 1, Episode 1, September 7, 1985.

47. *Small Wonder*, "Everybody into the Pool," directed by Bob Claver, written by Judith Bustany and Dawn Aldredge, Season 3, Episode 2, September 19, 1987.

48. Seth McAvoy, *Not Quite Human #1: Batteries Not Included* (New York: An Archway Paperback, 1985).

49. *Not Quite Human*, directed by Steven Hilliard Stern, written by Alan Ormsby (Walt Disney Television, June 19, 1987).

50. Lorraine Kerslake, "The Iron Man," *The Ted Hughes Society*, http://thetedhughessociety.org/ironman/.

51. Ted Hughes, *The Iron Man: A Story in Five Nights* (London: Faber and Faber, 1968), chap. 5.

52. *The Iron Giant*, directed by Brad Bird, written by Tim McCanlies (screenplay) and Brad Bird (story) (Warner Bros. Animation, 1999).

53. John M. Miller, "The Iron Giant," *Turner Classic Movies*, http://www.tcm.com/this-month/article/148012%7C0/The-Iron-Giant.html.

54. *Robots*, directed by Chris Wedge and Carlos Saldanha, written by David Lindsay-Abaire, Lowell Ganz, and Babaloo Mandel (screenplay) and Ron Mita, Jim McClain, and David Lindsay-Abaire (story) (Twentieth Century Fox Animation, 2005).

55. *Luxo Jr.*, directed and written by John Lasseter, animated short (Pixar Animation Studios, 1986).

56. *WALL-E*, directed by Andrew Stanton, written by Andrew Stanton and Jim Reardon (screenplay) and Andrew Stanton and Pete Docter (story) (Pixar Animation Studios, 2008).

57. Scott Lobdell and Gus Vazquez, "Sunfire & Big Hero 6," *Sunfire & Big Hero 6* #1 (September 1998).

58. Chris Claremont and David Nakayama, "Big Hero 6 in Brave New Heroes!" *Big Hero 6* #1 (November 2008).

59. *Big Hero 6*, directed by Don Hall and Chris Williams, written by Jordan Roberts, Robert L. Baird, and Daniel Gerson (Walt Disney Animation Studios, 2014).

60. Janet and Isaac Asimov, *Norby, the Mixed-Up Robot* (New York: Walker, 1983).

61. Nancy Larrick, "The All-White World of Children's Books," *Saturday Review*, September 11, 1965, 63–65, 84–85.

62. Jon Scieszka, *Frank Einstein and the Antimatter Robot* (New York: Amulet Books, 2014).

63. James Patterson and Chris Grabenstein, *House of Robots* (New York and Boston: Little, Brown, 2014).

64. Carolyn Thompson, "My Classmate the Robot: New York Pupil Attends Remotely," *Elwood [IN] Call-Leader*, February 15, 2013, 3.

65. "Science Bob" Pflugfelder and Steve Hockensmith, *Nick and Tesla's Robot Army Rampage* (Philadelphia: Quirk Books, 2014).

66. Dan Pilkey, *Ricky Ricotta's Mighty Robot* (New York: Scholastic, 2000).

67. Chris Gall, *Awesome Dawson* (New York and Boston: Little, Brown, 2013).

68. Steve Foxe, *Transformers Rescue Bots: Meet Blur* (New York and Boston: Little, Brown, 2016); Brandon T. Snider, *Transformers Rescue Bots: Meet Quickshadow* (New York and Boston: Little, Brown, 2017).

69. Dan Yaccarino, *Doug Unplugged* (New York: Alfred A. Knopf, 2013).

70. Dan Yaccarino, *If I Had a Robot* (New York: Viking, 1996).

71. Ame Dyckman, *Boy + Bot* (New York: Alfred A. Knopf, 2012).

72. Peter Brown, "The Wild Robot Lives!" *Peter Brown Studio*, March 10, 2016, http://www.peterbrownstudio.com/uncategorized/the-wild-robot/.

73. Peter Brown, *The Wild Robot* (New York and Boston: Little, Brown, 2016).

Chapter 13

1. *Space Patrol*, "The Androids of Algol," directed by Dick Darley (as Dik Darley), written by Norman Jolley, Season 5, Episode 4, January 22, 1955; "Double Trouble," directed by Dick Darley (as Dik Darley), written by Norman Jolley, Season 5, Episode 5, January 29, 1955; "The Android Invasion," directed by Dick Darley (as Dik Darley), written by Norman Jolley, Season 5, Episode 6, February 5, 1955.

2. *The Twilight Zone*, "The Lonely," directed by Jack Smight, written by Rod Serling, Season 1, Episode 7, November 13, 1959.

3. *The Twilight Zone*, "The Lateness of the Hour," directed by Jack Smight, written by Rod Serling, Season 2, Episode 8, December 2, 1960.

4. *The Twilight Zone*, "I Sing the Body Electric," directed by William F. Claxton (as William Claxton) and James Sheldon, written by Ray Bradbury, Season 3, Episode 32, May 18, 1962.

5. *My Living Doll*, "Boy Meets Girl," directed by Lawrence Dobkin, written by Bill Kelsay and Al Martin, Season 1, Episode 1, September 27, 1964.

6. Noel Brown, *The Hollywood Family Film: A History, from Shirley Temple to Harry Potter* (London: I. B. Taurus, 2012), 111–12.

7. Paul Loukides and Linda K. Fuller, eds., *Beyond the Stars: Studies in American Popular Film*, Volume 4: *Locales in American Popular Film* (Bowling Green, OH: Bowling Green State University Popular Press, 1993), 124.

8. *Dr. Goldfoot and the Bikini Machine* advertisement, *Greenville [SC] News*, November 24, 1965, 9.

9. *Get Smart*, "Back to the Old Drawing Board," directed by Bruce Bilson, written by Gary Clarke (as C. F. L'Amoreaux), Season 1, Episode 19, January 29, 1966.

10. *The Man from U.N.C.L.E.*, "The Sort of Do-It-

Yourself Dreadful Affair," directed by E. Darrell Hallenbeck, written by Harlan Ellison, Season 3, Episode 2, September 23, 1966.

11. *Batman*, "The Joker's Last Laugh" and "The Joker's Epitaph," directed by Oscar Rudolph, written by Peter Rabe (story) and Lorenzo Semple, Jr. (teleplay), Season 2, Episodes 47–48, February 15–16, 1967.

12. *Star Trek*, "What Are Little Girls Made Of?" directed by James Goldstone, written by Robert Bloch, Season 1, Episode 7, October 20, 1966.

13. Nick Thomas, "Make Room for Sherry Jackson," *The Spectrum*, March 31, 2016, http://www.thespectrum.com/story/entertainment/2016/03/31/make-room-sherry-jackson/82343114/.

14. *Star Trek*, "I, Mudd," directed by Marc Daniels, written by Stephen Kandel, Season 2, Episode 8, November 3, 1967.

15. *Star Trek*, "Requiem for Methuselah," directed by Murray Golden, written by Jerome Bixby, Season 3, Episode 19, February 14, 1969.

16. *The Wild, Wild West*, "The Night of Miguelito's Revenge," directed by James B. Clark, written by Jerry Thomas, Season 4, Episode 12, December 12, 1968.

17. *Mission: Impossible*, "Robot," directed by Reza Badiyi (as Reza S. Badiyi), written by Bruce Geller, Season 4, Episode 9, November 30, 1969.

18. Philip K. Dick, "Imposter," *Astounding Science Fiction* (June 1953), 62.

19. Algis Budrys, *Who?* (New York: Pyramid Books, 1958).

20. J. T. McIntosh, "Made in U.S.A.," *Galaxy Science Fiction* (April 1953).

21. Bill Warren, *Keep Watching the Skies! American Science Fiction Movies of the Fifties: The 21st Century Edition* (Jefferson, NC: McFarland, 2009), 169.

22. *The Creation of the Humanoids*, directed by Wesley Barry (as Wesley E. Barry), written by Jay Simms (Genie Productions, 1962).

23. Ira Levin, *The Stepford Wives* (New York: Random House, 1972).

24. Lynn Hayes, "Slaves to the Scrub Bucket," *Baltimore Sun*, October 15, 1972, 61; Virgil Miller Newton, Jr., "Suspense for Late Fall," *Tampa Tribune*, December 10, 1972, 51.

25. *The Stepford Wives*, directed by Bryan Forbes, written by William Goldman (Palomar Pictures, 1975); *The Stepford Wives*, directed by Frank Oz, written by Paul Rudnick (Paramount Pictures, 2004).

26. *The Bionic Woman*, "Kill Oscar," directed by Alan Crosland, Jr. (as Alan Crosland), written by Arthur Rowe and Oliver Crawford (story) and Arthur Rowe (teleplay), Season 2, Episode 5, October 27, 1976; *The Six Million Dollar Man*, "Kill Oscar, Part 2," directed by Barry Crane, written by Arthur Rowe and Oliver Crawford (story) and W. T. Zacha (as William T. Zacha) (teleplay), Season 4, Episode 6, October 31, 1976; *The Bionic Woman*, "Kill Oscar, Part 3," directed by Alan Crosland, Jr. (as Alan Crosland), written by Arthur Rowe and Oliver Crawford (story) and Arthur Rowe (teleplay), Season 2, Episode 6, November 3, 1976.

27. *Holmes and Yoyo*, "Pilot," directed by Jackie Cooper, written by Lee Hewitt, Jack Sher, and Leonard Stern, Season 1, Episode 1, September 25, 1976.

28. Gaby Woods, *Edison's Eve: A Magical History of the Quest for Mechanical Life* (New York: Alfred A. Knopf, 2002), 157.

29. Gregg Rickman, "Philip K. Dick on *Blade Runner*: 'They Did Sight Stimulation on My Brain,'" in *Retrofitting* Blade Runner: *Issues in Ridley Scott's* Blade Runner *and Philip K. Dick's* Do Androids Dream of Electric Sheep?, edited by Judith Kerman (Bowling Green, OH: Bowling Green State University Press, 1991), 107.

30. Rickman, "Philip K. Dick on *Blade Runner*," 105.

31. William M. Kolb, "Script to Screen: *Blade Runner* in Perspective," in *Retrofitting* Blade Runner: *Issues in Ridley Scott's* Blade Runner *and Philip K. Dick's* Do Androids Dream of Electric Sheep?, edited by Judith Kerman (Bowling Green, OH: Bowling Green State University Press, 1991), 144.

32. *Blade Runner*, directed by Ridley Scott, written by Hampton Fancher and David Webb Peoples (as David Peoples) (Warner Bros., 1982); *Blade Runner: The Final Cut* (Warner Bros., 2007).

33. Steve Carper, "Subverting the Disaffected City: Cityscape in *Blade Runner*," in *Retrofitting* Blade Runner: *Issues in Ridley Scott's* Blade Runner *and Philip K. Dick's* Do Androids Dream of Electric Sheep?, edited by Judith Kerman (Bowling Green, OH: Bowling Green State University Press, 1991), 185.

34. Joseph Francavilla, "The Android as *Doppelgänger*," in *Retrofitting* Blade Runner: *Issues in Ridley Scott's* Blade Runner *and Philip K. Dick's* Do Androids Dream of Electric Sheep?, edited by Judith Kerman (Bowling Green, OH: Bowling Green State University Press, 1991), 8.

35. *Blade Runner 2049*, directed by Denis Villeneuve, written by Hampton Fancher and Michael Green (Alcon Entertainment, 2017).

36. *Android*, directed by Aaron Lipstadt, written by James Reigle and Don Keith Opper (as Don Opper) (New World Pictures, 1982).

37. *Star Trek: The Next Generation*, "The Naked Now," directed by Paul Lynch, written by John D. F. Black and D. C. Fontana (as J. Michael Bingham) (story) and D. C. Fontana (as J. Michael Bingham) (teleplay), Season 1, Episode 2, October 3, 1987.

38. *Star Trek: The Next Generation*, "The Measure of a Man," directed by Robert Scheerer, written by Melinda M. Snodgrass, Season 2, Episode 9, February 11, 1989.

39. Rachel Abramowitz, "Regarding Stanley," *Los Angeles Times*, May 6, 2001.

40. *A.I. Artificial Intelligence*, directed by Steven Spielberg, written by Steven Spielberg (screenplay) and Ian Watson (screen story) (Amblin Entertainment, 2001).

41. *Ex Machina*, directed and written by Alex Garland (Universal Pictures International [UPI], United Kingdom, 2015).

Chapter 14

1. Lenny Bruce, "The Berkeley Concert (1965)—Full Transcript," *Scraps from the Loft*, August 8, 2017, http://scrapsfromtheloft.com/2017/08/25/lenny-bruce-berkeley-concert-1965-full-transcript/.

2. Andrew Ferguson, *The Sex Doll: A History* (Jefferson City, NC: McFarland, 2010), 16.

3. Iwan Bloch, *The Sexual Life of Our Time in Its*

Relation to Modern Civilization, translated from the sixth German edition by M. Eden Paul (London: Rebman, 1908), 648.

4. Bloch, *The Sexual Life of Our Time*, 647.

5. Ferguson, *The Sex Doll*, 10; Mark D. Griffiths, "Droidian Slips: A Brief Look at Robot Fetishism," *Psychology Today*, June 13, 2014, https://www.psychologytoday.com/us/blog/in-excess/201406/droidianslips.

6. Ferguson, *The Sex Doll*, 19.

7. Madame B*** [Alphonse Momas], *La Femme Endormie* (trans. unknown) (Melbourne [Paris: J. Renold], 1899).

8. Patrick J. Kearney, "Notes Towards a Bibliography of Alphonse Momas (1846–1933)," *Scissors and Paste Bibliographies*, http://www.scissors-and-paste.net/Momas.html.

9. Madame B***, *La Femme Endormie*.

10. Brad Ricca, *Super Boys: The Amazing Adventures of Jerry Siegel and Joe Shuster—The Creators of Superman* (New York: St. Martin's Press, 2013), 29–30.

11. Ray Bradbury, "Changeling," *Super Science Stories* (July 1949), 103.

12. William F. Nolan, "The Joy of Living," *If* (August 1954), 38.

13. Fritz Leiber, *The Silver Eggheads* (New York: Ballantine Books, 1962), 104.

14. Toad the Wet Sprocket, "Fall Down," written by Glen Phillips, Todd Nichols and Toad the Wet Sprocket, on *Dulcinea* (1994).

15. Tanith Lee, *The Silver Metal Lover* (New York: Nelson Doubleday, 1981).

16. Steven Popkes, "The Birds of Isla Mujares," *The Magazine of Fantasy and Science Fiction* (January 2003).

17. Benjamin Rosenbaum, "Droplet," *The Magazine of Fantasy and Science Fiction* (July 2002).

18. John Clute, "Evan Hunter," *The Encyclopedia of Science Fiction*, October 19, 2017, http://www.sf-encyclopedia.com/entry/hunter_evan; "The Robot Lovers," *Internet Speculative Fiction Database*, http://www.isfdb.org/cgi-bin/title.cgi?1750057.

19. Dean Hudson [Evan Hunter], *The Robot Lovers* (San Diego, CA: Corinth Publications, 1966).

20. Raymond E. Banks [Ralph Burch], *Duplicate Lovers* (Encino, CA: World-Wide Publishing/Hustler, 1980).

21. *Westworld*, directed and written by Michael Crichton (MGM, 1973).

22. Brian Tallerico, "The Long, Weird History of the Westworld Franchise," *Vulture*, September 30, 2016, http://www.vulture.com/2016/09/westworld-franchise-long-weird-history.html.

23. *SexWorld*, directed by Anthony Spinelli, written by Dean Rogers and Anthony Spinelli (screenplay) and Anthony Spinelli (story) (Essex Pictures Company, 1978).

24. *Westworld*, directed by Jonathan Nolan, written by Jonathan Nolan, Lisa Joy, and Michael Crichton (story) and Jonathan Nolan and Lisa Joy (teleplay), Season 1, 2016.

25. Garry Vander Voort, "The Androbot B.O.B.," *Retroist*, August 28, 2014, https://www.retroist.com/2014/08/28/the-androbot-b-o-b/.

26. Daniel J. Ruby, "Computerized Personalized Robots," *Popular Science* (May 1983), 100.

27. "Androbot B. O. B. and Topo," https://youtu.be/jkOctWWsj-A.

28. *Amber Aroused*, directed and written by Mark Davis (Caballero Home Video, 1985).

29. *A.I.: Artificial Intelligence* (Video Series 2016), directed by Stills by Alan.

30. "Made of the Future," *Weird Science #5* (January 1951).

31. Home page, Scifidreamgirls.com.

32. *Scifidreamgirls* (Video Series 2013–2014), directed by Jack Kona, Scifidreamgirls.com.

33. Jason Lee, *Sex Robots: The Future of Desire* (London: Palgrave Macmillan, 2017), 8.

34. *La Femme-objet* (*Programmed for Pleasure*), directed and written by Claude Mulot (as Frédéric Lansac) (Alpha France, France, 1981).

35. *Cybersex*, directed by Brad Armstrong (Vivid, 1996).

36. *2040*, directed and written by Brad Armstrong (Wicked Pictures, 2009).

37. *Sexbots: Programmed for Pleasure*, directed by Brad Armstrong (Wicked Pictures, 2016).

38. *Gigahoes* (Video Series 2014–2015), directed by David Wright, written by Kevin Ryss Gilligan and Adam Lash.

39. RealDoll Shop, https://www.realdoll.com/shop/.

40. Albert Goldman, from the journalism of Lawrence Schiller, *Ladies and Gentlemen—Lenny Bruce!!* (New York: Random House, 1974), 47.

Chapter 15

1. "Ford Plant Worker Dies in Accident," *Detroit Free Press*, January 26, 1979, 11.

2. Tim Kiska, "Robot Victim's Kin Win $10 Million," *Detroit Free Press*, August 10, 1983, 1A, 9A.

3. Milton Lesser, "Slaves to the Metal Horde," *Imagination* (June 1954).

4. Walter M. Miller, Jr., "I Made You," *Astounding Science Fiction* (March 1954).

5. Harry Harrison, "War with the Robots," *Science Fiction Adventures* (June 1962).

6. *The Avengers*, "The Cybernauts," directed by Sidney Hayers, written by Philip Levene, Season 4, Episode 3, March 28, 1966.

7. *The Avengers*, "Return of the Cybernauts," directed by Robert Day, written by Philip Levene, Season 6, Episode 1, February 21, 1968.

8. *The New Avengers*, "The Last of the Cybernauts…?" directed by Sidney Hayers, written by Brian Clemens, Season 1, Episode 3, March 9, 1979.

9. *Doctor Who*, "Robot: Part One–Robot: Part Four," directed by Christopher Barry, written by Terrance Dicks, Season 12, Episodes 1–4, December 28, 1974, and January 4, 11, and 18, 1975.

10. Steve Ryfle, *Japan's Favorite Mon-star: The Unauthorized Biography of the Big G* (Toronto: ECW Press, 1998), 196.

11. *Gojira tai Mekagojira* (*Godzilla vs. Mechagodzilla*), directed by Jun Fukuda, written by Jun Fukuda and Hiroyasu Yamamura (screenplay) and Masami Fukushima and Shin'ichi Sekizawa (story) (Toho, 1974).

12. "Mechagodzilla (GvMGII)," *Godzilla Wikia*, https://godzilla.fandom.com/wiki/Mechagodzilla_(GvMGII).

13. *Gojira vs. Mekagojira* (*Godzilla vs. Mechagodzilla II*), directed by Takao Okawara, written by Wataru Mimura (Toho, 1993).

14. *The Terminator*, directed by James Cameron, written by James Cameron and Gale Anne Hurd (Hemdale Film Corporation, 1984).

15. *Terminator 2: Judgment Day*, directed by James Cameron, written by James Cameron and William Wisher (Carolco Pictures, 1991).

16. Russ Burlingame, "Decoding Convergence with Jeff King: The Finale," *Comicbook*, April 8, 2015, http://comicbook.com/2015/05/27/decoding-convergence-with-jeff-king-the-finale/.

17. Karen Walker, "Ultron: The Black Sheep of the Avengers Family," *Back Issue!* (February 2010), 24.

18. Steve Englehart, "Avengers," *Avengers* #134 (April 1975); Steve Englehart, "Avengers," *Avengers* #135 (May 1975).

19. Roy Thomas, "Avengers," *Avengers* #66 (July 1969).

20. *Avengers* #135 (May 1975).

21. *Avengers: Age of Ultron*, directed and written by Joss Whedon (Marvel Studios, 2015).

22. *X-Men: Days of Future Past*, directed by Bryan Singer, written by Simon Kinberg (screenplay) and Jane Goldman, Simon Kinberg, and Matthew Vaughn (story) (20th Century Fox, 2014).

23. *Battlestar Galactica* (TV Miniseries), directed by Michael Rymer, written by Glen A. Larson (as Christopher Eric James) and Ronald D. Moore, December 8–9, 2003.

24. *Battlestar Galactica*, "No Exit," directed by Gwyneth Horder-Payton, written by Ryan Mottesheard, Season 4, Episode 15, February 13, 2009.

25. Robert Joustra and Alissa Wilkinson, *How to Survive the Apocalypse: Zombies, Cylons, Faith, and Politics at the End of the World* (Grand Rapids, MI: William B. Eerdmans, 2016), 67.

26. Joustra and Wilkinson, *How to Survive the Apocalypse*, 69.

Chapter 16

1. Sue Fleiss, *Robots, Robots Everywhere!* (New York: Golden Books, 2013), 7, 8.

2. Jerry Osborne, "The Continuing Interview with Connie Francis," *DISCoveries* (September 1991), posted July 2006, http://www.freewebs.com/conniefrancis/connie10.htm.

3. Neil Strauss, "Hardly a Pocket Calculator: Kraftwerk's Studio Goes on Tour," *New York Times*, June 11, 1998; "Kraftwerk—The Robots (live)," 1978, https://youtu.be/okhQtoQFG5s.

4. Marissa R. Moss, "Frank Zappa's Raunchy Rock Opera 'Joe's Garage' Debuts in L.A.," *Rolling Stone* (September 29, 2008).

5. Charles McNulty, "Review: 'Yoshimi Battles the Pink Robots' Sounds Thrilling, at Least," *Los Angeles Times*, November 19, 2012.

6. Jason Schreurs, "Robot Band Recruits Human Member to Write Songs, Hilarity Ensues," *Vice* (March 13, 2015), https://noisey.vice.com/en_ca/article/6vm4yb/compressorhead-nomeansno-johnwright-interview.

7. Lanre Bakare, "Meet Z-Machines, Squarepusher's New Robot Band," *The Guardian*, April 4, 2014.

8. John Apfelbaum, "How Many People Collect Stamps," May 23, 2013, https://www.apfelbauminc.com/blog/how-many-people-collect-stamps.

9. Steve Carper, "Robots on Stamps," *Flying Cars and Food Pills*, https://www.flyingcarsandfoodpills.com/robots-on-stamps.

10. "U.S. Orbits 'Robot Astronaut' and Brings It Back Successfully," *Philadelphia Daily News*, September 13, 1961, 2.

11. Tariq Malik, "Japan Launches Talking 'Robot Astronaut' Kirobo into Space," *Space.com*, August 3, 2013, https://www.space.com/22235-japan-launches-talking-space-robot-astronaut.html.

12. Forrest Wickman, "Star Wars Is a Postmodern Masterpiece," *Slate*, December 13, 2015, http://www.slate.com/articles/arts/cover_story/2015/12/star_wars_is_a_pastiche_how_george_lucas_combined_flash_gordon_westerns.html.

13. Dale Pollack, *Skywalking: The Life and Films of George Lucas* (New York: Crown Publishing, 1983), 142.

14. Pollack, *Skywalking*, 324.

15. Alex Raymond, *Flash Gordon*, February 18, 1934.

16. Douglas Adams, *The Hitchhiker's Guide to the Galaxy* (New York: Harmony Books, 1980), 94.

17. Douglas Adams, *The Restaurant at the End of the Universe* (New York: Harmony Books, 1981), 137.

18. Steve Carper, "The Look of the Future—Robots," *Flying Cars and Food Pills*, https://www.flyingcarsandfoodpills.com/the-look-of-the-future---robots.

19. "Inductees," *Robot Hall of Fame*, http://www.robothalloffame.org/inductees.html.

20. Adam Aston, "How Robots Lost Their Way," *Bloomberg Businessweek*, November 30, 2003, https://www.bloomberg.com/news/articles/2003-11-30/how-robots-lost-their-way.

21. Jonathan Hennessey, *The Comic Book Story of Video Games: The Incredible History of the Electronic Gaming Revolution* (Berkeley, CA, and New York: Ten Speed Press, 2017), 123.

22. Cory Doctorow, "I, Robot," *The Infinite Matrix* (February 15, 2005).

23. Mary Robinette Kowal, "Kiss Me Twice," *Asimov's Science Fiction* (June 2011).

24. Martin L. Shoemaker, "Today I Am Paul," *Clarkesworld Magazine* (August 2015).

25. Rachel Swirsky, "Eros, Philia, Agape," *Tor.com* (March 3, 2009).

26. Rachel Swirsky, "Grand Jeté (The Great Leap)," *Subterranean Magazine* (Summer 2014).

27. Ian McDonald, "The Djinn's Wife," in *More Human Than Human: Stories of Androids, Robots, and Manufactured Humanity*, edited by Neil Clarke (New York: Night Shade Books, 2014).

28. Catherynne M. Valente, "Silently and Very Fast," in *More Human Than Human: Stories of Androids, Robots, and Manufactured Humanity*, edited by Neil Clarke (New York: Night Shade Books, 2014), 505.

29. Valente, "Silently and Very Fast," 512.

30. Valente, "Silently and Very Fast," 488.

Bibliography

Big Little Books

Link, Stanley. *Tiny Tim and the Mechanical Men*. Racine, WI: Whitman Publishing Company, 1937.

Nowlan, Phil, and Lt. Dick Calkins. *Buck Rogers: 25th Century A.D. vs. the Fiend of Space*. Racine, WI: Whitman Publishing Company, 1940.

Winterbotham, R. R. *Maximo the Amazing Superman and the Supermachine*. Racine, WI: Whitman Publishing Company, 1941.

Comic Books

"Adam Link in Business." *Weird Science-Fantasy* #29 (June 1955).

"The Adventures of 'Sparagus and Chubby." *The Funnies* #1 (October 1936).

"The Adventures of 'Spargus and Chubby." *The Funnies* #2 (November 1936); #3 (December 1936).

"All for Love." *Space Adventures* #8 (September 1953).

"Air-Sub, DX." *Amazing Mystery Funnies*, vol. 2, #8 (August 1939).

"The Amazing Man." *Amazing Man Comics* #21 (March 1941).

"Atom the Cat." *Atom the Cat* #14 (January 1959).

"Atomic Mouse." *Atomic Mouse* #10 (October 1954).

"Atomictot." *All Humor Comics* #8 (Winter 1947).

"Auro Lord of Jupiter." *Planet Comics* #46 (January 1947).

"The Avenger." *The Avenger* #3 (June–July 1955).

"Avengers." *Avengers* #55 (August 1968); #57 (October 1968); #58 (November 1968); #66 (July 1969); #67 (August 1969); #68 (September 1969); #134 (April 1975); #135 (May 1975).

"Axle and Cam on the Planet Meco." *Popeye* #30 (September–November 1954).

"Beauty and the Beast." *Mister Mystery* #11 (May–June 1953).

"Big Hero 6 in Brave New Heroes!" *Big Hero 6* #1 (November 2008).

"The Black Terror." *Exciting Comics* #25 (February 1943).

"The Blue Blaze." *Mystic Comics* #3 (June 1940).

"The Boy King and the Giant." *Clue Comics* #1 (Feb-

ruary 1943); #4 (June 1943); #5 (August 1943); #9 (Winter 1944).

"Bozo the Robot." *Smash Comics* #13 (August 1940).

"Break-Up!" *Tomb of Terror* #15 (May 1954).

Bridwell, E. Nelson. "In Memoriam: Otto Oscar Binder." *The Amazing World of DC Comics* #3 (November 1974).

"Captain Future." *Startling Comics* #1 (June 1940).

"Captain Marvel." *Captain Marvel Adventures* #78 (November 1947).

"Captain Marvel, Jr." *Captain Marvel, Jr.* #93 (January 1951).

"Captain Midnight." *Captain Midnight* #9 (June 1943).

"Captain Nelson Cole of the Solar Force." *Planet Comics* #8 (August 1940).

"Captain Science." *Captain Science* #3 (April 1951).

"Captain Venture and the Planet Princess." *Nickel Comics* #7 (August 9, 1940).

"Captain Video." *Captain Video* #3 (June 1951).

"Challengers of the Unknown." *Showcase* #7 (April 1957).

"Convergence." *Convergence* #0–8 (2015).

"Cosmo Cat." *Cosmo Cat* #10 (October 1947).

"Dick Cole, Wonder Boy." *Blue Bolt* #11 (April 1941).

"Dickie Dean, the Boy Inventor." *Silver Streak Comics* #3 (March 1940); #5 (June 1940); #6 (September 1940); #8 (March 1941); #10 (May 1941); #14 (September 1941); #15 (October 1941).

"Dr. Diamond." *Cat-Man Comics* #8 (July 1941).

"Dr. Mortal." *Weird Comics* #3 (June 1940).

"Donald Duck." *Donald Duck* #28 (March–April 1953).

"Doom Patrol." *My Greatest Adventure* #80 (June 1963).

"Dynamic Man." *Dynamic Comics* #1 (July 1941).

"Edison Bell, Young Inventor." *Blue Bolt* #1 (June 1940); #2 (July 1940); #3 (August 1940).

"Electro." *Marvel Mystery Comics* #4 (February 1940).

"The Fantom of the Fair." *Amazing Mystery Funnies*, vol. 2, #11 (November 1939).

"Federal Men." *New Comics* #9 (October 1936); #10 (November 1936).

"The First Man to Reach the Moon." *Lost Worlds* #6 (December 1952).

"Flexo the Rubber Man." *Mystic Comics* #1 (March 1940).

"Futura." *Planet Comics* #43 (July 1946); #46 (January 1947).

"The Giant Robots of Kilgor." *Fantastic Comics* #4 (March 1940).

"The Green Claw." *Silver Streak Comics* #6 (September 1940).

"The Hangman." *Hangman Comics* #4 (Fall 1942).

"Hugh Hazzard and His Iron Man." *Smash Comics* #1 (August 1939); #2 (September 1939).

"The Human Torch." *All-Winners Comics*, vol. 2, #1 (August 1948).

"The Human Torch." *Human Torch* #2 (Fall 1940).

"The Human Torch." *Marvel Comics* #1 (October 1939).

"The Human Torch." *Marvel Mystery Comics* #2 (December 1939); #6 (April 1940); #7 (May 1940).

"I, Robot." *Weird Science-Fantasy* #27 (January–February 1955).

"Inspector Dayton." *Jumbo Comics* #17 (July 1940).

"Invasion of the Love Robots." *Amazing Adventures* #4 (July 1951).

"Iron Skull." *Amazing-Man Comics* #5 (September 1939); #7 (November 1939); #8 (February 1940); #11 (April 1940); #15 (August 1940); #21 (March 1941).

"Jet Powers." *Jet Powers* #1 (January 1951).

"Judgment Day." *Incredible Science Fiction* #33 (February 1956).

"Judgment Day." *Weird Fantasy* #18 (March–April 1953).

"The Justice Society of America." *All Star Comics* #26 (Fall 1945).

"Landor, Maker of Monsters." *Speed Comics* #4 (January 1940).

"The Lost World." *Planet Comics* #46 (January 1947).

"Made of the Future." *Weird Science* #5 (January 1951).

"Magnus, Robot Fighter." *Magnus, Robot Fighter* #1 (February 1963); #2 (May 1963); #3 (August 1963); #7 (August 1964).

"Major Mars." *Exciting Comics* #1 (April 1940).

"Man O'Metal." *Reg'lar Fellers Heroic Comics* #7 (July 1941).

"Manowar, the White Streak." *Target Comics* #1 (February 1940).

"The Marvel Family." *The Marvel Family* #5 (October 1946); #25 (May 1949); #89 (January 1954).

"Marvex, the Super Robot." *Daring Mystery Comics* #3 (April 1940); #4 (May 1940).

"Mech Cadet Yu." *Mech Cadet Yu* #1–8 (2017–2018).

"Mekano." *Wonder Comics* #1 (May 1944).

"The Metal Men." *Showcase* #37 (March–April 1962); #39 (August 1962).

"The Monster Doll." *Forbidden Worlds* #1 (July–August 1951).

"Mysta of the Moon." *Planet Comics* #35 (March 1945); #43 (July 1946); #46 (January 1947).

"The Mystery of the Metal Menace." *All Star Comics* #26 (Fall 1945).

"The Omnipotent Robot." *Space Adventures* #36 (October 1960).

"Phantom Lady." *Phantom Lady* #13 (August 1947); #17 (April 1948).

"A Pound of Flesh." *Mysterious Adventures* #16 (October 1953).

"Power Nelson, Futureman." *Futureman* #6 (August 1940).

"Revolt of the Robots." *Space Adventures* #2 (September 1952).

"Revolt of the Robots." *Space Detective* #3 (February 1952).

"Robert Robot." *Ding Dong* #1 (1946); #5 (1947).

"Robert the Robot." *Crazy* #7 (June 1954).

"Robotman." *Detective Comics* #141 (November 1948).

"Robotman." *Star Spangled Comics* #7 (April 1942).

"Robotmen of the Lost Planet." *Robotmen of the Lost Planet* #1 (January 1952).

"Samson" (on cover). *Fantastic Comics* #3 (February 1940).

"Sergeant Spook." *Blue Bolt* #35 (April 1943).

"The Shadow." *The Shadow Battles the Robot Master* (1945).

"The Shield" (on cover). *Pep Comics* #1 (January 1940).

"The Silent One." *Men's Adventures* #21 (May 1953).

"Silver Streak." *Silver Streak Comics* #10 (May 1941).

"Steel Sterling." *Zip Comics* #1 (February 1940).

"Sunfire & Big Hero 6." *Sunfire & Big Hero 6* #1 (September 1998).

"Superman." *Action Comics* #242 (July 1958); #544 (June 1983); #545 (July 1983).

"Supermouse." *Supermouse, the Big Cheese* #39 (May 1957).

"The Surrogates." *The Surrogates* #1–5 (2015–2016).

"Survival of the Fittest," *Weird Tales of the Future* #1, May 1952.

"Three Comrades." *Thrilling Comics*, vol. 2, #2 (June 1940).

"Tom Kerry, District Attorney." *Big Shot Comics* #8 (December 1940).

"Tommy Tinkle." *Hit Comics* #6 (December 1940).

"The Trial of Adam Link." *Weird Science-Fantasy* #28 (April 1955).

"The Wizard." *Shield-Wizard Comics* #13 (Spring 1944).

"Wonderman." *Wonder Comics* #20 (October 1948).

"X-Men." *X-Men* #14 (November 1965); #57 (June 1969).

Comic Strips

- Robot appearance dates are given when known; when two names are given, the first is the writer, the second the artist.

Barnaby (daily). Ted Ferro and Jack Morely (November 11–December 24, 1947).

Beyond Mars (Sunday). Jack Williamson and Lee Elias (1953).

Brick Bradford (daily). William Ritt and Clarence Gray (February 13–March 16, 1940; September 1–October 18, 1952).

Buck Rogers in the 25th Century (daily). Phil Nowlan and Dick Calkins (July 30–August 9, 1929; December 12–28, 1929; January 24–27, 1930; July 25–September 14, 1935); Dick Calkins (April 25–September 6, 1938; July 16–August 10, 1946).

Buck Rogers in the 25th Century (Sunday). Phil Nowlan and Russell Keaton (November 2–30, 1930); Phil Nowlan and Rick Yager (October 28, 1934–January 6, 1935); Rick Yager (June 12–September 18, 1938).

Dash Dixon (weekly). Dean Carr [Larry Antonette] (March 26–May 21, 1936 [no strip May 14]).

Dinky Dinkerton, Secret Agent 6 7/8 (Sunday). Art Huhta (September 19–October 18, 1941).

Don Dixon and the Hidden Empire (Sunday). Bob Moore and Carl Pfeufer (January 1–May 14, 1939).

Ella Cinders (daily). Charlie Plumb and Fred Fox (June 20–July 9, 1949).

Felix the Cat (daily). Otto Messmer (March 24–May 31, 1941).

Flash Gordon (daily). Alex Raymond. February 18, 1934.

Flash Gordon (daily). Harry Harrison and Dan Berry (December 25, 1961–March 16, 1962).

Freckles and His Friends (daily). Merrill Blosser (November 9, 1958).

Fritz von Blitz the Kaiser's Hoodoo (Sunday). H[arry]. C[ornell]. Greening (August 18, 1918–February 23, 1919).

Hairbreadth Harry (daily). C[harles]. W[illiam]. Kahles (February 14–17, 1927; September 30–October 12, 1929; May 16–June 20, 1931).

Invisible Scarlet O'Neil (daily). Russell Stamm (March 23–May 5, 1942; September 15–November 5, 1943).

Joe Palooka (daily). Mo Leff (January 3–18, 1956).

John Carter of Mars (Sunday). John Coleman Burroughs (April 26–September 6, 1942).

Majah Moovie (Sunday). H[arry]. C[ornell]. Greening (August 15, 1915–January 16, 1916).

Mandrake the Magician (daily). Lee Falk and Phil Davis (July 2–October 13, 1945).

Mickey Mouse (daily). Bill Walsh and Floyd Gottfredson (July 31–November 11, 1944).

Nancy (Sunday). Ernie Bushmiller (March 21, 1943).

Oaky Doaks (daily). Ralph Briggs Fuller (November 7, 1955–January 7, 1946).

Percy (Sunday). H[arry]. C[ornell]. Greening (October 1, 1911–January 12, 1913; 1919–1920).

Percy in Stageland (Sunday). H[arry]. C[ornell]. Greening (1919).

Pogo (daily). Walt Kelly (February 13, 1952).

Professor Dodger and His Automatic Servant Girl (weekly). Hans Horina (November 10–December 1, 1907 [no strip November 24]).

Superman (daily). Curt Swan and Wayne Boring (December 15, 1958–April 2, 1959).

Superman (Sunday). Jerry Siegel and Leo Nowak (October 27–December 15, 1940).

Thimble Theater (daily). Forrest "Bud" Sagendorf (May 9–July 27, 1963).

Tillie the Toiler (daily). Russ Westover (June 20–November 4, 1933).

Tim Tyler's Luck (daily). Lyman Young (November 2–December 5, 1942).

Fiction

- Collected story series are referred to by date of book publication followed by the range of original publication dates.

- *Astounding* changed its name multiple times, from *Astounding Stories of Super-Science* to *Astounding Stories* to *Astounding Stories of Super-Science* to *Astounding Stories* to *Astounding Science-Fiction* to *Astounding Science Fiction*. *Amazing Stories* temporarily changed its name to *Amazing Science Fiction Stories* from 1958 to 1960.

Ackerman, Forrest J., ed. *The Gernsback Awards 1926*. Volume 1. Los Angeles: Triton Books, 1982.

Adams, Douglas. *The Hitchhiker's Guide to the Galaxy*. New York: Harmony Books, 1980.

_____. *The Restaurant at the End of the Universe*. New York: Harmony Books, 1981.

Aldiss, Brian. "Super-Toys Last All Summer Long." *Harper's Bazaar* (December 1969).

Anderson, Poul. "Quixote and the Windmill." *Astounding Science Fiction* (November 1950).

Anonymous. *Big Hero 6*. New York: Golden Books, 2013.

Anonymous. "The Clericomotor." *Detroit Free Press*, July 20, 1884.

Anonymous. "Mosco's Automaton." *Chambers's Journal*, July 17, 1869.

Appleton, Victor, II [Richard Sklar]. *Tom Swift and His Giant Robot*. New York: Grosset and Dunlap, 1954.

Asimov, Isaac, ed. *Before the Golden Age*. Garden City, NY: Doubleday, 1974.

_____. "The Bicentennial Man." In *Stellar #2*, edited by Judy-Lynn del Rey. New York: Ballantine Books, 1976.

_____. *The Caves of Steel*. Garden City, NY: Doubleday, 1954.

_____. *I, Robot* (collection). New York: Gnome Press, 1950. (Original stories published from 1939 to 1950.)

_____. *The Rest of the Robots* (collection). Garden City, NY: Doubleday, 1964. (Original stories published from 1942 to 1957.)

_____. *Robots and Empire*. Garden City, NY: Doubleday, 1985.

_____. *The Robots of Dawn*. Garden City, NY: Doubleday, 1983.

Asimov, Isaac, and Robert Silverberg. *The Positronic Man*. New York: Doubleday Foundation, 1992.

Asimov, Janet, and Isaac Asimov. *Norby, the Mixed-Up Robot*. New York: Walker, 1983.

Bates, Harry. "Farewell to the Master." *Astounding Science-Fiction* (October 1940).

Baum, L. Frank. *Ozma of Oz*. Chicago: Reilly & Britton Co., 1907.

_____. *The Road to Oz*. Chicago: Reilly & Lee Co., 1909.

_____. *The Surprising Adventures of the Magical Monarch of Mo and His People*. Chicago: M. A. Donohue, 1903.

_____. *Tik-Tok of Oz*. Chicago: Reilly & Britton Co., 1914.

_____. *The Tin Woodman of Oz*. Chicago: Reilly & Lee Co., 1918.

Bear, Elizabeth. "Tideline." *Asimov's Science Fiction* (June 2007).

Beckwith, O. "The Robot Master." *Air Wonder Stories* (October 1929).

Bellamy, Elizabeth W. "Ely's Automatic Housemaid." *The Black Cat* (December 1899).

Benét, Stephen Vincent. "Nightmare Number Three." *New Yorker* (July 27, 1935).

Bennett, Charles H. *The Surprising, Unheard of and Never-to-Be-Surpassed Adventures of Young Munchausen*. London: Routledge, Warne, and Routledge, 1865.

Binder, Eando. *Adam Link—Robot* (collection). New York: Warner Paperback Library, 1965. (Original stories published from 1939 to 1941.)

_____. "I, Robot." *Amazing Stories* (January 1939).

Bradbury, Ray. "Changeling." *Super Science Stories* (July 1949).

Brown, Peter. *The Wild Robot*. New York and Boston: Little, Brown, 2016.

Budrys, Algis. "Dream of Victory." *Amazing Stories* (August/September 1953).

_____. *Who?* New York: Pyramid Books, 1958.

Burch, Ralph. [Banks, Raymond E.] *Duplicate Lovers*. Encino, CA: World-Wide Publishing/Hustler, 1980.

Caidin, Martin. *Cyborg*. New York: Arbor House, 1972.

Campbell, John W. "The Last Evolution." *Amazing Stories* (August 1932).

_____. "When the Atoms Failed." *Amazing Stories* (January 1930).

Campbell, M. L. "The Automatic Maid-of-All Work: A Possible Tale of the Near Future." *Canadian Magazine* (July 1893).

Clarke, Neil, ed. *More Human Than Human: Stories of Androids, Robots, and Manufactured Humanity*. San Francisco: Night Shade Books, 2014.

Conner, Jeff, ed. *Zombies vs. Robots: This Means War!* San Diego: IDW Publishing, 2012.

Cook, William Wallace. *A Round Trip to the Year 2000; Or, A Flight Through Time*. The Argosy (July–November 1903).

Crichton, Michael. *The Terminal Man*. New York: Alfred A. Knopf, 1972.

Cummings, Ray. "The Exile of Time." *Astounding Stories* (April–July 1931).

del Rey, Lester. "Helen O'Loy." *Astounding Science-Fiction* (December 1938).

_____. "A Pound of Cure." In *Star Science Fiction 2*, edited by Frederik Pohl. New York: Ballantine Books, 1953.

Dick, Philip K. "The Defenders." *Galaxy Science Fiction* (January 1953).

_____. *Do Androids Dream of Electric Sheep?* Garden City, NY: Doubleday, 1968.

_____. "Imposter." *Astounding Science Fiction* (June 1953).

_____. "James P. Crow." *Planet Stories* (May 1954).

_____. "Second Variety." *Space Science Fiction* (May 1953).

_____. *We Can Build You*. New York: DAW Books, 1972.

Dickson, Gordon R. "Robots Are Nice?" *Galaxy Science Fiction* (October 1957).

Doctorow, Cory. "I, Robot." *The Infinite Matrix* (February 15, 2005).

Dyckman, Ame. *Boy + Bot*. New York: Alfred A. Knopf, 2012.

Ellis, Edwin Sylvester. *The Steam Man of the Prairies*. New York: Beadle's American Novels 45, 1868. Released as *The Huge Hunter; or, The Steam Man of the Prairies*. New York: Pocket Novels 40, 1876. Posted March 21, 2009. http://www.gutenberg.org/files/7506/7506-h/7506-h.htm.

Fezandié, Clement. "The Secret of the Tel-Automaton." *Science and Invention* (June 1922).

Flagg, Francis. "The Synthetic Monster." *Wonder Stories* (December 1931).

Foxe, Steve. *Transformers Rescue Bots: Meet Blur*. New York and Boston: Little, Brown, 2016.

Fuller, Alice W. "A Wife Manufactured to Order." *The Arena* (July 1895).

G. H. P. [George Haven Putnam]. *The Artificial Mother: A Marital Fantasy*. New York: G. P. Putnam's Sons, 1894.

Gall, Chris. *Awesome Dawson*. New York and Boston: Little, Brown, 2013.

Galula, Abner J. "Automaton." *Amazing Stories* (November 1931).

Gault, William Campbell. "Made to Measure." *Galaxy Science Fiction* (January 1951).

Gernsback, Hugo. "Ralph 124C 41+." *Modern Electrics* (April–July 1911, September 1911, December 1911, January–March 1912).

Greenberg, Martin, ed. *The Robot and the Man*. New York: Gnome Press, 1953.

Gunn, James. "Little Orphan Android." *Galaxy Science Fiction* (September 1955).

Hamilton, Edmond. "The Metal Giants." *Weird Tales* (December 1926).

_____. "The Reign of the Robots." *Wonder Stories* (December 1931).

Harrison, Harry. "War with the Robots." *Science Fiction Adventures* (June 1962).

Hatke, Ben. *Little Robot*. New York: First Second Books, 2015.

Hawthorne, Julian. "An Automatic Enigma." *Belgravia* (May 1878).

Hoffmann, E. T. A. "The Sandman." Translated by

J. T. Bealby. In *The Best Tales of Hoffmann*, edited by E. F. Bleiler. New York: Dover, 1967.

Holmes, H. H. [Anthony Boucher]. "Q.U.R." *Astounding Science-Fiction* (March 1943).

_____. "Robinc." *Astounding Science-Fiction* (September 1943).

Horton, Rich, and Sean Wallace, eds. *Robots: The Recent A.I.* Gaithersburg, MD: Prime Books, 2012.

Hudson, Dean [Evan Hunter]. *The Robot Lovers*. San Diego, CA: Corinth Publications, 1966.

Huff, Melbourne. "The Robot Terror." *Scientific Detective Monthly* (March 1930).

Hughes, Ted. *The Iron Man: A Story in Five Nights*. London: Faber & Faber, 1968.

Jerome, Jerome K. *Novel Notes*. New York: Henry Holt, 1893.

Jones, Neil R. "The Jameson Satellite." *Amazing Stories* (July 1931).

Kelleam, Joseph E. "Rust." *Astounding Science-Fiction* (October 1939).

Keller, David H., MD. "Air Lines." *Amazing Stories* (January 1930).

_____. "The Psychophonic Nurse." *Amazing Stories* (November 1928).

_____. "The Threat of the Robot." *Science Wonder Stories* (June 1929).

Kellett, Ernest Edward. "A New Frankenstein." In *A Corner in Sleep and Other Impossibilities*. London: Jarrold & Sons, 1900.

Keyes, Daniel. "Robot—Unwanted." *Other Worlds Science Stories* (June 1952).

Kline, Otis Adelbert. "The Revenge of the Robot." *Thrilling Wonder Stories* (August 1936).

Kowal, Mary Robinette. "Kiss Me Twice." *Asimov's Science Fiction* (June 2011).

Laumer, Keith. *The Compleat Bolo* (collection). New York: Baen Books, 1990. (Original stories published from 1963 to 1986.)

Lee, Tanith. *The Silver Metal Lover*. New York: Nelson Doubleday, 1981.

Leiber, Fritz. "The Creature from Cleveland Depths." *Galaxy Science Fiction* (December 1962).

_____. *The Silver Eggheads*. New York: Ballantine Books, 1962.

Lesser, Milton. "Slaves to the Metal Horde." *Imagination* (June 1954).

Levin, Ira. *The Stepford Wives*. New York: Random House, 1972.

Liddell, C. H. [Henry Kuttner and C. L. Moore]. "Android." *The Magazine of Fantasy and Science Fiction* (June 1951).

Luper, Eric. *The Mysterious Moonstone*. New York: Scholastic, 2000.

Madame B*** [Alphonse Momas]. *La Femme Endormie*. Trans. unknown. Melbourne [Paris: J. Renold], 1899.

McAvoy, Seth. *Not Quite Human #1: Batteries Not Included*. New York: An Archway Paperback, 1985.

McDonald, Ian. "The Djinn's Wife." *Asimov's Science Fiction* (July 2006).

McGivern, William P. "Sidney, the Screwloose Robot." *Fantastic Adventures* (June 1941).

McIntosh, J. T. "Made in U.S.A." *Galaxy Science Fiction* (April 1953).

Merritt, A[braham]. *The Metal Monster*. Argosy All-Story Weekly (August 7–September 25, 1920).

Miller, Walter M., Jr. "I Made You." *Astounding Science Fiction* (March 1954).

Moore, C. L. "No Woman Born." *Astounding Science-Fiction* (December 1944).

Nolan, William F. "The Joy of Living." *If* (August 1954).

"Noname" [Harry Enton]. *The Steam Man of the Plains; or, The Terror of the West*. The Five Cent Wide Awake Library, January 24, 1883.

"Noname" [Luis Senarens]. *The Electric Man: or, Frank Reade, Jr., in Australia*. Boys of New York, October 10, 1886.

_____. *Frank Reade, Jr., and His New Steam Man; or, The Young Inventor's Trip to the Far West*. Frank Reade Library, September 24, 1892.

_____. *Frank Reade, Jr., and His Steam Wonder*. Boys of New York, February 4–April 29, 1882.

Nowlan, Philip Francis. "Armageddon—2419 A.D." *Amazing Stories* (August 1928).

O'Connor, William Douglas. "The Brazen Android." *Atlantic Monthly* (April–May 1891).

Oliver, Chad. "The Life Game." *Thrilling Wonder Stories* (June 1953).

Padgett, Lewis [Henry Kuttner]. "The Proud Robot." *Astounding Science-Fiction* (October 1943).

Padgett, Lewis [Henry Kuttner and C. L. Moore]. "Open Secret." *Astounding Science-Fiction* (April 1943).

_____. "The Twonky." *Astounding Science-Fiction* (September 1942).

Parisien, Dominik, and Navah Wolfe, eds. *Robots vs. Fairies*. New York: Saga Press, 2018.

Patterson, James, and Chris Grabenstein. *House of Robots*. New York and Boston: Little, Brown, 2014.

Perkins, Frederick Beecher. "The Man-ufactory." In *Devil-Puzzlers and Other Studies*. New York: G. P. Putnam's Sons, 1877.

Pflugfelder, "Science Bob," and Steve Hockensmith. *Nick and Tesla's Robot Army Rampage*. Philadelphia: Quirk Books, 2014.

Pilkey, Dan. *Ricky Ricotta's Mighty Robot*. New York: Scholastic, 2000.

Pohl, Frederik. "The Midas Plague." *Galaxy Science Fiction* (April 1954).

Popkes, Steven. "The Birds of Isla Mujares." *The Magazine of Fantasy and Science Fiction* (January 2003).

Rayer, F. G. "Deus Ex Machina." *New Worlds* (Winter 1950).

Reade, Philip. *Tom Edison Jr's Electric Mule; or, The Snorting Wonder of the Plains*. The Nugget Library #128, January 14, 1892.

_____. *Tom Edison Jr's Sky-Scraping Trip; or, Over*

the Wild West Like a Flying Squirrel. The Nugget Library #102, July 16, 1891.

Reeve, Arthur B., and John W. Grey. The Master Mystery. New York: Grosset & Dunlap, 1919.

Rideaux, Charles de Balzac. "The Rebel Robots." Scoops (February 10, 1934).

Robeson, Kenneth [Ryerson Johnson]. "The Fantastic Island." Doc Savage (December 1935).

Rosenbaum, Benjamin. "Droplet." The Magazine of Fantasy and Science Fiction (July 2002).

Russell, Eric Frank. "Jay Score." Astounding Science-Fiction (May 1941).

Saberhagen, Fred. Berserker (collection). New York: Ballantine Books, 1967. (Original stories published from 1963 to 1966.)

Schlossel, J[oseph]. "To the Moon by Proxy." Amazing Stories (October 1928).

Scieszka, Jon. Frank Einstein and the Antimatter Robot. New York: Amulet Books, 2014.

Seabright, Idris [Margaret St. Clair]. "Short in the Chest." Fantastic Universe (July 1954).

Shoemaker, Martin L. "Today I Am Paul." Clarkesworld Magazine (August 2015).

Simak, Clifford D. City (collection). New York: Gnome Press, 1952. (Original stories published from 1944 to 1951.)

Snider, Brandon T. Transformers Rescue Bots: Meet Quickshadow. New York and Boston: Little, Brown, 2017.

Stannard, W. W. "Mr. Corndropper's Hired Man." The Black Cat (October 1900).

Steffanson, Con [Ron Goulart]. The Lion Men of Mongo. New York: Avon Books, 1974.

Stuart, Don A. [John Campbell]. "Night." Astounding Stories (October 1935).

_____. "Twilight." Astounding Stories (November 1934).

Swirsky, Rachel. "Eros, Philia, Agape." Tor.com (March 3, 2009).

_____. "Grand Jeté (The Great Leap)." Subterranean Magazine (Summer 2014).

Tofte, Arthur. "Revolt of the Robots." Fantastic Adventures (May 1939).

Toombs, Robert. Electric Bob's Big Black Ostrich; or, Lost on the Desert. New York Five Cent Library #55, August 26, 1893.

Valente, Catherynne M. "Silently and Very Fast." Clarkesworld Magazine (October–December 2011).

VanderMeer, Jeff, and Ann VanderMeer, eds. The Big Book of Science Fiction. New York: Vintage, 2016.

Vanny, Jim. "The Radium Master." Wonder Stories (August 1930).

Villiers de L'Isle Adam. The Future Eve [1886]. Translated by Florence Crewe-Jones. Argosy All-Story Weekly (December 18, 1926–January 22, 1927).

_____. The Future Eve [1886]. Translated by Florence Crewe-Jones. New York: Baen Books, 2013.

_____. Tomorrow's Eve [1886]. Translated by Robert Martin Adams. Urbana: University of Illinois Press, 2011.

Vincent, Harl. "Rex." Astounding Stories (June 1934).

_____. "Terrors Unseen." Astounding Stories (March 1931).

Vonnegut, Kurt, Jr. Utopia 14 [Player Piano]. New York: Bantam Books, 1954.

White, Wade Albert. The Adventurer's Guide to Successful Escapes. New York and Boston: Little, Brown, 2016.

Williams, Robert Moore. "Robot's Return." Astounding Science-Fiction (September 1938).

Williamson, Jack. "After Worlds End." Marvel Science Stories (February 1939).

_____. "With Folded Hands…" Astounding Science Fiction (July 1947).

Wilson, Daniel H., and John Joseph Adams, eds. Robot Uprisings. New York: Vintage Books, 2014.

Wolf, Mari. "Robots of the World! Arise!" If: Worlds of Science Fiction (July 1952).

Wolfe, Gary K., ed. American Science Fiction: Nine Classic Novels of the 1950s (two-volume boxed set). New York: Library of America, 2012.

Yaccarino, Dan. Doug Unplugged. New York: Alfred A. Knopf, 2013.

_____. If I Had a Robot. New York: Viking, 1996.

Young, Robert F. "Doll-Friend." Amazing Science Fiction Stories (July 1959).

Journals

Jentsch, Ernst. "On the Psychology of the Uncanny" [1906]. Translated by Roy Sellars. Angelaki 2, no. 1 (1995): 7–16.

Klass, Philip. "'The Lady Automaton' by E. E. Kellett: A Pygmalion Source?" SHAW: The Annual of Bernard Shaw Studies 2 (1982): 75–100.

Landon, Brooks. "Slipstream Then, Slipstream Now: The Curious Connections between William Douglas O'Connor's 'The Brazen Android' and Michael Cunningham's Specimen Days." Science Fiction Studies 38, no. 1 (March 2011).

Oancea, Ana. "Edison's Modern Legend in Villiers' L'Eve Future." Nordlit 28 (2011): 173–87.

Movies (U.S. feature-length film, except as indicated)

A.I. Artificial Intelligence. Directed by Steven Spielberg. Written by Steven Spielberg (screenplay) and Ian Watson (screen story). Amblin Entertainment. 2001.

Amber Aroused. Directed and written by Mark Davis. Caballero Home Video. 1985.

Android. Directed by Aaron Lipstadt. Written by James Reigle and Don Keith Opper (as Don Opper). New World Pictures. 1982.

An Animated Doll. Produced by George Spoor and G. M. Anderson. Animated short. Essanay. 1908.

Astro Boy. Directed by David Bowers. Written by

Timothy Harris (as Timothy Hyde Harris) and David Bowers. Imagi Animation Studios. 2009.

Avengers: Age of Ultron. Directed and written by Joss Whedon. Marvel Studios. 2015.

The Bicentennial Man. Directed by Chris Columbus. Written by Nicholas Kazan. Columbia Pictures. 1999.

Big Hero 6. Directed by Don Hall and Chris Williams. Written by Jordan Roberts, Robert L. Baird, and Daniel Gerson. Walt Disney Animation Studios. 2014.

Blade Runner. Directed by Ridley Scott. Written by Hampton Fancher and David Webb Peoples (as David Peoples). Warner Bros. 1982; *Blade Runner: The Final Cut*. Warner Bros. 2007.

Blade Runner 2049. Directed by Denis Villeneuve. Written by Hampton Fancher and Michael Green. Alcon Entertainment. 2017.

Captain Video: Master of the Stratosphere. Directed by Spencer Gordon Bennet and Wallace Grissell. Screenplay by Royal K. Cole, Sherman L. Lowe, and Joseph F. Poland. Serial. Columbia. 1951.

Chikyu Boeigun (*The Mysterians*). Directed by Inoshiro Honda. Written by Takeshi Kimura (screenplay) and Jôjirô Okami and Shigeru Kayama (story). Toho. Japan. 1957.

A Clever Dummy. Directed by Herman C. Raymaker (as H. Raymaker). Short. Keystone. 1917.

The Colossus of New York. Directed by Eugène Lourié (as Eugene Lourie). Written by Thelma Schnee (screenplay) and Willis Goldbeck (story). William Alland Productions. 1958.

The Creation of the Humanoids. Directed by Wesley Barry (as Wesley E. Barry). Written by Jay Simms. Genie Productions. 1962.

Cybersex. Directed by Brad Armstrong. Vivid. 1996.

D.A.R.Y.L. Directed by Simon Wincer. Written by David Ambrose, Allan Scott, and Jeffrey Ellis. Paramount Pictures. 1985.

The Day the Earth Stood Still. Directed by Robert Wise. Written by Edmund H. North. Twentieth Century Fox. 1951.

Devil Girl from Mars. Directed by David MacDonald. Written by James Eastwood. Danziger Productions. United Kingdom. 1954.

Dr. Goldfoot and the Bikini Machine. Directed by Norman Taurog. Written by Elwood Ullman and Robert Kaufman (screenplay) and James H. Nicholson (as James Hartford) (story). American International Pictures. 1965.

Dr. Smith's Automaton. Short. Pathé Frères. France. 1910.

The Electric Leg. Directed by Percy Snow. Short. Clarendon. United Kingdom. 1912.

Ex Machina. Directed and written by Alex Garland. Universal Pictures International (UPI). United Kingdom. 2015.

The Fairylogue and Radio-Plays. Directed by Francis Boggs and Otis Turner. Written by L. Frank Baum. Short. Radio Play Corporation of America. 1908.

Forbidden Planet. Directed by Fred M. Wilcox (as Fred McLeod Wilcox). Written by Cyril Hume. MGM. 1956.

Gigahoes (Video Series). Directed by David Wright. Written by Kevin Ryss Gilligan and Adam Lash. 2014–2015.

Gog. Directed by Herbert L. Strock. Written by Tom Taggart (screenplay), Richard G. Taylor (additional dialogue), and Ivan Tors (story). Ivan Tors Productions. 1954.

Gojira tai Mekagojira (*Godzilla vs. Mechagodzilla*). Directed by Jun Fukuda. Written by Jun Fukuda and Hiroyasu Yamamura (screenplay) and Masami Fukushima and Shin'ichi Sekizawa (story). Toho. 1974.

Gojira vs. Mekagojira (*Godzilla vs. Mechagodzilla II*). Directed by Takao Okawara. Written by Wataru Mimura. Toho. 1993.

The Hitchhiker's Guide to the Galaxy. Directed by Garth Jennings. Written by Douglas Adams and Karey Kirkpatrick. 2005.

I, Robot. Directed by Alex Proyas. Written by Jeff Vintar and Akiva Goldsman (screenplay) and Jeff Vintar (screen story). Twentieth Century Fox. 2004.

The Invisible Boy. Directed by Herman Hoffman. Written by Cyril Hume. MGM. 1957.

The Iron Giant. Directed by Brad Bird. Written by Tim McCanlies (screenplay) and Brad Bird (story). Warner Bros. Animation. 1999.

Kronos. Directed by Kurt Neumann. Regal. 1957.

La Femme-objet (*Programmed for Pleasure*). Directed and written by Claude Mulot (as Frédéric Lansac). Alpha France. France. 1981.

La momia azteca contra el robot humano (*The Aztec Mummy Against the Humanoid Robot*). Directed by Rafael Portillo. Written by Guillermo Calderón (as Guillermo Calderon S.) and Alfredo Salazar. Mexico. 1958.

L'Uomo Meccanico (*The Mechanical Man*). Directed by André Deed. Milano Film. Italy. 1921.

Luxo Jr. Directed and written by John Lasseter. Animated short. Pixar Animation Studios. 1986.

The Master Mystery. Directed by Burton L. King (as Burton King). Written by Arthur B. Reeve and Charles Logue (as Charles A. Logue). Serial. Octagon Films. 1920.

The Mechanical Monsters. Directed by Dave Fleischer. Written by Izzy Sparber (as Isidore Sparber) and Seymour Kneitel. Animated short. Fleischer Studios. 1941.

Mekagojira no gyakushu (*Terror of Mechagodzilla*). Directed by Ishirô Honda. Written by Yukiko Takayama. Toho. 1978.

Metropolis. Directed by Fritz Lang. Written by Thea von Harbou. UFA. Germany. 1927.

Mother Riley Meets the Vampire (a.k.a. *Vampire Over London*). Directed by John Gilling. Written by Val Valentine. Fernwood Productions. United Kingdom. 1952.

The Phantom Empire. Directed by Otto Brower and B. Reeves Eason (as Breezy Eason). Written by Wallace MacDonald, Gerald Geraghty, and Hy Freedman (as H. Freedman). Serial. Mascot. 1935.

"Robby the Robot: Engineering a Sci-Fi Icon." Produced by Jonathan Strailey. On *Forbidden Planet: 50th Anniversary Edition* DVD. Warner Brothers. 2006.

Robot & Frank. Directed by Jake Schreier. Written by Christopher D. Ford. Stage 6 Films. 2012.

Robots. Directed by Chris Wedge and Carlos Saldanha. Written by David Lindsay-Abaire, Lowell Ganz, and Babaloo Mandel (screenplay) and Ron Mita, Jim McClain, and David Lindsay-Abaire (story). Twentieth Century Fox Animation, 2005.

Robot Stories. Directed and written by Greg Pak. Pak Film. 2003.

The Rubber Man. Produced by Siegmund Lubin. Short. Lubin. 1909.

Sammy's Automaton. Short. Eclipse. France. 1914.

Sexbots: Programmed for Pleasure. Directed by Brad Armstrong. Wicked Pictures. 2016.

Sex Kittens Go to College. Directed by Albert Zugsmith. Written by Robert Hill (screenplay) and Albert Zugsmith (story). Allied Artists Pictures. 1960.

SexWorld. Directed by Anthony Spinelli. Written by Dean Rogers and Anthony Spinelli (screenplay) and Anthony Spinelli (story). Essex Pictures Company. 1978.

Star Wars. Directed and written by George Lucas. Lucasfilm. 1977.

The Stepford Wives. Directed by Bryan Forbes. Written by William Goldman. Palomar Pictures. 1975.

The Stepford Wives. Directed by Frank Oz. Written by Paul Rudnick. Paramount Pictures. 2004.

Surrogates. Directed by Jonathan Mostow. Written by John Brancato and Michael Ferris. Touchstone Pictures. 2009.

Target Earth. Directed by Sherman A. Rose. Written by William Raynor (as Bill Raynor) (screenplay) and James H. Nicholson (as James Nicholson) and Wyott Ordung (original screenplay). Abtcon Pictures. 1954.

The Terminator. Directed by James Cameron. Written by James Cameron and Gale Anne Hurd. Hemdale Film Corporation. 1984.

Terminator 2: Judgment Day. Directed by James Cameron. Written by James Cameron and William Wisher. Carolco Pictures. 1991.

THX 1138. Directed and written by George Lucas. American Zoetrope. 1971.

Tobor the Great. Directed by Lee Sholem. Written by Philip MacDonald (screenplay) and Carl Dudley (story). Dudley Pictures. 1954.

2040. Directed and written by Brad Armstrong. Wicked Pictures. 2009.

21–87. Directed by Arthur Lipsett. Short. National Film Board of Canada. 1963.

Undersea Kingdom. Directed by B. Reeves Eason and Joseph Kane. Written by John Rathmell, Maurice Geraghty, and Oliver Drake (screenplay) and Tracy Knight and John Rathmell (original story). Serial. Republic. 1936.

WALL-E. Directed by Andrew Stanton. Written by Andrew Stanton and Jim Reardon (screenplay) and Andrew Stanton and Pete Docter (story). Pixar Animation Studios. 2008.

Westworld. Directed and written by Michael Crichton. MGM. 1973.

X-Men: Days of Future Past. Directed by Bryan Singer. Written by Simon Kinberg (screenplay) and Jane Goldman, Simon Kinberg, and Matthew Vaughn (story). 20th Century Fox. 2014.

Music

Albarn, Damon. "Everyday Robots." Written by Damon Albarn. On *Everyday Robots.* 2014.

B-52s. "Love in the Year 3000." Written by Keith Strickland, Kate Pierson, Fred Schneider, and Cindy Wilson. On *Funplex.* 2008.

Big Boy Bloater and the Limits. "Robot Girlfriend." Written by Big Boy Bloater. On *Luxury Hobo.* 2016.

BlöödHag. "Isaac Asimov." Written by BlöödHag. On *Necrotic Bibliophilia.* 2001.

The Buggles. "I Love You (Miss Robot)." Written by the Buggles. On *The Age of Plastic.* 1980.

Cage The Elephant. "Tiny Little Robots" Written by Cage The Elephant. On *Cage The Elephant.* 2008.

Compressorhead. "My Girlfriend's a Robot." Written by Tom Holliston. On *Party Machine.* 2017.

Cyrus, Miley. "Robot." Written by Miley Cyrus and John Shanks. On *Can't Be Tamed.* 2010.

Daft Punk. "Robot Rock." Written by Thomas Bangalter, Guy-Manuel de Homem-Christo, and Kae Williams. On *Human After All.* 2005.

Dance Gavin Dance. "Young Robot." Written by Jon Mess, Matt Mingus, Will Swan, Tim Feerick, and Tilian Pearson. On *Mothership.* 2016.

Duran Duran. "Electric Barbarella." Written by Nick Rhodes, Warren Cuccurullo, and Simon Le Bon. On *Medazzaland.* 1997.

Electric Light Orchestra (as ELO). "Yours Truly 2095." Written by Jeff Lynne. On *Time.* 1981.

Eminem. "Rap God." Written by Marshall Mathers, Bigram Zayas, Matthew "Filthy" Delgiorno, Stephen Hacker, Douglas Davis, Richard Walters, Dania Birks, Juana Burns, Juanita Lee, Fatima Shaheed, and Kim Nazel. On *The Marshall Mathers LP2.* 2013.

Fantastic Plastic Machine. "Electric Lady Land." Written by Tomoyuki Tanaka and Masaki Tsurugi. On *Luxury.* 1998.

The Flaming Lips. "Fight Test." Written by the Flaming Lips, Dave Fridmann, and Cat Stevens; "One More Robot/Sympathy 3000–21," "Yoshimi Battles the Pink Robots, Pt. 1," and "Yoshimi Battles the Pink Robots, Pt. 2." Written by the Flaming Lips. On *Yoshimi Battles the Pink Robots.* 2002.

Flight of the Conchords. "Robots." Written by Jemaine Clement and Bret McKenzie. On *Flight of the Conchords.* 2008.

49th Octave. "Bloody Expensive (Three Laws of

Robotics)." Music by 49th Octave. Three Laws of Robotics quoted from Isaac Asimov and from David Langford. Digital single. 2017.

Francis, Connie. "Robot Man." Written by George Goehring and Sylvia Dee. UK single. 1960.

Free Parking! "My Girlfriend Is a Robot." Written by Cody Gunther, Trevor Gilleece, Jesse Nebling, Brendan Steere, and Bogdan Niemoczynski. On *Asuka*. 2011.

Freezepop. "Robotron 2000." Written by the Duke of Candied Apples and Liz Enthusiasm. On *The Purple EP*. 2000.

Garraud, Joachim, featuring Ze Rebelle. "My Boyfriend Is a Robot." Written by Joachim Garraud and Sarah Zainal Abidin. Digital single. 2012.

Hanson Brothers. "My Girlfriend's a Robot." Written by Tom Holliston. On *Gross Misconduct*. 1992.

Hawkwind. "Robot." Written by Dave Brock and Robert Newton Calvert. On *PXR5*. 1979.

Hawthorne, Mayer. "Robot Love." Written by Kid Harpoon, John Hill, Bill Curtis, and Mayer Hawthorne. On *Where Does This Door Go*. 2013.

Horton, Jamie. "Robot Man." Written by George Goehring and Sylvia Dee. U.S. single. 1960.

Jay-Z, featuring Babyface and Foxy Brown. "(Always Be My) Sunshine." Written by Sean Carter, Daven Vanderpool, Daryl Barksdale, Bobby Robinson, James Harris III, and Terry Lewis. On *In My Lifetime, Vol. 1*. 1997.

Kabaal klankbaan. "Robot Girlfriend." Written by Floris Groenewald. Digital single. 2016.

Kraftwerk. "The Robots." Written by Ralf Hütter, Karl Bartos, and Florian Schneider-Esleben. On *Man-Machine*. 1978.

Kravitz, Lenny. "Black Velveteen." Written by Lenny Kravitz. On *5*. 1998.

Lee, Trip. "Robot." Written by William M. Barefield III and Kenneth Chris Mackey. On *The Good Life*. 2012.

Linkin Park. "Robot Boy." Written by Linkin Park. On *A Thousand Suns*. 2010.

Lords of Acid. "Robot Love." Written by Cornelia Anita Van Lierop, Maurice Joseph Engelen, and Olivier Jean Jacques Adams. On *Heaven Is an Orgasm*. 1998.

Marina and the Diamonds. "I Am Not a Robot." Written by Marina Diamandis and Liam Howe. On *The Family Jewels*. 2009.

Mr. Bungle. "None of Them Knew They Were Robots." Written by Trey Spruance, Mike Patton, and Danny Heifetz. On *California*. 1999.

Never Shout Never. "Robot." Written by Christofer Drew Ingle. On *Time Travel*. 2011.

Pet Shop Boys. "Sad Robot World." Written by Neil Tennant and Chris Lowe. On *Super*. 2016.

Radiohead. "Paranoid Android." Written by Thom Yorke, Jonny Greenwood, Philip Selway, Ed O'Brien, and Colin Greenwood. On *OK Computer*. 1997.

Reagan Youth. "Brave New World." Written by Dave Rubinstein and Paul Bakija. On *Volume 2*. 1990.

Reed, Lou. "No Money Down." Written by Lou Reed. On *Mistrial*. 1986.

Rhett and Link. "Robot Girlfriend." Written by Rhett and Link. On *Up to This Point*. 2010.

Robyn. "Fembot." Written by Robyn and Klas Åhlund. On *Body Talk Pt. 1*. 2010.

Röyksopp, featuring Robyn. "The Girl and the Robot." Written by Svein Berge, Torbjorn Brundtland, and Robyn. On *Junior*. 2009.

Servotron. "3 Laws (Abolished)." On *No Room for Humans*. 1996.

Show Business Giants. "My Girlfriend's a Robot." Written by Tom Holliston. On *Gold Love*. 1988.

Styx. "Mr. Roboto." Written by Dennis DeYoung. On *Kilroy Was Here*. 1983.

t.A.T.u. "Robot." Written by Valeriy Polienko and Aleksander Voitinskiy. On *Against the Traffic*. 2001.

They Might Be Giants. "Become a Robot." Written by They Might Be Giants. On *Demo Tape*. 1985.

_____. "Robot Parade." Written by They Might Be Giants. On *No!* 2002.

Toad the Wet Sprocket. "Fall Down." Written by Glen Phillips, Todd Nichols and Toad the Wet Sprocket. On *Dulcinea*. 1994.

Traxman. "The Robots." Written by Cornelius Ferguson. On *Da Mind of Traxman*. Japanese special edition, 2012.

Was (Not Was). "Robot Girl." Written by David Was and Don Was. On *What Up, Dog?* 1988.

Weezer. "I'm a Robot." Written by Rivers Cuomo. On *Death to False Metal*. 2010.

Zappa, Frank. "A Token of My Extreme" and "Sy Borg." Written by Frank Zappa. On *Joe's Garage Acts II & III*. 1979.

Zru Vogue. "Atomic Robot Man vs. The Beautiful People." Written by Andrew L. Jackson and Max Tyrell. On *Unlimited Enjoyment Instant Gratification*. 1998

Music Videos

Branson. Richie. "I, for One, Welcome Our Robot Overlords." 2017. https://youtu.be/_0A04gwKBdY.

"Daft Punk's Electroma 2006." 2006. https://youtu.be/puYIvZ2umio.

"Kraftwerk—The Robots (live)." 1978. https://youtu.be/okhQtoQFG5s.

"Lou Reed—No Money Down." 1986. https://youtu.be/XiyX70ZqsVQ.

"Macy Gray—B.O.B (Official Music Video)." 2016. https://youtu.be/bfmuUf-47M4.

"Styx—Mr. Roboto (Relaid Audio)." 1983. https://youtu.be/uc6f_2nPSX8.

Newspapers and Magazines

"2 Great Days to Shop and Enjoy Chester." *Delaware County Daily Times* (Chester, PA), June 6, 1963.

"13-Year-Old Boy Creates Robot." *Sayre [PA] Evening Times*, July 30, 1957.

"$8,091,700 of Loan Quota Is Subscribed." *St. Louis Star and Times*, April 24, 1919.

Abramowitz, Rachel. "Regarding Stanley." *Los Angeles Times*, May 6, 2001.

"Academy Notes." *St. Johnsbury [VT] Caledonian*, November 25, 1908.

"'Aladdin' To-Night." *Wheeling Daily Intelligencer*, January 9, 1888.

"Almost Human." *Cincinnati Enquirer*, August 31, 1911.

"Alpha Loses Her Temper." *Lancashire Evening Post*, September 15, 1934.

"Amateur Vaudeville by One Thousand Club." *Galveston [TX] Daily News*, February 23, 1912.

"Among Best Shows." *Phoenix Republic*, September 23, 1962.

"Amusements." *Daily Arkansas Gazette* (Little Rock), October 1, 1907.

"Amusements." *New Orleans Times–Picayune*, March 4, 1887.

Atchison [KS] Daily Champion, April 13, 1869.

"At the Family." *Scranton Republican*, February 24, 1907.

"At the Mechanic Theatre." *The Times* (London), December 22, 1795.

"At the Salt Palace." *Salt Lake City Herald*, July 15, 1900.

"Austin & Stone's." *Boston Sunday Globe*, April 26, 1896.

"Automated Santa." *Pittsburgh Press Sunday Magazine*, December 16, 1962.

"An Automatic Ethiopian Minstrel Troupe." *Brooklyn Daily Eagle*, December 1, 1895.

"Automatic Minstrels." *Oakland Tribune*, November 4, 1906.

"Automation Puzzles World Magicians." *Sandusky [OH] Register*, December 21, 1919.

"The Automaton Hugger." *Richmond [IN] Item*, August 12, 1896.

"Automaton of Woman Sings, Talks, and Walks." *Washington Times*, August 30, 1911.

"Automaton Woman Walks." *Frederick [MD] News*, September 11, 1911.

"The Avenue Theater." *Pittsburgh Daily Post*, October 5, 1902.

"Avenue Theater—'Jolly Grass Widows.'" *Detroit Free Press*, December 31, 1906.

Babb, Stanley E. "The Grand Tour." *Galveston Daily News*, January 6, 1924.

Bakare, Lanre. "Meet Z-Machines, Squarepusher's New Robot Band." *The Guardian*, April 4, 2014.

Barnes, Mike. "Robert Kinoshita, Robot Designer for 'Forbidden Planet' and 'Lost in Space,' Dies at 100." *Hollywood Reporter*, January 13, 2015.

Barton, Olive Roberts. "Televox and House Work Is Made Easy." *Appleton [WI] Post-Crescent*, March 13, 1928.

"Bell Theater Has Weird Act on This Week's Bill." *Oakland Tribune*, February 1, 1909.

"Big Looie." *Detroit Free Press Sunday Graphic*, July 2, 1950.

"Big Looie Plays a Solo." *Rochester Democrat and Chronicle*, February 20, 1938.

Bingay, Malcom W. "Good Morning." *Detroit Free Press*, April 5, 1935.

"Binghamton Man Invented One of First Autos Ever Constructed in This Country." *Binghamton [NY] Press and Sun-Bulletin*, June 19, 1909.

"Blue Monday." *Albuquerque Journal*, July 11, 1938.

"Boneless, Bloodless, Fleshless Man May Replace Human Beings in the Industry of the World." *Evansville [IN] Press*, June 13, 1914.

Boone, Andrew R. "Garco, Indestructible 'Worker,' Can Do Dangerous Jobs for Humans." *Lafayette [IN] Journal and Courier*, March 13, 1954.

_____. "Plug-In Built in 90 Days." *Popular Science* (December 1953).

"Boy, 13, Builds His Own Robot—and It Works." *Hutchinson [KS] News*, October 25, 1930.

"Boy Builds Robot." *Newport [RI] Daily News*, March 24, 1954.

"Boy with a £13,000 Toy." *Sydney Morning Herald*, December 5, 1954.

"Brainy Machines." *Elmira [NY] Star-Gazette*, April 13, 1943.

Brooklyn Daily Eagle, January 24, 1868.

"Brother of Televox Is Operated by Light; New Mechanical Servant Shown in Action." *New York Times*, April 18, 1929.

Broun, Heywood. "Seeing Things at Night." *Pittsburgh Daily Post*, October 15, 1922.

"Buck Rogers." *New Yorker* (December 22, 1934).

Butler, Samuel. "Darwin among the Machines." Letter to the editor, *Christchurch, NZ Press*, June 13, 1863.

Campbell, John W. "Robots." *Astounding Science-Fiction* (November 1939).

"Changes in Vaudeville." *Chicago Inter-Ocean*, August 25, 1902.

"Chicago Opera House." *Chicago Inter-Ocean*, March 4, 1888.

"Chip Off the Old Block." *Time* (January 4, 1954).

"City and Provincial News." *Manitoba Free Press* (Winnipeg), April 10, 1878.

"Claudo the 'Mechanical Man.'" *Camden [NJ] Post*, December 6, 1956.

"A Clever Mechanical and Electrical Automaton." *Scientific American* (January 13, 1907).

"A Clockwork Man That Talks and Sings." *San Antonio Light and Gazette*, April 23, 1911.

"A Coal-Burning Steam Buggy." *St. Joseph [MO] Observer*, July 3, 1931.

Colligan, Douglas. "The Robots Are Coming." *New York Magazine* (July 30, 1979).

"Constructor of First Electric Chair in World to Exhibit 'Enigmarelle,' Famed Automaton at Home-Coming." *New Philadelphia [OH] Daily Times*, August 17, 1910.

Cook, Max B. "Aero's 'Mechanical Brains.'" *Honolulu Advertiser*, July 31, 1944.

Correspondent (Branch No. 11, Davenport, Iowa). Letter to the editor. *Leather Workers' Journal* (July 1914).

Correspondent (Ryegate Local No. 153). Letter to the editor. *Paper Makers Journal* (February 1912).

Courtney, William B. "Matinee Ladies." *Bristol [PA] Daily Courier*, June 13, 1927.

Crocker, Herbert C. "The Electric Man." *Illustrated World* (December 1916).

"Current Topics." *Burlington Weekly Free Press*, February 21, 1868.

Curtis, Olga. "Toy Industry Unveils 1954 Line Featuring 'Robert the Robot' and Streamlined Autos." *Lubbock Evening Journal*, March 9, 1932.

"Daniel Lambert." *Pittsburgh Daily Commercial*, January 15, 1868.

Danver, Charles F. "Origin of Sparko." *Pittsburgh Post-Gazette*, January 14, 1941.

Daw, Richard. "Youth's Resourceful Robot Helps Sell Boys on Science." *Vineland [NJ] Daily Journal*, December 11, 1958.

Dean, James W. "The Screen." *Springfield [MO] Leader and Press*, December 20, 1922.

"A Declaration of Independence." *New York Times*, September 28, 1930.

Dorman, Marjorie. "'Noname' Talks about Himself." *Brooklyn Daily Eagle*, February 25, 1923.

"Down-in-Four." *Time* (September 23, 1930).

Dr. Goldfoot and the Bikini Machine advertisement. *Greenville [SC] News*, November 24, 1965.

Driscoll, Charles B. "The World and All." *Lansing State Journal*, May 31, 1937.

Du Brown, Rick. "'I Spy' and Western Off to a Good Start." *Santa Rosa [CA] Press Democrat*, September 16, 1965.

Durand, Theodosia. "World Fairs and Affairs." *Santa Rosa [CA] Press Democrat*, June 21, 1936.

Eames, Lucy N. "Clearer Diagnosis and Simpler Treatment." *Michigan State Medical Journal* (June 1912).

"Easy to Convince Public, According to 'Electrical Man,' Who Has Done Stunt." *Salt Lake City Inter-Mountain Republican*, November 19, 1906.

Elam, F. Leland. "Strange Abode." *Los Angeles Times Sunday Magazine*, June 12, 1939.

"Electric Frankenstein." *Washington Evening Star*, August 6, 1890.

"The Electric Man." *Greensboro [NC] Patriot*, September 4, 1895.

"An Electric Man." *Monongahela [PA] Daily Republican*, October 13, 1890.

Elliott, Carl F. "Sees Robot Age in Near Future." *Brooklyn Daily Eagle*, October 22, 1933.

Engels, F[rederick]. "Obituary of Helena Demuth." *The People's Press*, November 11, 1890.

"Englewood Theatre." *Suburban Economist* (Chicago), December 14, 1917.

"English Inventor Fears Three-Ton Frankenstein He Worked On 14 Years." *Minneapolis Star Tribune*, September 18, 1932.

"English Robot Makes Lecture." *Altoona [PA] Mirror*, October 12, 1928.

"Ephemeris." *Pittsburgh Weekly Gazette*, January 9, 1869.

"Eric Robot Tours Phila. Shrines as Guest of Inquirer." *Philadelphia Inquirer*, March 12, 1929.

"Eric Robot's Debut Is Eagerly Awaited by Philadelphians." *Philadelphia Inquirer*, March 9, 1929.

Eufalua [OK] Republican, February 7, 1913.

"Events of Importance." *Winnipeg Tribune*, January 20, 1906.

"Exhibiting Mechanical Man." *Alton [IL] Evening Telegraph*, June 25, 1914.

"Exhibition at the Union Hotel." *New York Evening Post*, December 6, 1803.

"Famous Robot Man Will Appear Here." *Berkeley Daily Gazette*, February 20, 1936.

"Faranta's New Theatre." *New Orleans Picayune*, November 16, 1884.

Feck, Luke. "Average—A Compliment." *Cincinnati Enquirer*, November 23, 1962.

"Fifth Avenue Theater." *New York Times*, October 17, 1882.

"The First Automobile in America, Built in 1868." *Montpelier [ID] Examiner*, October 29, 1909.

"First Prize in Contest Goes to McKeesport Man." *Pittsburgh Post-Gazette*, March 2, 1929.

"Ford Plant Worker Dies in Accident." *Detroit Free Press*, January 26, 1979.

Fox, Helen. "Automatic Device Saves Labor for Control Systems." *Munster [IN] Times*, June 26, 1924.

"'Frankenstein' Had Nothing on This Robot." *Santa Cruz [CA] Sentinel*, September 20, 1932.

"Free-Lance Robot Hunts Job at Fair." *New York Times*, July 23, 1939.

Fremont [OH] Weekly Journal, March 20, 1868.

Fuehrer, Sherwood. "Gismo and I." *Boy's Life* (June 1956).

Fyles, Franklin. "Crane and Kelcey in Two New Plays." *Washington Post*, September 25, 1904.

Gaver, Jack. "Surgeon, Ex-Showman Turn Out 'Rod Brown.'" *Louisville Courier-Journal*, May 16, 1954.

Gernsback, Hugo. Editor response to letter. *Science Wonder Quarterly* (September 1929).

Gilbert, Douglas. "Odd Jobs." *Pittsburgh Press*, December 11, 1940.

Gill, Alan. "In the Age of Dee-Ah-Tump." *Marion [OH] Star*, August 24, 1962.

Gilmore, Eddy. "British Scientist Predicts Automatic Housewife." *Racine [WI] Journal Times*, September 25, 1962.

"'Gloomy Harris.'" *Amarillo [TX] Globe-Times*, April 18, 1934.

"Gloomy Harris Draws Crowds." *Greenwood [SC] Index-Journal*, September 22, 1935.

"Gossip from Paris." *Philadelphia Inquirer*, July 6, 1890.

"Gossip of the Stage Folk." *Washington Post*, October 28, 1906.

Grautsky, Amara. "Yankees–Red Sox to Play Two Games at Olympic Stadium in London in June 2019." *New York Daily News*, May 8, 2018.

"A Great Entertainment." *Fort Worth Daily Gazette*, March 31, 1887.

"Great Attraction. Ventriloquism." *Nashville Tennessean*, May 16, 1840.

H. L. "Things Wise and Otherwise." *New Orleans Times–Democrat*, March 19, 1893.

"Haddock's Exhibition of Androides." *Hartford Courant*, May 26, 1829.

"Haddock's Mechanical Exhibition of Androides." *New York Evening Post*, May 26, 1820.

Hagy, Robert R., Jr. "Tough Mechanical Man Scorned by Girls Who Boss Him Around." *Pittsburgh Post-Gazette*, January 3, 1941.

"The Harris Theater." *Pittsburgh Daily Post*, January 14, 1912.

"Has Built a New Percy." *Alton [IL] Evening Telegraph*, June 26, 1913.

"Have You Seen 'Willie Vocalite.'" *Los Angeles Times*, May 19, 1932.

Hayes, Lynn. "Slaves to the Scrub Bucket." *Baltimore Sun*, October 15, 1972.

"He Knows All the Answers." *Port Huron [MI] Times Herald*, October 29, 1934.

"Hercules, the Wonder." *Scranton Republican*, January 24, 1893.

"He Was a Deadhead." *Logan [OH] Hocking Sentinel*, February 8, 1900.

"The Hippodrome." *The Times* (London), October 12, 1904.

"The Home of the Steam Man." *Philadelphia Times*, January 22, 1883.

"Hopkins' Theaters." *Chicago Inter-Ocean*, December 7, 1896.

"How Moto Girl Plan Originated." *Arkansas Democrat* (Little Rock), December 2, 1910.

Huhn, Austin. "Clarence—Radio Robot." *Radio Craft* (October 1939).

"Hum of City Streets." *Philadelphia Times*, January 9, 1902.

"The Human Doll." *Columbus Daily Advocate*, April 22, 1912.

"A Human Wax Figure." *Philadelphia Times*, November 6, 1898.

"Husband Hunters, Wait Awhile! See If Robot Hits Mansfield." *Mansfield [OH] News-Journal*, January 5, 1929.

"Icicle Girl and Mechanical Man Boost Chevrolet." *Wichita Daily Times*, November 2, 1935.

"In the Automobile World." *Washington Post*, May 29, 1909.

"In the World of Electricity." *New York Times*, August 4, 1895.

"In 2028 … Games Take Longer Than Ever." *ESPN The Magazine* (April 23, 2018).

"In 2028 … NFL Stadiums Shrink." *ESPN The Magazine* (April 23, 2018).

"Industrial." *Plymouth [IN] Weekly Republican*, March 26, 1868.

"Iowa City Man Scores Big Hit." *Iowa City Press-Citizen*, June 8, 1909.

"The Iron Man." *Bristol [IN] Banner*, October 23, 1908.

"Iron Monster Turns Traitor." *Detroit Free Press*, September 19, 1932.

"Is It a Man or a Mechanical Figure?" *Detroit Free Press*, April 17, 1902.

"It Walks and Talks." *St. Louis Republic*, September 2, 1900.

"Jay W. Winton." *Rochester Democrat and Chronicle*, October 15, 1907.

"Just a Human Robot." Letter to the editor. *Wisconsin State Journal* (Madison), August 29, 1933.

Kaempffert, Waldemar. "Science Produces the 'Electrical Man.'" *New York Times*, October 23, 1927.

"Katrina von Televox." *Hagerstown [MD] Morning Herald*, April 22, 1931.

Kaufman, Stan. "Even Fem Flasher Can't Faze Battle Creek's Resident 'Robot.'" *Battle Creek [MI] Enquirer*, January 26, 1977.

Kiska, Tim. "Robot Victim's Kin Win $10 Million." *Detroit Free Press*, August 10, 1983.

"The Largest Toy Ever Brought to America." *Brooklyn Daily Eagle*. December 1, 1895.

Larrick, Nancy. "The All-White World of Children's Books." *Saturday Review*, September 11, 1965.

"The Latest Wonder." *Fort Wayne Daily Gazette*, January 15, 1868.

Lear, John. "Can a Mechanical Brain Replace *You*?" *Collier's* (April 3, 1953).

"Life of Robot's Master Menaced by Steel Monster." *Benton Harbor [MI] News-Palladium*, September 19, 1932.

"Lifelike Robot Speaks, Smokes, and Drinks." *Popular Science* (October 1935).

"Like Frankenstein, Robot Man Turns on Inventor." *Des Moines Register*, September 19, 1932.

"Local and Police Paragraphs and Minor Items of Metropolitan News." *New York Herald*, April 7, 1870.

Logan, Clarence A. "Is Sir Oliver Lodge Being Fooled by Wireless?" *San Francisco Chronicle*, May 23, 1920.

"London Wins Great Game." *Pittsburgh Press*, January 1, 1905.

"'Looie' Jealous Dummy Stooge." *Lansing State Journal*, January 20, 1938.

Lovece, Frank. "'Jetsons' Live-Action Remake from ABC in the Works." *Newsday*, August 17, 2017.

"Machine Made Milk." *Lebanon [PA] Daily News*, October 25, 1912.

"Machine-Man Amusing Roof-Garden Patrons in New York." *Indianapolis News*, June 3, 1902.

"Makes Giant Figure Walk." *Yale [MI] Expositor*, September 7, 1900.

Maloney, Russell. "Toy Fair." *New Yorker* (April 13, 1935).

"A Man of Wood Really Walks." *New York World*, June 28, 1896.

Mantle, Burns. "Super-Robot Results." *Salt Lake Tribune*, October 22, 1922.

Martin, Robert E. "Mechanical Men Walk and Talk." *Popular Science Monthly* (December 1928).

"The Master Mind." *Rocky Mount [NC] Telegram*, August 10, 1944.

McAdam, Wayne. Letter to the editor. *Science Wonder Quarterly* (September 1929).

McGuire, Edward. "Santa'll Have More Talky Toys This Year." *Portsmouth [OH] Times*, September 22, 1955.

McNulty, Charles. "Review: 'Yoshimi Battles the Pink Robots' Sounds Thrilling, At Least." *Los Angeles Times*, November 19, 2012.

"Mechanical? Maybe! But He Blushes Says 13,000th Visitor." *Franklin [IN] Evening Star*, September 22, 1934.

"A Mechanical Giant." *Charlotte News*, September 14, 1900.

"A Mechanical Lady Turns Out to Be Real." *Minneapolis Journal*, March 25, 1905.

"Mechanical Maid to Do Her Stunts." *Philadelphia Inquirer*, September 23, 1930.

"Mechanical Man." *Aberdeen Press and Journal*, December 31, 1928.

"A Mechanical Man." *London Globe*, April 26, 1919.

"Mechanical Man and Inventor Are Here This Week." *The [MIT] Tech*, February 25, 1929.

"Mechanical Man Breaks Down on Tryout." *Alton [IL] Evening Telegraph*, April 13, 1918.

"'Mechanical Man' Comes to Defy Wisecrackers, Jokers." *Akron Beacon Journal*, September 22, 1936.

"Mechanical Man Electric Marvel Put on Exhibit." *Altoona [PA] Mirror*, April 11, 1939.

"Mechanical Man Gets Pain in His 'Tummy.'" *Pittsburgh Press*, March 5, 1929.

"Mechanical Man Here Today." *Pittsburgh Press*, May 3, 1913.

"'Mechanical' Man to Give Performances at Block's." *Indianapolis Star*, February 13, 1935.

"Mechanical Man to Walk City Streets." *Fort Wayne Journal-Gazette*, June 5, 1914.

"Mechanical Man Visioned as Servant for Housewife." *Des Moines Tribune*, February 22, 1928.

"Mechanical Man Walks Down the Street." *Alton [IL] Evening Telegraph*, June 22, 1914.

"Mechanical Man Will Visit Pittsburgh Soon." *Pittsburgh Press*, May 2, 1913.

"Mechanical Man's Talk Startles Gotham Audience." *Miami News*, October 25, 1934.

"Mechanical Men That Excel Any Human Being." *San Antonio Light*, September 9, 1931.

"'Mechanical Men' Who Answer Vocal Orders Give Promise of Being Big Help around Home." *Brooklyn Daily Eagle*, October 14, 1927.

"Mental Trouble in the Making." *Monthly Bulletin of the Massachusetts Society for Mental Hygiene* (January 1923).

"'Metropolis' Held Hypnotic Picture." *Minneapolis Star*, October 22, 1927.

"'Metropolis' Spectacular Film Scheduled for State." *Reading [PA] Times*, August 22, 1927.

"Millionaire Dentist Has Million-Melody Home." *Allentown [PA] Morning Call*, May 11, 1937.

"Miss Automaton, She Talks." *New York Observer*, September 14, 1911.

"Miss Katrina von Televox." *Altoona [PA] Mirror*, October 3, 1930.

"Miss Televox to Be in Elyria Saturday." *Elyria [OH] Chronicle Telegram*, June 18, 1931.

"Mr. William Robot Attends Party." *Munster [IN] Times*, March 19, 1934.

"'Monsieur X' Shows Girlish Blonde Hair." *Beatrice [NB] Daily Sun*, August 2, 1936.

"Monzello's Mechanical Minstrels." *Variety* (August 29, 1908).

Moore, Charles. "Mechanical Man Alpha Ends Career after Owner Hurt." *San Bernardino County [CA] Sun*, February 18, 1936.

Moss, Marissa R. "Frank Zappa's Raunchy Rock Opera 'Joe's Garage' Debuts in L.A." *Rolling Stone* (September 29, 2008).

"Most Modern of Men—He's Mechanical!" *Asbury Park [NJ] Press*, February 22, 1928.

"Motor Show Is Opened." *New York Times*, November 4, 1900.

Muralt, Ted. "Making 'Mr. X'—A Radio Robot." *Radio-Craft* (December 1935).

"Musical Matters." *Chicago Tribune*, January 23, 1887.

"N.Y. Aquarium." *New York Herald*, November 19, 1879.

"A New Auto If You Can Tickle Mechanical Man." *Ames [IA] Daily Tribune*, July 17, 1935.

"New Developments in Electricity." *Modern Mechanics* (March 1931).

"New Mechanical Models—Mechanical Man." *Meccano* (January 1931).

"A 'New' Steam Man." *American Artisan and Illustrated Journal of Popular Science* (March 10, 1869).

"A New Steam Man." *Black Hills Daily Times* (Deadwood, SD), May 18, 1883.

"The Newark Steam Man." *Washington Evening Star*, March 7, 1868.

"News and Personal." *Louisville Daily Courier*, April 25, 1868.

"News of the Day." *Charleston Daily News*, March 11, 1870.

Newton, Virgil Miller, Jr. "Suspense for Late Fall." *Tampa Tribune*, December 10, 1972.

"'Nickle' Novels Were Good Prophets." *Wilmington [DE] News Journal*, August 24, 1920.

Northrop, W. B. "An Electric Man." *The Strand* (November 1900).

"Odd Inventions Shown." *New York Tribune*, November 4, 1900.

"Oh, Kiss Me, You Mechanical Fool." *Ogden [UT] Standard Examiner*, September 21, 1962.

"Olympic." *New York Evening Post*, October 2, 1837.

"Only Westinghouse Could Build Willie Vocalite." *Coshocton [OH] Tribune*, March 25, 1934.

"Oregon's New School Law Held Dangerous." *Woodland [CA] Daily Democrat*, January 3, 1923.

Orpheus. "Mimes and Music." *New Zealand Evening Post*, December 3, 1904.

"Owner Finds Robot Is Frankenstein." *Carlisle [PA] Sentinel*, October 20, 1932.

"Paavo Nurmi, Finn, Is Athletic Monk." *New Castle News*, August 15, 1924.

"Page Frankenstein: Robot Shoots Man." *Hamilton [OH] Journal News*, September 20, 1932.

Palmer, Raymond A. "The Observatory by the Editor." *Amazing Stories* (August 1938).

"Paris Stage Novelties." *New York Times*, February 27, 1881.

Parker, Dorothy. "A Mash Note to Crockett Johnson." *PM*, October 3, 1943.

Parsons, William Barclay. "Real Robots That Do Work for Man." *New York Times*, September 15, 1929.

"Party Looked On while Zutka Did 'Stunts.'" *Pittsburgh Press*, May 14, 1905.

"Passing the Camera." *St. Louis Republic*, September 9, 1900.

"Patent No. 4175." *Canadian Mechanics' Magazine and Patent Office Record*, Volumes 3–6 (January 1875).

Pawlenty, Tim. "We're at the Dawn of the Fourth Industrial Revolution." *Minneapolis Star Tribune*, June 5, 2017.

"Peale's Museum." *New York Evening Post*, December 22, 1840.

"Peale's Museum & Gallery of the Fine Arts." *New York Evening Post*, February 10, 1835.

"'Percy the Mechanical Man,' (Bray)." *Moving Picture World* (December 23, 1916).

"Perfect Robot on First Exhibition Complains of Pain; Screwdriver Aids." *Indianapolis Star*, May 26, 1934.

"The Perfect Woman." *Birmingham [UK] Daily Gazette*, June 16, 1949.

Perlman, William J. Letter to the editor. *New York Times*, January 25, 1923.

"Personal." *Wheeling Daily Intelligencer*, June 4, 1868.

"Personal and Pertinent." *Wilmington Evening Journal*, June 12, 1909.

Phillips, H. I. "The Once Over." *Oakland Tribune*, October 25, 1927.

Phillips, H. L. "Trying Out the Televox, or Mechanical Man." *Harrisburg Evening News*, March 5, 1928.

Pittsburgh Daily Commercial, March 13, 1869.

"Police Methods Will Be Shown." *Harrisburg Evening News*, October 18, 1935.

"Popular Entertainments." *Philadelphia Inquirer*, January 28, 1887.

Potter, Mark. "A Robot Who Reads the Newspapers." *Leeds Mercury*, August 24, 1932.

Prager, Arthur. "Bless My Collar Button If It Isn't Tom Swift." *American Heritage* (December 1976).

"Professor Wyman, Ventriloquist and Wizard." *Washington Evening Star*, January 20, 1862.

"Psycho Revealed." *New Orleans Times–Democrat*, December 11, 1910.

"Putting the Mechanical Man to Work." *Zanesville [OH] Sunday Times Signal*, December 9, 1928.

"Queens Boy, 13, Constructs Robotron the Robot." *Bridgeport Post*, July 30, 1957.

"R.U.R." *Oakland Tribune*, March 11, 1923.

Rau, Herb. "The Things That Happen Here." *Miami News*, January 18, 1956.

"Redmond's London Marionettes." *Brooklyn Daily Eagle*, September 1, 1877.

"A Remarkable Automaton." *Syracuse Herald*, June 7, 1909.

"Reporter Gets Interview." *Bakersfield Californian*, March 1, 1955.

Richards, William C. "World to Greet New Wonders in Next 100 Years." *Detroit Free Press* (centennial section), May 10, 1931.

"Roaming Robot." *Minneapolis Star Tribune*, August 11, 1955.

"Robert the Robot Walks, Talks." *Popular Science* (December 1954).

Robinson, Murray. "Planet Parenthood." *Collier's* (January 5, 1952).

"Robot." *Time* (November 5, 1934).

"'Robota' Gives Mystifying Act Here." *Ithaca Journal*, July 18, 1930.

"Robot Gives Girls New Thrill." *Akron Beacon Journal*, December 8, 1931.

"Robot Goes Wild, Knocks Out Man." *Des Moines Register*, June 28, 1935.

"Robot in Revolt." *Dundee Courier and Advertiser*, September 19, 1932.

"Robot Reader." *Brownsville [TX] Herald*, March 24, 1954.

"A Robot Romance." *Mexia [TX] Evening News*, January 11, 1923.

"Robot Spurns Method Acting." *Pasadena [CA] Independent Star-News*, March 13, 1966.

"Robot Turns Frankenstein, Shoots Creator." *Arizona Republic* (Phoenix), September 19, 1932.

"Robot Utters a Warning, Then Shoots Creator." *Chicago Tribune*, September 19, 1932.

"Robot Woman Socks Escort from Behind." *Elmira [NY] Star-Gazette*, October 4, 1934.

"Rock 'Em Sock 'Em Robots." *Kingston [NY] Daily Freeman*, October 30, 1964.

Rogers, Jude. "Why Kraftwerk Are Still the World's Most Influential Band." *The Guardian*, January 27, 2013.

"Romantic Old Maids Can Hear the Words of Love They Long For." *San Antonio Light*, July 1, 1928.

"Rowe a Humane Man." *Delaware County Daily Times* (Chester, PA), January 22, 1883.

Ruby, Daniel J. "Computerized Personalized Robots." *Popular Science* (May 1983).

"St. Charles Orpheum." *New Orleans Times–Democrat*, March 11, 1902.

Sandburg, Carl. Letter to the editor. *New York Times*, January 28, 1923.

"'San-Velo.'" *Los Angeles Times*, September 29, 1957.

Sarrazin, Gabriel. "France." *The Athenæum*, January 1, 1887.

Saunders, Hortense. "Mr. Televox, Plain but Useful, and in All a Live-Wire Fellow." *Pittsburgh Press*, February 23, 1928.

Scheffaur, Herman G. "An Impression of the German Film 'Metropolis.'" *New York Times*, March 6, 1927.

"Science's New Mechanical Man." *Richmond [IN] Item*, June 27, 1930.

"Scientific News." *English Mechanic and World of Science*, no. 585 (June 9, 1876).

"Scientist Predicts Robots of Future Will Do

Drudgery." *Burlington [VT] Free Press*, February 21, 1935.

"See 'Claudo' the Mechanical Man." *Baltimore Sun*, October 31, 1926.

"See 'Gloomy Gus' the Frozen Faced Man." *Alexandria [LA] Town Talk*, March 2, 1933.

"See Nerv-o." *Lansing [MI] State Journal*, August 10, 1939.

"See Waxo—The Mechanical Man." *Longview [TX] News-Journal*, October 26, 1933.

"Shades of Brick Bradford! 'Elektro' Is Almost Human." *Munster [IN] Times*, April 12, 1939.

Shelley, Percy Bysshe. "On 'Frankenstein.'" *The Athenæum*, November 10, 1832.

Shippey, Lee. "The Courtship of Miles Televox." *Los Angeles Times*, June 20, 1928.

"Shot by the 'Monster' of His Own Creation." *Ogden [UT] Standard-Examiner*, October 23, 1932.

Sime [Silverman]. "Araco." *Variety* (April 14, 1906).

_____. "Reded and Hadley." *Variety* (September 15, 1906).

_____. Untitled. *Variety* (April 6, 1906).

_____. "William Gane's 'Automatic Minstrels.'" *Variety* (August 22, 1908).

"Singing Light Tower at Fair Blends Music and Color." *Popular Mechanics* (June 1939).

Smith, H. Allen. "Writer Marvels When He Asks Robot How He Feels and Later Reports Severe Pain in Side." *Great Falls Tribune*, May 27, 1934.

Smith, John. "Stop the Villain!" *North-Carolina Star* (Raleigh), November 27, 1812.

"Smuggling a Passenger." *Wilmington [NC] Tri-Weekly Commercial*, February 26, 1853.

"Space Patrol Conquers Kids." *Life* (September 1, 1952).

"Sparko, the Perfect Pup." *Galveston [TX] Daily News*, June 10, 1940.

"A Steam Man." *Cincinnati Enquirer*, January 17, 1883.

"The Steam Man." *Lexington [MO] Weekly Caucasian*, December 12, 1868.

"Steam Man." *Orange [VA] Native Virginian*, January 24, 1868.

"The Steam Man." *Scientific American* (March 28, 1868).

"The Steam Man." *Scientific American* (May 21, 1870).

"The Steam Man." *Scientific American* (April 15, 1893).

"Steam Man." *Warrenton [MO] Banner*, December 15, 1868.

Steam Man advertisement. *Boston Post*, April 19, 1896.

"The Steam Man—His First Appearance on the Street." *New York Tribune*, January 24, 1868.

"The Steam Man Coming." *Harrisburg Daily Independent*, September 18, 1880.

"The Steam Man's Demise." *Cincinnati Enquirer*, January 19, 1883.

"Stock Companies." *Rochester Democrat and Chronicle*, February 4, 1900.

Strauss, Neil. "Call Them the Beatles of Electronic Dance Music." *New York Times*, June 15, 1997.

_____. "Hardly a Pocket Calculator: Kraftwerk's Studio Goes on Tour." *New York Times*, June 11, 1998.

Sullivan, Kay. "A Robot in the Family." *Jackson [MI] Clarion-Ledger* (*Parade* magazine supplement), October 17, 1954.

Sullivan, Paul Barrett. "Motoring." *Brooklyn Life*, May 29, 1909.

Sydney Morning Herald, June 23, 1868.

"Tag Along." *Reno Gazette-Journal*, October 7, 1954.

Talbot, Margaret. "A Stepford for Our Times." *Atlantic* (December 2003).

"Talking Robot Grows Mature." *Greenville [MS] Delta Democrat-Times*, October 3, 1940.

Tazewell [VA] Clinch Valley News, June 19, 1891.

"Teaching New Dog Old Tricks." *Brainerd [MN] Daily Dispatch*, May 4, 1940.

"'The Tempters' at Majestic." *Pittston [PA] Gazette*, March 21, 1918.

"Theater Memoranda." *Pittsburgh Daily Post*, January 23, 1917.

"Theaters and Music." *Brooklyn Daily Eagle*, September 2, 1888.

"Theatrical Gossip." *Chicago Inter-Ocean*, November 6, 1892.

"This Automaton Will Walk, Dance and Write." *Brooklyn Daily Eagle*, August 24, 1904.

"Thodar the Robot to Parade on Landis Ave. Saturday." *Vineland [NJ] Daily Journal*, April 27, 1955.

Thompson, Carolyn. "My Classmate the Robot: New York Pupil Attends Remotely." *Elwood [IN] Call-Leader*, February 15, 2013.

"Three Big Hits at the Grand." *Altoona [PA] Tribune*, May 7, 1907.

"Tin Can with an Idea." *Life* (December 13, 1954).

"The Tivoli." *Oakland Tribune*, May 11, 1901.

Tom Swift advertisement. *Hartford Courant*, November 28, 1954.

"Trick of a Ventriloquist." *Wetumpka [AL] Argus*, September 9, 1840.

Tucker, Robert G. "Coming Theatrical Events Casting Pleasant Shadows." *Indianapolis Star*, November 19, 1922.

"U.S. Orbits 'Robot Astronaut' and Brings It Back Successfully." *Philadelphia Daily News*, September 13, 1961.

"Uncle Sam to Drive 'Fritz' around City." *St. Louis Post-Dispatch*, April 20, 1919.

"A Unique Exhibition." *Wilkes-Barre Times*, April 17, 1899.

Unsigned review of *Around the World in Eighty Days*, by Jules Verne. *The Galaxy: A Magazine of Entertaining Reading* (September 1873).

Unsigned review of "An Automatic Enigma," by Julian Hawthorne. *The Academy* (May 11, 1898).

Unsigned review of "An Automatic Enigma," by Julian Hawthorne. *The Library Table* (June 8, 1878).

Unsigned review of "A Wife Made to Order," by Alice W. Fuller. *Stone* (August 1895).

"Vaudeville." *Harrisburg Telegraph*, March 11, 1924.

"Ventriloqual Automatons." *Azizola [AZ] Oasis*, February 6, 1897.

"Ventriloquism!" *Mississippi Free Trader* (Natchez), January 17, 1838.

"A Ventriloquist's Trick." *Salt Lake City Tribune*, July 31, 1898.

"Ver Valin Ventriloquist." *Muncie [IN] Star Press*, January 28, 1923.

"Victims of the Machines." *Sioux Falls [SD] Argus-Leader*, February 4, 1930.

Walker, Karen. "Ultron: The Black Sheep of the Avengers Family." *Back Issue!* (February 2010).

"Walking Automation Is a Mechanical Wonder." *Buffalo Courier*, September 2, 1900.

"Walking Automaton Is a True Wonder." *Buffalo Express*, September 2, 1900.

"Walking Giant." *Cincinnati Enquirer*, August 4, 1901.

"The Walking Steam Man." *Reading [PA] Times*, August 5, 1878.

"Walks by Steam." *Wilmington [DE] News Journal*, July 6, 1878.

Washington Republican, February 8, 1868.

"'Wax Doll,' a Man Who Bears Scars by Curious Persons." *St. Louis Star and Times*, May 12, 1912.

"'Waxo' Is Featured at Rainbow Gardens." *Akron Beacon Journal*, March 2, 1926.

"'Waxo' the Mechanical Man." *Uniontown [PA] Morning Herald*, August 14, 1936.

"What May Happen When Robots Do All the Work." *Nashville Tennessean* (*American Weekly* supplement), August 5, 1938.

"What Mechanical Man May Do." *New Castle [PA] News*, February 24, 1928.

"Who Would Do the Nagging?" *Hutchinson [KS] News*, September 22, 1962.

"Will He Do for President?" *Nashville Union and Dispatch*, January 19, 1868.

"Willie Vocalite." *Bakersfield Californian*, May 7, 1932.

"Willie Vocalite." *Decatur [IL] Herald*, February 10, 1932.

"Willie Vocalite Is 'Some' Robot; Does Everything but Pay Taxes." *Racine [WI] Journal Times*, April 7, 1932.

"Willie Will Attend Chicago Exhibit If He Learns to Smoke." *Muncie [IN] Evening Press*, August 7, 1933.

"A Wonderful Artificial Man." *Dundee Evening Telegraph and Post*, March 22, 1911.

"The Wonderful Steam Man!" *Nashville Tennessean*, October 19, 1869.

"Wonderland's Offerings." *Detroit Free Press*, February 18, 1896.

"Wonders Will Never Cease!" *New York Evening Post*, January 13, 1831.

"A Wooden Blonde That Will Smile, Breathe, Blush, Sing, Play, Wink." *Hamilton [OH] Evening Journal*, June 16, 1923.

"The World's Wonder Robot." *Van Wert [OH] Daily Bulletin*, December 3, 1931.

Wye, E. "At the Sign of the Cat and Fiddle." *Single Tax Review* (November–December 1922).

"Young Inventor." *Minneapolis Star Tribune* (*This Week* supplement), February 27, 1955.

"Zadock Pratt Dederick." *Houston Post*, June 30, 1921.

Nonfiction

Anderson, Vicki. *The Dime Novel in Children's Literature*. Jefferson, NC: McFarland, 2004.

Bacon, Roger. *Friar Bacon, His Discovery of the Miracles of Art, Nature, and Magick, Faithfully translated out of Dr Dees own Copy, by T.M. and never before in English*. London: Simon Miller, 1659. Transcribed, printed and published privately by Dr. Alan R. Young, PhD. Caen, France (September 1993). http://www.sacred-texts.com/aor/bacon/miracle.htm.

Barnum, P. T. *Life of P. T. Barnum*. London: Sampson Row, Son, & Co., 1855.

Bassior, Jean-Noel. *Space Patrol: Missions of Daring in the Name of Early Television*. Jefferson, NC: McFarland, 2012.

Blackbeard, Bill, Dale Crain, and James Vance, eds. *100 Years of Comic Strips*. New York: Barnes & Noble Books, 2004.

Bloch, Iwan. *The Sexual Life of Our Time in Its Relation to Modern Civilization*. Translated from the sixth German edition by M. Eden Paul. London: Rebman, 1908.

Borden, Bill, with Steve Posner. *The Big Book of Big Little Books*. San Francisco: Chronicle Books, 1997.

Bouissou, Jean-Marie. "Manga: A Historical Overview." In *Manga: An Anthology of Global and Cultural Perspectives*, edited by Toni Johnson-Woods. London: Continuum, 2010.

Bradbury, Ray. "Buck Rogers in Apollo Year 1." In *The Collected Works of Buck Rogers in the 25th Century*, edited by Robert C. Dille. New York: Chelsea House, 1970.

Brown, Noel. *The Hollywood Family Film: A History, from Shirley Temple to Harry Potter*. London: I. B. Taurus, 2012.

Bunte, Jim, Dave Hallman, and Heinz Mueller. *Vintage Toys: Robots and Space Toys*. Iola, IA: Kraue Publications, 1999.

Carper, Steve. "Subverting the Disaffected City: Cityscape in *Blade Runner*." In *Retrofitting* Blade Runner: *Issues in Ridley Scott's* Blade Runner *and* Philip K. Dick's Do Androids Dream of Electric Sheep?, edited by Judith Kerman. Bowling Green, OH: Bowling Green State University Press, 1991.

Carter, Paul. *The Creation of Tomorrow*. New York: Columbia University Press, 1977.

Castleman, Harry, and Walter J. Podrazik. *Watching TV: Six Decades of American Television*. Second edition. Syracuse, NY: Syracuse University Press, 2003.

Cohen, John. *Human Robots in Myth and Science*. South Brunswick and New York: A. S. Barnes, 1967.

Davin, Eric Leif. *Partners in Wonder: Women and the Birth of Science Fiction, 1926–1965*. Lanham, MD: Lexington Books, 2006.

Edwards, John Milton [William Wallace Cook]. *The Fiction Factory*. Ridgewood, NJ: The Editor Company, 1912.

Emerson, Ralph Waldo. *Emerson in His Journals.* Edited by Joel Porte. Cambridge, MA: Belknap Press, 1982.

Ferguson, Andrew. *The Sex Doll: A History.* Jefferson City, NC: McFarland, 2010.

Fleiss, Sue. *Robots, Robots Everywhere!* New York: Golden Books, 2013.

Francavilla, Joseph. "The Android as *Doppelgänger.*" In *Retrofitting* Blade Runner: *Issues in Ridley Scott's* Blade Runner *and Philip K. Dick's* Do Androids Dream of Electric Sheep?, edited by Judith Kerman. Bowling Green, OH: Bowling Green State University Press, 1991.

Freud, Sigmund. "The Uncanny." In *The Standard Edition of the Complete Psychological Works of Sigmund Freud. XVII,* edited and translated by James Strachey et al. London: Hogarth Press and the Institute of Psycho-Analysis, 1955.

Gerrold, David. *The Trouble with Tribbles.* New York: Ballantine Books, 1973.

Goldman, Albert, from the journalism of Lawrence Schiller. *Ladies and Gentlemen—Lenny Bruce!!* New York: Random House, 1974.

Goulart, Ron. *Great History of Comic Books.* Chicago: Contemporary Books, 1986.

Gower, John. *Confessio Amantis: Book 4.* Edited by Russell A. Peck. Translated by Andrew Galloway. http://d.lib.rochester.edu/teams/text/peck-gower-confessio-amantis-book-4.

Graydon, Danny. *The Jetsons: The Official Guide to the Cartoon Classic.* Philadelphia: Running Press, 2011.

Green, Abel, and Joe Laurie, Jr. *Show Biz: From Vaude to Video.* New York: Henry Holt, 1953.

Hajdu, David. *The Ten-Cent Plague: The Great Comic Book Scare and How It Changed America.* New York: Farrar, Straus and Giroux, 2008.

Hardy, Phil. *The Encyclopedia of Science Fiction Movies.* Minneapolis: Woodbury Press, 1986.

Hatfield, H. Stafford. *Automaton, or The Future of the Mechanical Man.* London: Kegan Paul, Trench, Trubner & Co., 1928.

Heinlein, Robert A. Introduction to *Tomorrow, the Stars.* Edited by Robert A. Heinlein. New York: Signet, 1953.

Hennessey, Jonathan. *The Comic Book Story of Video Games: The Incredible History of the Electronic Gaming Revolution.* Berkeley, CA, and New York: Ten Speed Press, 2017.

Herman, Daniel. *Buck Rogers in the 25th Century: The Complete Newspaper Sundays.* Volume 1. New Castle, PA: Hermes Press, 2010.

Horn, Maurice, ed. *100 Years of American Newspaper Comics.* New York: Gramercy Books, 1996.

Hornyak, Timothy N. *Loving the Machine: The Art and Science of Japanese Robots.* Tokyo, New York, and London: Kodansha International, 2006.

Joustra, Robert, and Alissa Wilkinson. *How to Survive the Apocalypse: Zombies, Cylons, Faith, and Politics at the End of the World.* Grand Rapids, MI: William B. Eerdmans, 2016.

Klíma, Ivan. *Karel Čapek: Life and Work.* Translated by Norma Comrada. North Haven, CT: Catbird Press, 2001.

Kolb, William M. "Script to Screen: *Blade Runner* in Perspective." In *Retrofitting* Blade Runner: *Issues in Ridley Scott's* Blade Runner *and Philip K. Dick's* Do Androids Dream of Electric Sheep?, edited by Judith Kerman. Bowling Green, OH: Bowling Green State University Press, 1991.

Ladd, Fred, with Harvey Deneroff. *Astro Boy and Anime Come to the Americas: An Insider's View of the Birth of a Pop Culture Phenomenon.* Jefferson, NC: McFarland, 2009.

Lee, Jason. *Sex Robots: The Future of Desire.* London: Palgrave Macmillan, 2017.

Loukides, Paul, and Linda K. Fuller, eds. *Beyond the Stars: Studies in American Popular Film.* Volume 4: *Locales in American Popular Film.* Bowling Green, OH: Bowling Green State University Popular Press, 1993.

Loyd, Lady Mary, trans. *Villiers de L'Isle Adam: His Life and Works, from the French of Vicomte Robert du Pontavice de Heussey.* London: William Heinemann, 1894.

Macleod, Henry Duncan. *The Elements of Political Economy.* London: Longman, Brown, Green, Longmans, and Roberts, 1858.

Metzger, Th. [*sic*]. *Blood and Volts: Edison, Tesla, & the Electric Chair.* Brooklyn: Autonomedia, 1996.

Miller, Cynthia. "Domesticating Space: Science Fiction Series Come Home." In *Science Fiction Film, Television, and Adaptation Across the Screens,* edited by J. P. Telotte and Gerald Duchovnay. New York: Routledge, 2012.

Mills, Robert. "Talking Heads, or, A Tale of Two Clerics." In *Disembodied Heads in Medieval and Early Modern Culture,* edited by Catrien Santing, Barbara Baert, and Anita Traninger. Leiden: Brill, 2013.

Monroe, Paul. Section introduction to *Lost in Space 25th Anniversary Tribute Book* by James Van Hise. Las Vegas: Pioneer Books, 1990.

Moore, C. L. Introduction to *Robots Have No Tails* by Henry Kuttner. New York: Lancer, 1973.

Moskowitz, Sam. *Seekers of Tomorrow.* New York: Ballantine Books, 1967.

Peel, John. *The Lost in Space Files.* Granada Hills, CA: Schuster and Schuster, 1987.

Pollack, Dale. *Skywalking: The Life and Films of George Lucas.* New York: Crown Publishing, 1983.

Rainone, Joseph. *The Art & History of American Popular Series.* Volume 1. CreateSpace Independent Publishing Platform, 2013.

Reginald, Robert. *Contemporary Science Fiction Authors.* New York: Arno Press, 1974.

Ricca, Brad. *Super Boys: The Amazing Adventures of Jerry Siegel and Joe Shuster—The Creators of Superman.* New York: St. Martin's Press, 2013.

Richmond, Simon. *The Rough Guide to Anime: Japan's Finest from Ghibli to Gankutsuō.* London: Rough Guides, 2009.

Rickman, Gregg. "Philip K. Dick on *Blade Runner:* 'They Did Sight Stimulation on My Brain.'" In

Retrofitting Blade Runner: *Issues in Ridley Scott's* Blade Runner *and Philip K. Dick's* Do Androids Dream of Electric Sheep?, edited by Judith Kerman. Bowling Green, OH: Bowling Green State University Press, 1991.

Riley, Michael O. *Oz and Beyond: The Fantasy World of L. Frank Baum.* Lawrence: University Press of Kansas, 1997.

Rogers, Katharine M. *L. Frank Baum: Creator of Oz.* New York: St. Martin's Press, 2002.

Ryfle, Steve. *Japan's Favorite Mon-star: The Unauthorized Biography of the Big G.* Toronto: ECW Press, 1998.

Schaut, Scott. *Robots of Westinghouse, 1924–Today.* Mansfield, OH: Scott Schautt, Mansfield Memorial Museum, 2006.

Shaw, George Bernard. *The Collected Works of George Bernard Shaw: Plays, Novels, Articles, Letters, and Essays.* e-artnow, 2015.

Smith, Johanna M., ed. *Mary Shelley, Frankenstein: Complete, Authoritative Text with Biographical and Historical Contexts, Critical History, and Essays from Five Contemporary Critical Perspectives.* Boston: Bedford Books of St. Martin's Press, 1992.

Sunderland, Luke. *Old French Narrative Cycles: Heroism between Ethics and Morality.* Woodbridge, UK, and Rochester, NY: Boydell and Brewer, 2010.

Warren, Bill. *Keep Watching the Skies! American Science Fiction Movies of the Fifties: The 21st Century Edition.* Jefferson, NC: McFarland, 2009.

Warrick, Patricia S. *The Cybernetic Imagination in Science Fiction.* Cambridge, MA, and London: MIT Press, 1980.

Wertham, Fredric. *Seduction of the Innocent.* New York: Rinehart & Company, 1954.

Woods, Gaby. *Edison's Eve: A Magical History of the Quest for Mechanical Life.* New York: Alfred A. Knopf, 2002.

Radio

The Hitchhiker's Guide to the Galaxy (BBC Radio Series 1978, 1980). Written by Douglas Adams and John Lloyd.

Tom Corbett, Space Cadet (Radio Series 1952). "The Giants of Mercury." Season 1, Episodes 20–21, March 11 and 13, 1952.

Stage Presentations

Barbier, Jules, and Michel Carré. *Les Contes d'Hoffmann* (1851 opera).

Čapek, Karel. *R.U.R. (Rossum's Universal Robots)* [1921]. Translated by David Wyllie, 2006. https://archive.org/stream/R.U.R./R.U.R._djvu.txt.

Offenbach, Jacques. *The Tales of Hoffmann* (1881 opera).

Saint-Léon, Arthur, and Léo Delibes. *Coppélia* (1870 ballet).

Television

A.I.: Artificial Intelligence (Video Series 2016). Directed by Stills by Alan.

The Addams Family (TV Series 1964–1966). "Lurch's Little Helper." Directed by Sidney Lanfield. Written by Phil Leslie. Season 2, Episode 27, March 18, 1966.

Almost Human (TV Series 2013–2014). Created by J. H. Wyman.

Astroboy (TV Series). "The Birth of Astroboy." Directed by Noboru Ishiguro and Osamu Tezuka. Written by Osamu Tezuka. Episode 1, Season 1, September 7, 1963.

The Avengers (TV Series, 1961–1969). "The Cybernauts." Directed by Sidney Hayers. Written by Philip Levene. Season 4, Episode 3, March 28, 1966.

The Avengers (TV Series, 1961–1969). "Return of the Cybernauts." Directed by Robert Day. Written by Philip Levene. Season 6, Episode 1, February 21, 1968.

Batman (TV Series 1966–1968). "The Joker's Last Laugh" and "The Joker's Epitaph." Directed by Oscar Rudolph. Written by Peter Rabe (story) and Lorenzo Semple, Jr. (teleplay). Season 2, Episodes 47–48, February 15–16, 1967.

Battlestar Galactica (TV Miniseries). Directed by Michael Rymer. Written by Glen A. Larson (as Christopher Eric James) and Ronald D. Moore. December 8–9, 2003.

Battlestar Galactica (TV Series 2004–2009). "No Exit." Directed by Gwyneth Horder-Payton. Written by Ryan Mottesheard. Season 4, Episode 15, February 13, 2009.

The Bionic Woman (TV Series 1976–1978). "Fembots in Las Vegas." Directed by Michael Preece. Written by Arthur Rowe. Season 3, Episode 3, September 24, 1977.

The Bionic Woman (TV Series 1976–1978). "Fembots in Las Vegas, Part 2." Directed by Michael Preece. Written by Arthur Rowe. Season 3, Episode 4, October 1, 1977.

The Bionic Woman (TV Series 1976–1978). "Kill Oscar." Directed by Alan Crosland, Jr. (as Alan Crosland). Written by Arthur Rowe and Oliver Crawford (story) and Arthur Rowe (teleplay). Season 2, Episode 5, October 27, 1976.

The Bionic Woman (TV Series 1976–1978). "Kill Oscar, Part 3." Directed by Alan Crosland, Jr. (as Alan Crosland). Written by Arthur Rowe and Oliver Crawford (story) and Arthur Rowe (teleplay). Season 2, Episode 6, November 3, 1976.

Camp Camp (TV Series 2016–2017). "Anti-Social Network." Directed by Jordan Cwierz. Written by Gray G. Haddock. Season 2, Episode 2, June 16, 2017.

Captain Video and His Video Rangers (TV Series 1949–1955). "I, Tobor." Written by Isaac Asimov. November 2–December 7, 1953.

Captain Video and His Video Rangers (TV Series 1949–1955). "The Return of Tobor." Written by

Carey Wilber. Season 6, Episode 15, December 6, 1954.

Captain Z-Ro (TV Series 1955–1956). "Roger the Robot." Directed by David Butler. Written by Roy Steffens. Season 1, Episode 9, February 12, 1956.

Captain Z-Ro (TV Series 1955–1956). "The Great Pyramid of Giza." Directed by David Butler. Written by Roy Steffens. Season 1, Episode 22, May 13, 1956.

Doctor Who (TV Series 1963–1989). "Robot: Part One–Robot: Part Four." Directed by Christopher Barry. Written by Terrance Dicks. Season 12, Episodes 1–4, December 28, 1974, and January 4, 11, and 18, 1975.

Futurama (TV Series 1999–2013). Created by Matt Groening and David X. Cohen.

Future Cop (TV Series 1976–1977). "Pilot." Directed by Jud Taylor. Written by Allen S. Epstein (as Allen Epstein) and Anthony Wilson. May 1, 1976.

Get Smart (TV Series 1965–1970). "Anatomy of a Lover." Directed by Bruce Bilson. Written by Gary Clarke (as C. F. L'Amoreaux). Season 2, Episode 1, September 9, 1967.

Get Smart (TV Series 1965–1970). "Back to the Old Drawing Board." Directed by Bruce Bilson. Written by Gary Clarke (as C. F. L'Amoreaux). Season 1, Episode 19, January 29, 1966.

Get Smart (TV Series 1965–1970). "It Takes One to Know One." Directed by Earl Bellamy. Written by Gary Clarke (as C. F. L'Amoreaux). Season 1, Episode 16, January 7, 1967.

Get Smart (TV Series 1965–1970). "Run, Robot, Run." Directed by Bruce Bilson. Written by Gary Clarke (as C. F. L'Amoreaux). Season 3, Episode 23, March 16, 1968.

Get Smart (TV Series 1965–1970). "When Good Fellows Get Together." Directed by Sidney Miller. Written by Gary Clarke (as C. F. L'Amoreaux). Season 3, Episode 8, November 18, 1967.

Get Smart (TV Series 1965–1970). "The Worst Best Man." Directed by Gary Nelson. Written by Chris Hayward, Allan Burns, Arne Sultan, and Leonard Stern. Season 4, Episode 6, October 26, 1968.

Gilligan's Island (TV Series 1964–1967). "Gilligan's Living Doll." Directed by Leslie Goodwins. Written by Bob Stevens. Season 2, Episode 21, February 10, 1966.

Hazel (TV Series 1961–1966). "Hazel's Contract." Directed by William D. Russell. Written by Peggy Chantler Dick. Season 2, Episode 2, September 27, 1962.

The Hitchhiker's Guide to the Galaxy (TV Series 1981). Directed by Alan J. W. Bell. Written by Douglas Adams and John Lloyd.

Holmes and Yoyo (TV Series 1976). "Pilot." Directed by Jackie Cooper. Written by Lee Hewitt, Jack Sher, and Leonard Stern. Season 1, Episode 1, September 25, 1976.

The Honeymooners (TV Series 1955–1956). "TV or Not TV." Directed by Frank Satenstein. Written by Marvin Marx and Walter Stone. Season 1, Episode 1, October 1, 1955.

I've Got a Secret (TV Series 1952–1967). Hosted by Garry Moore. Season 6, Episode 38, March 28, 1958.

The Jetsons (TV Series 1962–1963). "The Coming of Astro." Directed by Joseph Barbera and William Hanna. Written by Tony Benedict (story and teleplay). Season 1, Episode 5, October 21, 1962.

The Jetsons (TV Series 1962–1963). "Dude Planet." Directed by Joseph Barbera and William Hanna. Written by Walter Black (story) and Tony Benedict (teleplay). Season 1, Episode 22, February 17, 1963.

The Jetsons (TV Series 1962–1963). "G. I. Jetson." Directed by Joseph Barbera and William Hanna. Written by Barry E. Blitzer (story) and Tony Benedict (teleplay). Season 1, Episode 19, January 27, 1963.

The Jetsons (TV Series 1962–1963). "Jetson's Nite Out." Directed by Joseph Barbera and William Hanna. Written by Harvey Bullock and R. S. Allen (story) and Tony Benedict (teleplay). Season 1, Episode 3, October 7, 1962.

The Jetsons (TV Series 1962–1963). "Las Venus." Directed by Joseph Barbera and William Hanna. Written by Barry E. Blitzer (story) and Tony Benedict (teleplay). Season 1, Episode 13, December 16, 1962.

The Jetsons (TV Series 1962–1963). "Rosie's Boyfriend." Directed by Joseph Barbera and William Hanna. Written by Walter Black (story) and Tony Benedict (teleplay). Season 1, Episode 8, November 11, 1962.

The Jetsons (TV Series 1962–1963). "Rosie the Robot." Directed by Joseph Barbera and William Hanna. Written by Larry Markes (story) and Tony Benedict (teleplay). Season 1, Episode 1, September 23, 1962.

The Jetsons (TV Series 1962–1963). "Uniblab." Directed by Joseph Barbera and William Hanna. Written by Barry E. Blitzer (story) and Tony Benedict (teleplay). Season 1, Episode 10, November 25, 1962.

Johnny Jupiter (TV Series 1953–1954). "Duckworth and the Professor." Directed by Howard Magwood. Written by Jerome Coopersmith (screenplay and story) and Sam Rockingham (story). Season 2, Episode 8, October 25, 1953.

Lancelot Link, Secret Chimp (TV Series 1970–1971). "The Reluctant Robot." Produced by Stan Burns and Mike Marmer. Season 1, Episode 9, February 26, 1970.

Lost in Space (TV Series 1965–1968). "Deadliest of the Species." Directed by Don Richardson. Written by Robert Hamner. Season 3, Episode 11, November 22, 1967.

Lost in Space (TV Series 1965–1968). "The Ghost Planet." Directed by Nathan Juran. Written by Peter Packer. Season 2, Episode 3, September 25, 1966.

Lost in Space (TV Series 1965–1968). "Junkyard in Space." Directed by Ezra Stone. Written by Barney Slater. Season 3, Episode 24, March 6, 1968.

Lost in Space (TV Series 1965–1968). "The Mechanical Men." Directed by Seymour Robbie. Written by Barney Slater. Season 2, Episode 28, April 5, 1967.

Lost in Space (TV Series 1965–1968). "The Reluctant Stowaway." Directed by Anton Leader (as Tony Leader). Written by Shimon Wincelberg (as S. Bar-David). Season 1, Episode 1, September 15, 1965.

Lost in Space (TV Series 1965–1968). "War of the Robots." Directed by Sobey Martin. Written by Barney Slater. Season 1, Episode 20, February 9, 1966.

The Man from U.N.C.L.E. (TV Series 1964–1968). "The Double Affair." Directed by John Newland. Written by Clyde Ware. Season 1, Episode 8, November 17, 1964.

The Man from U.N.C.L.E. (TV Series 1964–1968). "The Sort of Do-It-Yourself Dreadful Affair." Directed by E. Darrell Hallenbeck. Written by Harlan Ellison. Season 3, Episode 2, September 23, 1966.

Mann & Machine (TV Series 1992). "Prototype." Directed by Vern Gillum. Written by Robert De Laurentiis and Dick Wolf (story) and Robert De Laurentiis (teleplay). Season 1, Episode 1, April 5, 1992.

The Many Loves of Dobie Gillis (TV Series 1959–1963). "Beethoven, Presley and Me." Directed by Guy Scarpitta. Written by Dean Riesner. Season 4, Episode 24, March 13, 1963.

MGM Parade (TV Series 1955–1956). Hosted by Walter Pidgeon. Season 1, Episodes 32–33, March 14 and 21, 1956.

Mission: Impossible (TV Series 1966–1973). "Robot." Directed by Reza Badiyi (as Reza S. Badiyi). Written by Bruce Geller. Season 4, Episode 9, November 30, 1969.

My Living Doll (TV Series 1964–1965). "Boy Meets Girl." Directed by Lawrence Dobkin. Written by Bill Kelsay and Al Martin. Season 1, Episode 1, September 27, 1964.

The New Avengers (TV Series, 1976–1977). "The Last of the Cybernauts…?" Directed by Sidney Hayers. Written by Brian Clemens. Season 1, Episode 3, March 9, 1979.

Not Quite Human (TV Movie 1987). Directed by Steven Hilliard Stern. Written by Alan Ormsby. Walt Disney Television. June 19, 1987.

Perry Como's Kraft Music Hall (TV Series 1948–1967). Hosted by Perry Como. Season 8, Episode 23, February 18, 1956.

Planet Patrol (TV Series 1963–1968). "The Telepathic Robot." Directed by Frank Goulding. Written by Roberta Leigh. Season 2, Episode 4, July 24, 1966.

Quark (TV Series 1978). "May the Source Be with You." Directed by Hy Averback. Written by Steve Zacharias. Season 1, Episode 2, February 24, 1978.

Rocky Jones, Space Ranger (TV Series 1954). "Out of This World." Directed by Hollingsworth Morse.

Written by Arthur Hoerl. Season 1, Episodes 34–36, October 12, 19, and 26, 1954.

Rod Brown of the Rocket Rangers (TV Series 1953–1954). "The Robot Robber of Deimos." Directed by George Gould. Season 1, Episode 30, November 7, 1953.

The Six Million Dollar Man (TV Movie 1973). Directed by Richard Irving. Written by Terrence McDonnell. March 7, 1973.

The Six Million Dollar Man (TV Series 1974–1978). "The Bionic Woman." Directed by Dick Moder. Written by Kenneth Johnson. Season 2, Episode 19, March 16, 1975.

The Six Million Dollar Man (TV Series 1974–1978). "The Bionic Woman, Part 2." Directed by Dick Moder. Written by Kenneth Johnson. Season 2, Episode 20, March 23, 1975.

The Six Million Dollar Man (TV Series 1974–1978). "Kill Oscar, Part 2." Directed by Barry Crane. Written by Arthur Rowe and Oliver Crawford (story) and W. T. Zacha (as William T. Zacha) (teleplay). Season 4, Episode 6, October 31, 1976.

Small Wonder (TV Series 1985–1989). "Everybody into the Pool." Directed by Bob Claver. Written by Judith Bustany and Dawn Aldredge. Season 3, Episode 2, September 19, 1987.

Small Wonder (TV Series 1985–1989). "Vicki's Homecoming." Directed by John Bowab. Written by Howard Leeds. Season 1, Episode 1, September 7, 1985.

Space Patrol (TV Series 1950–1955). "The Android Invasion." Directed by Dick Darley (as Dik Darley). Written by Norman Jolley. Season 5, Episode 6, February 5, 1955.

Space Patrol (TV Series 1950–1955). "The Androids of Algol." Directed by Dick Darley (as Dik Darley). Written by Norman Jolley. Season 5, Episode 4, January 22, 1955.

Space Patrol (TV Series 1950–1955). "Double Trouble." Directed by Dick Darley (as Dik Darley). Written by Norman Jolley. Season 5, Episode 5, January 29, 1955.

Star Trek (TV Series 1966–1969). "I, Mudd." Directed by Marc Daniels. Written by Stephen Kandel. Season 2, Episode 8, November 3, 1967.

Star Trek (TV Series 1966–1969). "Requiem for Methuselah." Directed by Murray Golden. Written by Jerome Bixby. Season 3, Episode 19, February 14, 1969.

Star Trek (TV Series 1966–1969). "What Are Little Girls Made Of?" Directed by James Goldstone. Written by Robert Bloch. Season 1, Episode 7, October 20, 1966.

Star Trek: The Next Generation (TV Series 1987–1994). "Encounter at Farpoint." Directed by Corey Allen. Written by D. C. Fontana and Gene Roddenberry. Season 1, Episode 1, September 26, 1987.

Star Trek: The Next Generation (TV Series 1987–1994). "The Measure of a Man." Directed by Robert Scheerer. Written by Melinda M. Snodgrass. Season 2, Episode 9, February 11, 1989.

Star Trek: The Next Generation (TV Series 1987–1994). "The Naked Now." Directed by Paul Lynch. Written by John D. F. Black and D. C. Fontana (as J. Michael Bingham) (story) and D. C. Fontana (as J. Michael Bingham) (teleplay). Season 1, Episode 2, October 3, 1987.

Tales of Tomorrow (TV Series 1951–1953). "Read to Me, Herr Doktor." Directed by Don Medford. Written by Alvin Sapinsley. Season 2, Episode 30, March 20, 1953.

The Thin Man (TV Series 1957–1959). "Robot Client." Directed by Oscar Rudolph. Written by Devery Freeman. Season 1, Episode 23, February 28, 1958.

The Twilight Zone (TV Series 1959–1964). "The Brain Center at Whipple's." Directed by Richard Donner. Written by Rod Serling. Season 5, Episode 33, May 15, 1964.

The Twilight Zone (TV Series 1959–1964). "I Sing the Body Electric." Directed by William F. Claxton (as William Claxton) and James Sheldon. Written by Ray Bradbury. Season 3, Episode 32, May 18, 1962.

The Twilight Zone (TV Series 1959–1964). "The Lateness of the Hour." Directed by Jack Smight. Written by Rod Serling. Season 2, Episode 8, December 2, 1960.

The Twilight Zone (TV Series 1959–1964). "The Lonely." Directed by Jack Smight. Written by Rod Serling. Season 1, Episode 7, November 13, 1959.

The Twilight Zone (TV Series 1959–1964). "Uncle Simon." Directed by Don Siegel. Written by Rod Serling. Season 5, Episode 8, November 15, 1963.

Westworld (TV Series 2016, 2018). Directed by Jonathan Nolan. Written by Jonathan Nolan, Lisa Joy, and Michael Crichton (story) and Jonathan Nolan and Lisa Joy (teleplay).

The Wild, Wild West (TV Series 1965–1969). "The Night of Miguelito's Revenge." Directed by James B. Clark. Written by Jerry Thomas. Season 4, Episode 12, December 12, 1968.

Websites

"A. Merritt's *The Metal Monster.*" *Skulls in the Stars* (blog), February 2, 2009. https://skullsinthestars.com/2009/02/02/a-merritts-the-metal-monster/.

Adcock, John. "George & Norman Munro." *Punch in Canada* (blog), August 20, 2008. http://punch-incanada.blogspot.com/2008/08/rivals-from-nova-scotia.html.

Amaro, Richard. "San Diego Invites the World to Balboa Park a Second Time." *Journal of San Diego History* 31, no. 4 (Fall 1985). http://sandiegohistory.org/journal/1985/october/invites/.

"Androbot B. O. B. and Topo." https://youtu.be/jkOctWWsj-A.

Apfelbaum, John. "How Many People Collect Stamps." May 23, 2013. https://www.apfelbauminc.com/blog/how-many-people-collect-stamps.

Aston, Adam. "How Robots Lost Their Way." *Bloomberg Businessweek*, November 30, 2003. https://www.bloomberg.com/news/articles/2003-11-30/how-robots-lost-their-way.

"Beadle's Pocket Novels." *Beadle and Adams Dime Novel Digitization Project.* http://www.ulib.niu.edu/badndp/pn-a.html.

British Pathé. "The Face of Things—To Come! Alpha the Robot" (newsreel). June 12, 1934. https://www.britishpathe.com/video/the-face-of-things-to-come-alpha-the-robot/.

Brown, Peter. "The Wild Robot Lives!" *Peter Brown Studio*, March 10, 2016. http://www.peterbrownstudio.com/uncategorized/the-wild-robot/.

Bruce, Lenny. "The Berkeley Concert (1965)—Full Transcript." *Scraps from the Loft*, August 25, 2017. http://scrapsfromtheloft.com/2017/08/25/lenny-bruce-berkeley-concert-1965-full-transcript/.

Burlingame, Russ. "Decoding Convergence with Jeff King: The Finale." *Comicbook*, April 8, 2015. http://comicbook.com/2015/05/27/decoding-convergence-with-jeff-king-the-finale/.

Carper, Steve. "The Look of the Future—Robots." *Flying Cars and Food Pills.* https://www.flyingcarsandfoodpills.com/the-look-of-the-future---robots.

_____. "Robots on Stamps." *Flying Cars and Food Pills.* https://www.flyingcarsandfoodpills.com/robots-on-stamps.

Clute, John. "Edisonade." *The Encyclopedia of Science Fiction*, January 18, 2018. http://www.sf-encyclopedia.com/entry/edisonade.

_____. "Evan Hunter." *The Encyclopedia of Science Fiction*, October 19, 2017. http://www.sf-encyclopedia.com/entry/hunter_evan.

Coker, Rory. "Rod Brown of the Rocket Rangers." *The Space Hero Files.* http://216.75.63.68/space/text/index.phtml.

Cragston, Scott. "Robert the Robot." *Robotapedia.* http://www.attackingmartian.com/robert_the_robot_extra.html.

Dederick, Zadoc P., and Isaac Grass. "Improvement in Steam Carriage." U.S. Patent 75874, granted March 24, 1868. https://patents.google.com/patent/US75874.

"Doc Atomic." "Atomic Robot Man (Unknown/1949/Japan/5 inches)." *Astounding Artifacts*, June 16, 2009. http://astoundingartifacts.blogspot.com/2009/06/atomic-robot-man-unknown-1949.html.

"Edwin Sylvester Ellis." *Beadle and Adams Dime Novel Digitization Project.* http://www.ulib.niu.edu/badndp/ellis_edward.html.

Felchner, William J. "Ideal's Robot Commando: Collectible 1960s Toy." *Knoji.* https://vintages-antiques-collectibles.knoji.com/ideals-robot-commando-collectible-1960s-toy/.

"*Forbidden Planet* movie poster." *AllPosters.* https://www.allposters.com/-sp/Forbidden-Planet-Posters_i12191582_.htm.

Griffiths, Mark D. "Droidian Slips: A Brief Look at Robot Fetishism." *Psychology Today*, June 13, 2014. https://www.psychologytoday.com/us/blog/in-excess/201406/droidian-slips.

Handy, Jam. "Leave It to Roll-Oh." 1940. https://you tu.be/KSnJBNijsVU.

Hix, Lisa. "Attack of the Vintage Toy Robots! Justin Pinchot on Japan's Coolest Postwar Export." *Collector's Weekly*, November 18, 2010. https://www.collectorsweekly.com/articles/attack-of-the-vintage-toy-robots-justin-pinchot-on-japans-coolest-postwar-export/.

Hoggett, Reuben. "1874—Adam Ironsides—The Steam Man—C. C. Roe a.k.a. Capt. Rowe (Canadian)." *Cybernetic Zoo*. http://cyberneticzoo.com/steammen/1874-adam-ironsides-the-steam-man-c-c-roe-canadian/.

_____. "1875 'Psycho' the Whist-playing Automaton—Maskelyne & Clarke (British)." *Cybernetic Zoo*. http://cyberneticzoo.com/not-quite-robots/1875-psycho-the-whist-playing-automaton-maskelyne-clarke-british/.

_____. "1928—Eric Robot—Capt. Richards & A.H. Reffell (English)." *Cybernetic Zoo*. http://cyberneticzoo.com/robots/1928-eric-robot-capt-richards-english/.

_____. "1930—Alpha the Robot (American)." *Cybernetic Zoo*. http://cyberneticzoo.com/pseudo-automatons-and-robots/1930-alpha-the-robot-american/.

_____. "1932—Alpha the Robot—Harry May (English)." *Cybernetic Zoo*. http://cyberneticzoo.com/robots/1932-alpha-the-robot-harry-may-english/.

_____. "1932—George Robot—Capt. W.H. Richards (British)." *Cybernetic Zoo*. http://cyberneticzoo.com/robots/1932-%E2%80%93-george-robot-%E2%80%93-capt-w-h-richards-british/.

_____. "1954—'Gismo the Peaceful'—Sherwood Fuehrer (American)." *Cybernetic Zoo*. http://cyberneticzoo.com/robots/1954-gismo-the-peaceful-sherwood-fuehrer-american/.

_____. Anonymous broadsheet "The Great Steam King" (undated [probably March 1869]). *Cybernetic Zoo*. http://cyberneticzoo.com/steammen/1868-steam-man-eno-american/.

Holtz, Allan. "Ink-Slinger Profile: H. C. Greening." *Stripper's Guide* (blog), September 8, 2011. https://strippersguide.blogspot.com/2011/09/ink-slinger-profiles-hc-greening.html.

_____. "Obscurity of the Day: Professor Dodger and His Automatic Servant Girl." *Stripper's Guide* (blog), November 30, 2012. https://strippersguide.blogspot.com/2012/11/obscurity-of-day-professor-dodger-and.html.

"Inductees." *Robot Hall of Fame*. http://www.robothalloffame.org/inductees.html.

Jay, Alex. "Ink-Slinger Profiles by Alex Jay: Larry Antonette." *Stripper's Guide* (blog), June 11, 2014. http://strippersguide.blogspot.com/2014/06/ink-slinger-profiles-by-alex-jay-larry.html.

Kearney, Patrick J. "Notes Towards a Bibliography of Alphonse Momas (1846–1933)." *Scissors and Paste Bibliographies*. http://www.scissors-and-paste.net/Momas.html.

Kemerling, Garth. "Descartes: A New Approach." http://www.philosophypages.com/hy/4b.htm.

Kerslake, Lorraine. "The Iron Man." *The Ted Hughes Society*. http://thetedhughessociety.org/ironman/.

Kiger, Patrick J. "Rossum's Universal Robots." *How Stuff Works*. https://science.howstuffworks.com/10-evil-robots1.htm.

LaBruce, Bruce. "Notes on Camp/Anti Camp." July 7, 2015. http://brucelabruce.com/2015/07/07/notes-on-camp-anti-camp/.

"Making Learning Fun." *Baruch College Alumni Magazine* (November 3, 2015). https://blogs.baruch.cuny.edu/bcam/2015/11/03/making-learning-fun/.

Malik, Tariq. "Japan Launches Talking 'Robot Astronaut' Kirobo into Space." Spacewww, August 3, 2013. https://www.space.com/22235-japan-launches-talking-space-robot-astronaut.html.

Markham, Don. "Atomic Rabbit." *Toonopedia*. http://www.toonopedia.com/atomicrb.htm.

_____. "Felix the Cat." *Toonopedia*. http://www.toonopedia.com/felix.htm.

McIntyre, John T. *The House of Beadle & Adams*. "Chapter XIII: The Final Years, 1890 to 1897." *Beadle and Adams Dime Novel Digitization Project*. http://www.ulib.niu.edu/badndp/chap13.html.

"Mechagodzilla (GvMGII)." *Godzilla Wikia*. https://godzilla.fandom.com/wiki/Mechagodzilla_(GvMGII).

"The Middleton Family at the New York World's Fair." San Jose State University. http://www.sjsu.edu/faculty/wooda/middleton/middletonelektro.html.

Miller, John M. "The Iron Giant." *Turner Classic Movies*. http://www.tcm.com/this-month/article/148012%7C0/The-Iron-Giant.html.

Moore, George R. U.S. Patent No. 454570, granted June 23, 1891. https://www.google.com/patents/US454570.

Murray, Michael D. "Frederick Beecher Perkins: Library Pioneer and Curmudgeon." Master's thesis, San Diego State University, 2009. scholarworks.sjsu.edu/cgi/viewcontent.cgi?article=4973&context=etd_theses.

"*The Mysterians* movie poster." *AllPosters*. https://www.allposters.com/-sp/The-Mysterians-1959-Posters_i6250399_.htm.

"*The Mysterians* movie trailer." https://youtu.be/AHbL8a71Hoo.

Novak, Matt. "50 Years of the Jetsons: Why the Show Still Matters." *Paleofuture* (blog), September 19, 2012. http://blogs.smithsonianmag.com/paleofuture/2012/09/50-years-of-the-jetsons-why-the-show-still-matters/.

Obituary Record of Graduates of Yale University Deceased during the Academical Year Ending in June, 1899. mssa.library.yale.edu/obituary_record/1859_1924/1898-99.pdf.

Osborne, Jerry. "The Continuing Interview with Connie Francis." *DISCoveries* (September 1991). Posted July 2006. http://www.freewebs.com/conniefrancis/connie10.htm.

Perew, Louis Philip, and Joseph A. Dischinger. U.S.

Patent No. 949287, granted February 15, 1910. https://www.google.com/patents/US949287.

Phillips, Mark. "The History of *Lost in Space*." *Lost in Space Wikia*. http://lostinspace.wikia.com/wiki/The_History_of_Lost_In_Space.

RealDoll Shop. https://www.realdoll.com/shop/.

Repasky, Mark. "Thodar the Robot." March 25, 2018. http://www.innbythemill.com/thodar.htm.

"Ripley Believe-It-Or-Not Odditorium at a Century of Progress." *Postcardy* (blog), January 7, 2015. http://postcardy.blogspot.com/2015/01/ripley-believe-it-or-not-odditorium-at.html.

"Robert the Robot from 1950's Toy Vintage Robot." https://youtu.be/FDQEV43DeFM.

"The Robot Lovers." *Internet Speculative Fiction Database.* http://www.isfdb.org/cgi-bin/title.cgi?1750057.

Roe, C. C. Broadsheet (undated [probably 1878]). http://epe.lac-bac.gc.ca/100/205/301/ic/cdc/industrial/ironman.htm.

Rydell, Robert W. "Century of Progress Exposition." *Encyclopedia of Chicago.* http://www.encyclopedia.chicagohistory.org/pages/225.html.

Schreurs, Jason. "Robot Band Recruits Human Member to Write Songs, Hilarity Ensues." *Vice* (March 13, 2015). https://noisey.vice.com/en_ca/article/6vm4yb/compressorhead-nomeansnojohnwright-interview.

"Science Fiction Movies." *Ultimate Movie Rankings.* http://www.ultimatemovierankings.com/science-fiction-movies-2/.

Scifidreamgirls (Video Series 2013–2014). Directed by Jack Kona. Scifidreamgirls.com.

Sontag, Susan. "Notes on 'Camp.'" *The Partisan Review* (Fall 1964). https://faculty.georgetown.edu/irvinem/theory/Sontag-NotesOnCamp-1964.html.

Stradley, Don. "'Captain Video & His Video Rangers': This 1950s Sci-fi Adventure TV Show Was a Night Flight Fave." *Night Flight*, May 31, 2017. http://nightflight.com/captain-video-his-video-rangers-this-1950s-sci-fi-adventure-tv-show-was-a-night-flight-fave.

Szondy, David. "The Original Robby the Robot Goes Up for Auction." *New Atlas*, October 27, 2017. https://newatlas.com/robby-robot-bonhams/51922/.

Tallerico, Brian. "The Long, Weird History of the Westworld Franchise." *Vulture*, September 30, 2016. http://www.vulture.com/2016/09/westworld-franchise-long-weird-history.html.

Thomas, Bryon. "'Space Patrol': This Early 50s 'Space Opera' Set in the 30th Century Aired on 'Night Flight' in the Mid-80s." *Night Flight*, September 7, 2016. http://nightflight.com/space-patrol-this-early-50s-space-opera-set-in-the-30th-century-aired-on-night-flight-in-the-mid-80s/.

Thomas, Nick. "Make Room for Sherry Jackson." *The Spectrum*, March 31, 2016. http://www.thespectrum.com/story/entertainment/2016/03/31/make-room-sherry-jackson/82343114/.

Vander Voort, Garry. "The Androbot B.O.B." *Retroist*, August 28, 2014. https://www.retroist.com/2014/08/28/the-androbot-b-o-b/.

"Vintage and Antique Space Toys." *Collector's Weekly.* https://www.collectorsweekly.com/toys/space.

Wall, Bill. "Ajeeb the Chess-Playing Automaton." *Chessmaniac* (blog), January 17, 2015. http://www.chessmaniac.com/ajeeb-the-chess-playing-automaton/.

Westinghouse. "The Middleton Family at the New York World's Fair." 1939. https://www.youtube.com/watch?v=YF594h8KUXw.

Wickman, Forrest. "Star Wars Is a Postmodern Masterpiece." *Slate*, December 13, 2015. http://www.slate.com/articles/arts/cover_story/2015/12/star_wars_is_a_pastiche_how_george_lucas_combined_flash_gordon_westerns.html.

Zeman, Carrie R. "The Remarkable Story of Edward S. Ellis." October 15, 2012. https://athrillingnarrative.com/2012/10/15/the-remarkable-story-of-edward-s-ellis/.

Index

285

www.ingramcontent.com/pod-product-compliance
Lightning Source LLC
Chambersburg PA
CBHW081736270326
41932CB00020B/3294